MECHANISMS OF CELL DEATH AND OPPORTUNITIES FOR THERAPEUTIC DEVELOPMENT

Perspectives in Translational Cell Biology

MECHANISMS OF CELL DEATH AND OPPORTUNITIES FOR THERAPEUTIC DEVELOPMENT

Edited by

DAIQING LIAO

Department of Anatomy and Cell Biology,
College of Medicine, University of Florida, FL, United States

ELSEVIER

ACADEMIC PRESS

An imprint of Elsevier

Academic Press is an imprint of Elsevier
125 London Wall, London EC2Y 5AS, United Kingdom
525 B Street, Suite 1650, San Diego, CA 92101, United States
50 Hampshire Street, 5th Floor, Cambridge, MA 02139, United States
The Boulevard, Langford Lane, Kidlington, Oxford OX5 1GB, United Kingdom

Library of Congress Cataloging-in-Publication Data
A catalog record for this book is available from the Library of Congress

British Library Cataloguing-in-Publication Data
A catalogue record for this book is available from the British Library

ISBN: 978-0-12-814208-0

For information on all Academic Press publications visit our website at https://www.elsevier.com/books-and-journals

Publisher: Andre G. Wolff
Editorial Project Manager: Naomi Robertson
Production Project Manager: Maria Bernard
Cover Designer: Mark Rogers

Typeset by TNQ Technologies

Contents

List of contributors

Michael C. Chung
Department of Physics, University of Florida, Gainesville, FL, United States

Haiming Dai
Department of Pathology and Laboratory Medicine, Hefei Cancer Hospital, Chinese Academy of Sciences, Hefei, Anhui, China; Anhui Province Key Laboratory of Medical Physics and Technology, Institute of Health & Medical Technology, Hefei Institutes of Physical Science, Chinese Academy of Sciences, Hefei, Anhui, China

Juan Guan
Department of Physics, University of Florida, Gainesville, FL, United States; Department of Anatomy and Cell Biology, College of Medicine, University of Florida, Gainesville, FL, United States

Jia Jia
Department of Pathology and Laboratory Medicine, Hefei Cancer Hospital, Chinese Academy of Sciences, Hefei, Anhui, China; Anhui Province Key Laboratory of Medical Physics and Technology, Institute of Health & Medical Technology, Hefei Institutes of Physical Science, Chinese Academy of Sciences, Hefei, Anhui, China

Rui Kang
Department of Surgery, UT Southwestern Medical Center, Dallas, TX, United States

Scott H. Kaufmann
Department of Molecular Pharmacology and Experimental Therapeutics, Mayo Clinic, Rochester, MN, United States; Division of Oncology Research, Mayo Clinic, Rochester, MN, United States

Mary E. Law
Department of Pharmacology & Therapeutics, College of Medicine, University of Florida, Gainesville, FL, United States

Brian K. Law
Department of Pharmacology & Therapeutics, College of Medicine, University of Florida, Gainesville, FL, United States; UF-Health Cancer Center, University of Florida, Gainesville, FL, United States

Daiqing Liao
Department of Anatomy and Cell Biology, College of Medicine, University of Florida, Gainesville, FL, United States

X. Wei Meng
Department of Molecular Pharmacology and Experimental Therapeutics, Mayo Clinic, Rochester, MN, United States; Division of Oncology Research, Mayo Clinic, Rochester, MN, United States

Gautam Sethi
Department of Pharmacology, Yong Loo Lin School of Medicine, National University of Singapore, Singapore, Singapore

Muthu K. Shanmugam
Department of Pharmacology, Yong Loo Lin School of Medicine, National University of Singapore, Singapore, Singapore

Xiaokun Shu
Department of Pharmaceutical Chemistry and Cardiovascular Research Institute, University of California San Francisco, San Francisco, CA, United States

Daolin Tang
Department of Surgery, UT Southwestern Medical Center, Dallas, TX, United States

Tsz-Leung To
Broad Institute, Cambridge, MA, United States

Mengxiong Wang
Department of Radiation Oncology, College of Medicine, Stanford University, Palo Alto, CA, United States

Kaiqin Ye
Department of Pathology and Laboratory Medicine, Hefei Cancer Hospital, Chinese Academy of Sciences, Hefei, Anhui, China; Anhui Province Key Laboratory of Medical Physics and Technology, Institute of Health & Medical Technology, Hefei Institutes of Physical Science, Chinese Academy of Sciences, Hefei, Anhui, China

CHAPTER 1

Apoptosis, necroptosis, and pyroptosis in health and disease: an overview of molecular mechanisms, targets for therapeutic development, and known small molecule and biologic modulators

Daiqing Liao
Department of Anatomy and Cell Biology, College of Medicine, University of Florida, Gainesville, FL, United States

1. Introduction

Regulated cell death (RCD) occurs during development, tissue damage and repair, and pathogen infection. There are several forms of RCD, among which apoptosis, necroptosis, and pyroptosis are well understood in terms of physiological/pathophysiological triggers, precise signaling cascades, the activation and regulation of supramolecular complexes, and effectors/executioners of eventual cell demise [1–4]. Apoptosis was initially described as "shrinkage necrosis" [5,6], and is generally considered a non-lytic form of cell death, whereas cell lysis occurs in necroptotic and pyroptotic cell death [3,7–12]. As such, these forms of RCD have distinct immunological consequences [4,13–16]. This chapter focuses on molecular mechanisms underlying apoptosis, necroptosis, and pyroptosis, their roles in diverse human diseases, and drug discovery efforts on modulating these three mechanisms of RCD. There are several other forms of RCD such as ferroptosis, entotic cell death, NETosis, parthanatos, lysosome-dependent cell death, autophagy-dependent cell death, alkaliptosis, and oxeiptosis, for which the underlying molecular mechanisms are relatively less well defined [1,4,17,18]. Ferroptosis is discussed in Chapter 8 of this volume, while other forms of RCD including apoptosis, autophagy, necroptosis, pyroptosis are covered in other chapters of this volume.

Mechanisms of Cell Death and Opportunities for Therapeutic Development
ISBN 978-0-12-814208-0
https://doi.org/10.1016/B978-0-12-814208-0.00008-7

2. The apoptotic pathways and major targets for drug development

Apoptosis plays a major role in embryonic development to remove unwanted epithelial cells for tissue organization and morphogenesis [2,19–21]. It also functions to eliminate superfluous and potentially dangerous cells (e.g., infected and damaged cells and extra lymphocytes) for the maintenance of homeostasis [22]. Apoptosis is a nonlytic form of cell death. During apoptotic cell death, caspases are orderly activated, resulting in the digestion of many intracellular substrates, chromatin fragmentation, and reorganization of cytoskeletons to induce membrane blebbing [23,24]. The dying cells disintegrate into small cellular debris known as apoptotic bodies with intact membrane on their surfaces without uncontrolled release of cellular contents. Apoptotic cells release certain metabolites to serve as "find-me" signals for engulfment by neighboring phagocytic cells such as macrophages [7–9,11,12]. Thus, no immunogenic contents are released. As such, apoptosis is considered as a nonimmunogenic or uninflammatory form of cell death [13].

A hallmark of cancer cells is resistance to apoptotic cell death [25]. Thus, sensitizing cancer cells to apoptosis represents the major focus of drug discovery efforts in the past 3 decades. The defined molecular functions and the availability of three-dimensional structures of key regulators of apoptosis have led to extensive efforts to develop small molecule agents to induce apoptosis in cancer cells as potential cancer therapeutics [26–38]. Agents that block functions of proteins in several steps of the apoptotic pathway including BCL-2 family proteins, and inhibitor of apoptosis proteins (IAPs) have been discovered and evaluated in the clinic for cancer therapy (Fig. 1.1). Biologics that mimic the proapoptotic functions of death ligands known collectively as proapoptotic receptor agonists (PARAs) have been tested in clinical studies for treating diverse cancer types [39,40] (Fig. 1.1).

2.1 An overview of the molecular mechanism of apoptosis

Apoptosis is generally categorized into extrinsic and intrinsic mechanisms, although both mechanisms are interconnected and converge on mitochondria-mediated events [41,42] (Fig. 1.1). The extrinsic apoptosis pathway is activated upon the engagement of death ligands with death receptors and the formation of cytoplasmic death-inducing signaling complex (DISC), which recruits the initiator caspase 8 (or 10), resulting in proximity-induced autoactivation of these initiator caspases. The activated caspase 8 then cleaves and activates the effector or executioner caspases (3, 6, or 7), resulting in the apoptotic cell death. Growth factor or nutrient

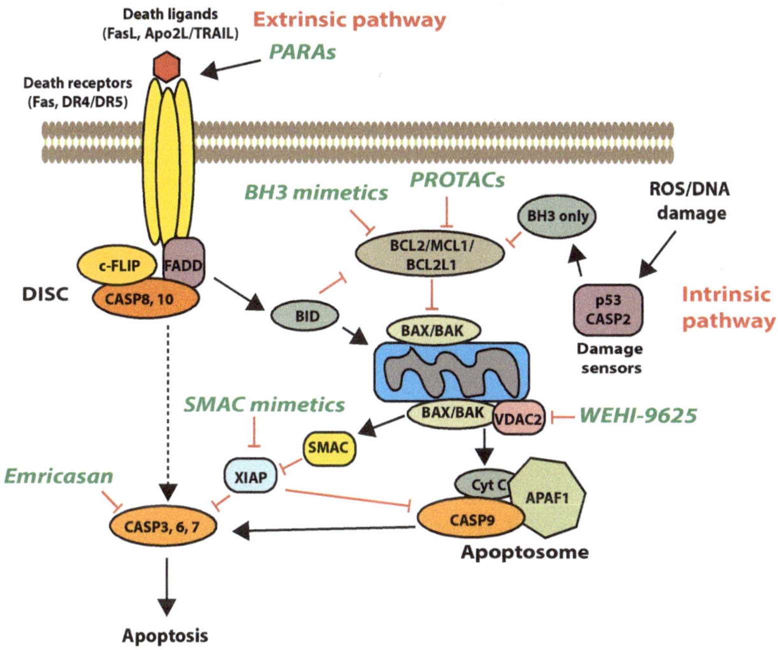

Figure 1.1 ***The apoptotic pathways and major targets for drug development.*** The extrinsic apoptosis pathway is initiated by binding of death ligands to death receptors and the formation of cytoplasmic death–inducing signaling complex (DISC) and the activation of the initiator caspase 8 (or 10), which activates the effector or executioner caspases (3, 6, or 7), resulting in the apoptotic cell death. Proapoptotic receptor agonists (PARAs) promote the extrinsic pathway of apoptosis and they have been tested in clinical trials for the treatment of cancer. Reactive oxygen species (ROS), DNA damage, or cytosolic calcium overload trigger intrinsic apoptosis pathway by activating damage sensors such as p53. p53 transactivates proapoptotic BCL-2 family members to promote apoptosis. The proapoptotic BH3-only proteins induce BAX/BAK pore formation to release cytochrome *c*, SMAC, and other apoptogenic proteins from mitochondria. Cytochrome *c* binds to APAF1 to trigger its conformational change to form an oligomer of a disc-like complex with sevenfold symmetry known as apoptosome. The initiator caspase 9 is recruited to and autoactivated in apoptosome. The activated form of caspase 9 cleaves and activates caspases 3, 6, or 7, resulting in apoptosis. Caspase 2 is activated by the PIDDosome, which comprises PIDD1, RAIDD, and caspase 2. Activated caspase 2 cleaves BID to t-BID to trigger the release of apoptogenic proteins from mitochondria and subsequent apoptosis. Agents targeting various proteins in the intrinsic apoptosis pathways are shown. Most of these agents including BH3 and SMAC mimetics induce apoptosis, while others such as the pan-caspase inhibitor emricasan and the VDAC2 binder WEHI-9625 block apoptosis.

withdrawal, loss of cell attachment, steroid hormone, developmental cues, exposure to cytotoxic agents such as chemotherapy drugs, reactive oxygen species (ROS), DNA damage, or cytosolic calcium overload trigger intrinsic apoptosis pathway by activating damage sensors such as p53. p53 transactivates proapoptotic BCL-2 family members BH3-only proteins such as BBC3 (also known as PUMA) to promote apoptosis. Transcriptional upregulation as well as posttranslational protein stabilization of BH3-only proteins increases their concentration. The BH3-only proteins then interact with the antiapoptotic BCL-2 family members known as multi-domain BCL-2 proteins such as BCL-2, BXL-xL, and MCL1, which prevent the activation of the proapoptotic multidomain BCL-2 proteins including BAX, BAK, and BOK [43]. When the intracellular antiapoptotic BCL-2 proteins are bound by BH3-only proteins, BAX/BAK are released from complexes with the antiapoptotic BCL-2 proteins, undergo conformational change, and oligomerize to form a pore on the mitochondrial outer membranes (MOMs), resulting in the release cytochrome *c*, SMAC, and other apoptogenic proteins from mitochondria [44—47]. Cytochrome *c* binds to APAF1 to trigger its conformational change to form an oligomer of a disc-like complex with sevenfold symmetry known as apoptosome [48—50]. The initiator caspase 9 is recruited to and activated by apoptosome. The activated form of caspase 9 cleaves and activates caspases 3, 6, or 7, resulting in apoptosis. Caspase 2 is activated by the PIDDosome, which comprises PIDD1, RAIDD, and caspase 2 [51,52]. Activated initiator caspases (caspases 2, 8, and 9) also cleave BID to produce a truncated form of BID known as t-BID, which triggers the release of apoptogenic proteins from mitochondria and subsequent apoptosis.

2.2 BCL-2 family proteins

As noted above, BCL-2 family members consist of both antiapoptotic proteins and proapoptotic proteins. The antiapoptotic BCL-2 proteins have four BCL-2 homology domains (BH1-4). The so-called multidomain proapoptotic BCL-2 proteins contain BH1-3, which includes BAX, BAK, and BOK. These proteins form pores in the MOM when activated by apoptotic signaling [44—47]. A group of eight BCL-2 family members (BID, BAD, BIM, BIK, BMF, NOXA, PUMA, and HRK) contain only BH3 domain. These proteins are collectively known as BH3-only BCL-2 proteins. These proteins share little sequence homology beyond the BH3 sequence. As discussed extensively in Chapter 7 in this volume, the detailed mechanistic understanding of how the BCL-2 family regulates apoptosis has led to the development of the BH3 mimetics that target the BH3-binding groove of the antiapoptotic BCL-2 proteins (Fig. 1.1). These agents include small molecules and short polypeptides [29—31,45]. The BCL-2 inhibitor

ABT-199/GDC-0199 (also known as venclexta and venetoclax) induces intrinsic apoptosis [53] and has been approved by the Food and Drug Administration (FDA) for treating CLL (chronic lymphocytic leukemia) and SLL (small lymphocytic lymphoma) in 2016 [54]. Degradation of antiapoptotic BCL-2 proteins using proteolysis targeting chimeras (PROTACs) has shown considerable promise for treating cancer in preclinical models and the lead drug candidate DT2216 has entered clinical trials [55—57].

2.3 Inhibitor of apoptosis proteins (IAPs)

IAPs bind and inhibit caspases [23]. There are eight IAPs in humans, all of which contain one to three baculovirus IAP repeat (BIR) domains. Five IAPs, cIAP1 (BIRC2), cIAP2 (BIRC3), XIAP (BIRC4), NL-IAP/Livin (BIRC7), and BIRC8 (ILP2), contain a RING domain in their C-terminus, which suggests they may have intrinsic E3 ubiquitin ligase activity [58—60]. Four of these proteins (cIAP1, cIAP2, XIAP, and ILP2) also have a ubiquitin-binding motif known as ubiquitin-associated domain [60,61]. Additionally, cIAP1 and cIAP2 harbor a caspase recruiting domain (CARD), which is involved in autoinhibition of their E3 ubiquitin ligase activity [62—64]. BIRC1 encodes a protein known as NAIP that is involved in inflammasome activation and pyroptosis as discussed below in this chapter. BIRC6 (also known as Bruce/Apollon) is a large protein with 4845 amino acid residues. It serves as both ubiquitin-conjugating enzyme and ubiquitin ligase and has been shown recently to inhibit autophagy [65].

IAPs are generally highly expressed in cancer cells and have been extensively investigated as targets of anticancer therapy [66,67]. The design of IAP antagonists is inspired by the observation that four hydrophobic residues (AVPI/F) known as IAP-binding motif (IBM) inserts into a deep hydrophobic pocket in most of the BIRs of IAPs and this interaction blocks IAP—caspase association, thereby relieving their inhibition of activated caspases to promote apoptosis [27,68—71]. Many mitochondrial proteins including SMAC (second mitochondria-derived activator of caspase, encoded by the DIABLO gene) contain an IBM, which emerges within a new N-terminus upon cleavage of the import signal peptide during their mitochondrial import process [48,72,73]. A structural study shows that SMAC homodimerizes via an extensive helical hydrophobic interface and thus is intrinsically bivalent [26]. The synthetic compounds that are modeled after IBM are known as SMAC mimetics (Fig. 1.1). Two types of SMAC mimetics, monovalent and bivalent, have been designed and synthesized [27,69–71]. Monovalent IAP antagonists have one BIR-binding ligand, whereas bivalent SMAC mimetics have two BIR-binding ligands separated by a linker. Bivalent SMAC mimetics (e.g., birinapant) are more potent (up to 1000-fold) than monovalent IAP antagonists [27,74–76]. Interestingly, binding of both monovalent and bivalent SMAC mimetics to cIAPs has been shown to relieve auto-inhibition of the E3 ubiquitin ligase activity of cIAPs, resulting in their autoubiquitination and degradation [62,63].

cIAPs ubiquitinate RIPK1 (receptor-interacting protein kinase 1), which is required for assembling signaling complexes for activating the canonical NF-κB pathway as well as mitogen-activated protein kinase (MAPK) signaling cascade [77–79] (Fig. 1.2). SMAC mimetics-mediated degradation of cIAPs therefore inhibits cell survival and proliferation signaling and can lead to apoptosis or necroptosis as discussed below (Fig. 1.2).

There are renewed interests in using SMAC mimetics for cancer therapy. Both monovalent and bivalent SMAC mimetics have been tested as monotherapy or in combination with other therapeutic modalities such as chemotherapy, immunotherapy, and radiation for treating leukemias/lymphomas and solid tumors [35,36,80]. Eight SMAC mimetics including five monovalent inhibitors (GDC-0152, CUDC-427 (GDC-0917), Debio 1143 (AT-406), LCL161, and BI 891065) and three bivalent analogs (birinapant (TL32711), APG-1387, and HGS1029 (AEG40826)) have been evaluated in the clinic for the treatment of hematological and solid malignancies [35]. A nonpeptidomimetic antagonist of cIAP1/2 and XIAP, ASTX66 (tolinapant), has entered phase I/II clinical trials for treating solid tumors and lymphoma [38,81] (ClinicalTrials.gov Identifier: NCT02503423). Whereas monotherapies with these IAP antagonists show good safety profiles and on-target activity, they exhibit only limited anticancer efficacy. Therefore, combination therapies with other agents and the screen for patients who may exhibit better response to SMAC mimetics have been suggested for further therapeutic development of these compounds [35].

2.4 PARAs

Recombinant forms of some death ligands such as tumor necrosis factor (TNF)–related apoptosis-inducing ligand TRAIL (also known as Apo2L or

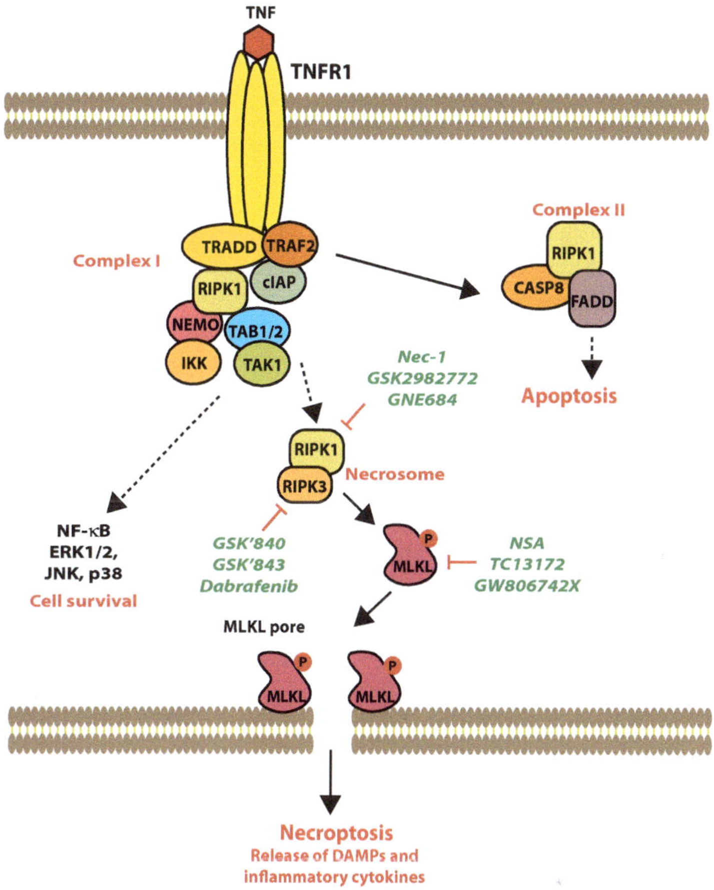

Figure 1.2 ***Mechanisms of necroptosis, major drug targets, and small molecule inhibitors.*** TNF/TNFR interaction promotes the assembly of the membrane-associated complex I. This signaling complex provides a platform for activating the NF-κB and MAPK pathways, which upregulate gene expression programs for cell survival and proliferation. Under certain conditions such as inhibition of cIAPs, RIPK1, FADD, and caspase 8 assemble the cytosolic complex II that activates the caspase cascade to induce apoptosis. When caspase 8 is inhibited, RIPK1 interacts with and phosphorylates RIPK3 to initiate the formation of an amyloid-like complex known as necrosome that recruits and phosphorylates MLKL. Phosphorylated MLKL assembles an oligomer which interacts with the cytosolic leaflet of the plasma membranes, resulting in membrane rapture, the release of cellular contents known as DAMPs, and necroptosis. Small molecule inhibitors targeting RIPK1, RIPK3, and MLKL are depicted in green.

TNFSF10) are potent inducers of extrinsic apoptosis and have good safety profiles in contrast to severe toxicity associated with TNF and Fas ligand to normal tissues [40,82–85]. The first-generation recombinant TRAIL is a soluble form of human TRAIL extracellular domain lacking the transmembrane domain and a short cytoplasmic tail (Apo2L/AMG-951/dulanermin) [82]. Several agonistic antibodies that selectively target DR4 (mapatumumab) [86] or DR5 (drozitumab [87], conatumumab/AMG-655 [88], lexatumumab [89], and tigatuzumab [90]) have been discovered and tested. Some of these biologics have been tested in clinical trials for treating hematological and solid tumors including nonsmall cell lung cancer, colon, pancreatic cancers and non-Hodgkin's lymphoma [40,85]. Although these agents are well tolerated, anticancer efficacy has not been demonstrated. Current strategies include combination regimens with other treatment modalities including chemotherapy and immunotherapy and identification of patients with biomarkers that predict good therapeutic response [40,85]. Additionally, some of the early generation TRAIL-based therapeutics do not have desirable pharmaceutical properties such as short plasma half-life and low binding affinity to death receptors [38,91]. These drawbacks have spurred efforts to engineer biologics with superior target-binding affinity and pharmaceutical properties. ABBV-621 (Eftozanermin alfa) is a second-generation TRAIL receptor agonist recombinant protein with an IgG1-Fc mutant backbone linked to two sets of TRAIL receptor binding domain monomers. Thus, ABBV-21 is a hexavalent agonist, which has been shown to bind to DR4 and DR5 death receptors with nanomolar affinity. It displays on-target induction of DISC formation and caspase 8 activation. It shows good preclinical efficacy in diverse cancer types without notable toxicity in nonrodent species [91]. ABBV-21 is now in clinical studies for treating leukemia/lymphoma and solid tumors [92,93] (ClinicalTrials.gov Identifier: NCT03082209).

2.5 Other targets of the apoptosis pathways

Fas-associated death domain (FADD)-like IL-1β—converting enzyme inhibitory protein (FLIP, also known as cFLIP, encoded by the CFLAR gene) is a caspase-like protein but lacks catalytic activity. It regulates apoptosis, necroptosis, autophagy, and inflammation [94,95]. Multiple FLIP isoforms including FLIP(L) and FLIP(S) are expressed. FLIP(L) can promote or inhibit caspase 8 activity, while FLIP(S) is an inhibitor of caspase 8 [96]. A recent study shows that FLIP(L) blocks pyroptosis by blocking caspase 8-mediated cleavage of gasdermin D (GSDMD) [97] (also see the pyroptosis section below). Inhibitors that block the interaction between FLIP and the DISC complex have been described [38]. Caspase inhibitors, such as the pan-caspase inhibitor emricasan (Fig. 1.1), have been reported and there are renewed interests in using such inhibitors for clinical applications [98–103]. Voltage-dependent anion channel (VDAC) of the MOM is a major metabolite channel and a target for drug discovery to modulate mitochondrial physiology [104]. A recent report identified a VDAC2 inhibitor, WEHI-9625, that blocks BAX/BAK-dependent apoptosis [103,105] (Fig. 1.1).

3. Necroptosis: mechanism, disease implications, and potential therapeutics

In contrast to programmed cell death or RCD, accidental cell death is referred to as cell death that is triggered by overwhelming physical or chemical stress [1]. Historically, necrosis has been considered as an accidental or uncontrollable mode of cell death. Necrosis is characterized by cell swelling, membrane rupture, and the ensuing release of cellular contents including inflammatory cytokines [106–109]. It is now clear that there are defined mechanisms underlying several different forms of necrosis, such as necroptosis, pyroptosis, and ferroptosis, that exhibit similar lytic features of cell demise.

3.1 Mechanism of necroptosis

Upon binding of TNF to its cognate receptors, primarily TNF receptor 1 (TNFR1 encoded by the TNFRSF1A gene), the cytoplasmic domain of TNFR1 recruits the adaptor protein TRADD, resulting in the assembly of a signaling complex known as complex I, consisting of several proteins including RIPK1, cIAP1/2, and TRAF2 (Fig. 1.2). In this complex, RIKP1 is ubiquitinated by cIAPs. The ubiquitin chain attached to RIPK1 facilitates the recruitment and activation of two kinase complexes (the IKK complex comprising NEMO, IKK1 and IKK2, and the TAB/TAK-1 complex), resulting in the activation of signaling cascades leading to NF-κB activation and the expression of inflammatory cytokines and other genes such as cFLIP (CFLAR) [78,79,109–111]. As noted above, cFLIP is an inhibitor of caspase 8 [96,112,113]. Complex I can also activate the MAPKs ERKs, JNK, and p38 to promote cell survival and proliferation [110]. Under certain conditions, such as cIAP inhibition (e.g., by SMAC mimetics), RIPK1 deubiquitination (e.g., by the deubiquitinase cylindromatosis—CYLD), or the absence of cFLIP, a complex comprising RIPK1, FADD, and caspase 8 (also known as complex II or IIa) forms, which triggers caspase 8 activation and apoptosis (Fig. 1.2). When caspase 8 activity is inhibited (e.g., by genetic inactivation or with the pan caspase inhibitor z-VAD-fmk), RIPK1 interacts with RIPK3, via the RIP homotypic interaction motif (RHIM) [114,115]. RIPK1 and RIPK3 form a large amyloid-like complex known as necrosome (also known as complex IIb). In cell cultures, certain cells undergo necroptotic cell death when exposed to TNFα, a cIAP inhibitor (e.g., SMAC mimetics) and a caspase inhibitor (e.g., z-VAD-fmk) [116,117]. RIPK3 is phosphorylated and activated by RIPK1. RIPK3 can also be activated by ZBP1 (Z-DNA binding protein 1, also known as DAI) in response to exogenous DNAs (e.g., viral DNA) in the cytosol [118,119]. Additionally, in response to bacterial and viral infection as well as certain endogenous stresses, RIPK3 is activated via interaction with TRIF (also known as TICAM1) that serves as an adaptor of Toll-like receptors (TLRs) 3 or 4 (TLR3 or TLR4)–mediated signal transduction pathway [111,119]. Activated RIPK3 then phosphorylates the pseudokinase MLKL (mixed lineage kinase domain-like) to promote its oligomerization, plasma membrane translocation, membrane permeabilization, and cell death [111,116,120–122]. The kinase activity of RIPK1 is essential for necrosome formation, as RIPK1 inhibitors such as Nec-1 prevent necroptosis [123,124] (Fig. 1.2). RIPK3 kinase activity is required for MLKL activation; genetic or pharmacological RIPK3 inhibition blocks necroptosis [117,125–128]. It should be noted that, in addition to TNFR1 and TLRs, other receptors such as the interferon α/β and γ receptors (IFNAR1 and INFGR1), can also trigger necroptosis [111,129–134].

3.2 Necroptosis in diseases

Necroptosis plays various roles in physiological and pathological processes. Given that mice lacking Ripk3 (Ripk3$^{-/-}$) [135] and Mlkl (Mlkl$^{-/-}$) [136] are viable, necroptosis per se may not be required for normal mouse development. Interestingly, embryonic lethality of Casp8 knockout (Casp8$^{-/-}$) can be rescued by concomitant Ripk3 knockout (Ripk3$^{-/-}$) in mice, suggesting that excessive necroptosis in Casp8 knockout mice appears to impair animal development [137–139]. Other similar genetic studies demonstrate the importance of cross-talk between apoptosis and necroptosis for normal mouse development (reviewed in Ref. [111]).

Deregulation of necroptosis is implicated in several pathological conditions. Necroptotic cells are immunogenic, which promotes antitumor immune response [140–142]. RIPK1/RIPK3-mediated necroptosis is an important mechanism that controls viral infection in vitro and in vivo [2,111,117,126,143–148]. Necroptosis may also have roles in controlling bacterial infection and etiology of infectious diseases [109,111,148].

RIPK3-dependent necroptosis contributes to the etiology of cardiovascular diseases as demonstrated in studies based on various cell and animal models [110,111,149–152]. The RIPK1 inhibitors afforded protection against various forms of neurological, liver, and kidney diseases in preclinical studies, implicating necroptosis as a potential cause for these pathologies [111,153–157].

Necroptosis is extensively linked to different forms of inflammatory diseases [111,158–161]. Recent studies show that necroptosis along with apoptosis and pyroptosis (PANoptosis) is implicated in SARS-CoV-2–triggered severe inflammatory responses and acute lung damages associated with patient mortality due to coronavirus disease 2019 (COVID-19) [162,163]. Significantly, homozygous loss of function of TANK-binding kinase 1 (TBK1) in humans causes systemic autoinflammation, not overt viral disease, due to TNF-induced and RIPK1-dependent necroptosis. Clinical disease of autoinflammation can be reversed with anti-TNF treatment [164].

Necroptosis displays various roles in cancer. As mentioned above, necroptotic tumor cells promote antitumor immunity. In studies using syngeneic mouse models, the RIPK1–RIPK3–MLKL necroptosis pathway was shown to be critical to the development of antitumor immune response. Damage-associated molecular patterns (DAMPs) released from necroptotic tumor cells promote the maturation of bone marrow–derived dendritic cells to cross-present tumor-associated antigens to CD8+ cytotoxic T lymphocytes [13,16,140–142,165–167]. According to the NCBI Gene portal, RIPK3 expression varies significantly in different tissues with relatively high expression in some organs such as the colon, small intestine, lung, and lymph node and low in others such as the liver, ovary, and pancreas, while RIPK1 is ubiquitously expressed. The expression levels of MLKL are also quite variable in different tissues with relatively low levels in liver, pancreas, and ovary. In tumors, studies have shown that RIPK3 is silenced by DNA methylation in breast cancer and other cancer types and its reactivation potentiates response to chemotherapy [168–170]. Consequently, most cancer cell lines are insensitive to agents that promote necroptotic cell death [168,171,172]. Downregulation of MLKL in several types of cancer has also been observed [122,170]. Nonetheless, RIPK1 has notable oncogenic functions [173,174], likely due to its role in promoting cell survival and proliferation (Fig. 1.2). Notably, tumor cell–induced necroptosis of endothelial cells appears to promote cancer metastasis [175]. Necroptotic signaling could also induce an immunosuppressive tumor microenvironment (TME) [176], thereby promoting tumor growth and progression.

3.3 Necroptosis-targeting agents

Inhibition or activation of necroptosis may have therapeutic applications for different human diseases. Inflammatory diseases can be treated with agents that block necroptosis, whereas compounds that promote necroptosis can be used to induce cancer cell death as potential cancer therapeutics. Many cancer therapeutics as well as components of traditional medicines have been shown to induce necroptosis [134,170]. Inhibitors targeting the major regulators of necroptosis namely RIPK1, RIPK3, and MLKL have been discovered [177] (Fig. 1.2). In a screen of ~15,000 compounds, necrostatin-1 (Nec-1) was identified to inhibit necroptosis of human monocytic U937 cells induced by TNFα and z-VAD-fmk [124]. Subsequent studies show Nec-1 and related compounds directly target RIPK1 by binding to an allosteric site outside of the ATP-binding pocket in the kinase domain of RIPK1 [123,178,179]. GNE684 is a potent RIPK1 inhibitor that binds to the same hydrophobic pocket as necrostatins [180]. Necrostatins and GNE684 are also known as type III kinase inhibitors [177]. In vitro, GNE684 inhibited TNF-induced necroptosis. Interestingly, GNE684 blocks inflammatory diseases but not tumor growth or metastasis in mouse models [180,181]. Other highly selective type III RIPK1 inhibitors have also been identified [172,182–184], among which GSK2982772 has undergone clinical trials for treating psoriasis, rheumatoid arthritis, and ulcerative colitis [183,185–187]. GSK2982772 is well tolerated with mostly mild adverse events (AEs) and no severe treatment-related AEs were observed [185–187]. GSK2982772 shows promising clinical results in patients with psoriasis [186]. However, in patients with moderate-to-severe rheumatoid arthritis, a regimen of GSK2982772 dosing did not show a meaningful clinical improvement of the disease [187]. Two other RIPK1 inhibitors GSK3145095 and DNL747 have undergone clinical trials in patients with advanced solid tumors and neurodegenerative diseases such as Alzheimer's disease (AD), respectively [177,188].

Potent and selective RIPK3 inhibitors have also been discovered [117,172,177,189,190] (Fig. 1.2). Dabrafenib, an inhibitor of the serine/threonine kinase $BRAF^{V600E}$ that has been approved by the FDA for treating melanomas, is an ATP-competitive RIPK3 inhibitor [172,177]. GSK′843 is another ATP-competitive RIPK3 inhibitor [177]. RIPK3 inhibitors block necroptosis in cell-based studies and systemic inflammation in vivo in mouse models [117,189]. RIPK3 inhibitors at a high concentration can induce apoptosis [117]. This observation along with embryonic lethality of mice with knockin of the $RIPK3^{D161N}$ kinase-dead mutant [128] raises safety concerns for in vivo use of RIPK3 inhibitors. As such, interests in clinical developments of RIPK3 inhibitors are not high [177,181]. No RIPK3 inhibitors have been tested in clinical trials thus far [177].

In a cell-based high throughput screen (HTS) of ~200,000 compounds, an MLKL inhibitor necrosulfonamide (NSA) was discovered. NSA binds covalently to Cys86 located in α-helix 4 of the MLKL N-terminal domain [116]. This HTS also identified an MLKL inhibitor with a xanthine core [116]. A series of related compounds including TC13172 with such a xanthine core potently and selectively inhibit necroptosis [191,192] (Fig. 1.2). Structural studies show that an inhibitor in this class specifically binds to a pocket in the executioner domain of MLKL that locks the protein in the autoinhibitory state, thus preventing oligomerization and necroptosis [192]. A screen of a library of 367 small molecules against the recombinant mouse MLKL pseudokinase domain using a thermal stability shift assay led to the identification of compound 1 (also known as GW806742X) as an MLKL inhibitor, which has similar potency of necroptosis inhibition as NSA [191,193] (Fig. 1.2). Thus far, no MLKL inhibitors have been advanced to the clinic. In addition to induction of necroptosis, emerging evidence suggests MLKL has roles in other biological processes including binding to certain lipid molecules and cross-talk with other cell death pathways [194]. Its roles in several pathologies including inflammatory diseases, type 2 diabetes, and cancer will attract more attentions to the development of MLKL inhibitors for clinical applications [122,194].

3.4 Agents that indirectly affect necroptosis

Inhibition of necroptosis is important for the resolution phase of inflammation in tissue repair. Inflammation in the intestine correlates with increased death of intestinal epithelial cells (IECs). In healthy gut tissues, initial inflammation and damaged epithelia also signal for inflammation cessation and tissue recovery. This recovery phase is compromised in chronic inflammatory bowel diseases, such as Crohn's disease and ulcerative colitis. In these disease conditions, inflammation is not properly resolved, triggering excessive IEC death and barrier dysfunction, perpetuating further inflammation and tissue damage [195]. In IECs, TNF signaling triggers necroptosis via the RIPK1–RIPK3–MLKL pathway [196]. It has been shown that metabolites derived from arachidonic acid such as prostaglandin E2 (PGE2) play important roles in inflammation resolution [195,197,198]. PGE2 is produced before other lipid metabolites important for resolving inflammation [198], suggesting that it has signaling function. A recent study shows that PGE2 suppresses inflammation through its receptor EP4 (encoded the PTGER4 gene). Mechanistically, PGE2-EP4 signaling inhibits necroptosis by suppressing RIPK1 phosphorylation, MLKL oligomerization, and membrane translocation in IECs [199]. A selective, orally bioavailable EP4 receptor agonist inhibits IEC necroptosis thereby promoting inflammation resolution [199]. EP4 agonists have also been shown to ameliorate other inflammatory diseases such as osteoarthritis [200] and tubulointerstitial injury in nephrotoxic serum nephritis [201].

4. Pyroptosis

Infection of host cells by pathogens such as bacteria initiates a form of inflammatory cell death known as pyroptosis to limit infection. Earlier studies show that lytic necrotic cell death occurs in macrophages infected with the human enteroinvasive pathogen *Shigella flexneri* [202,203]. Later studies show that activation of inflammatory caspases 1 (also known as IL-1 converting enzyme, ICE) and 5 in humans and 1 and 11 in mice is critical to this rapid lytic form of cell death [203–207]. In the canonical pyroptotic pathway, activation of caspase 1 occurs in disc-like oligomeric structures known as inflammasomes [208] (Fig. 1.3). When cells are exposed to (a) DAMPs (endogenous danger molecules released from damaged or dying cells such as high-mobility group box 1, S100 proteins, and heat shock proteins, interleukin-1α (IL-1α), and ATP), (b) pathogen-associated molecular patterns (PAMPs, e.g., bacterial lipopolysaccharides (LPSs), viral nucleic acids (double-stranded RNAs, dsRNA), and other microbe-derived molecules), (c) inert structures such as crystalline particulates, or (d) imbalances in intracellular potassium (K^+) concentration, several proteins assemble to form inflammasomes through specific protein–protein interactions [208–210]. Several proteins have been shown to mediate the formation of the inflammasomes. These include three members of the NLR

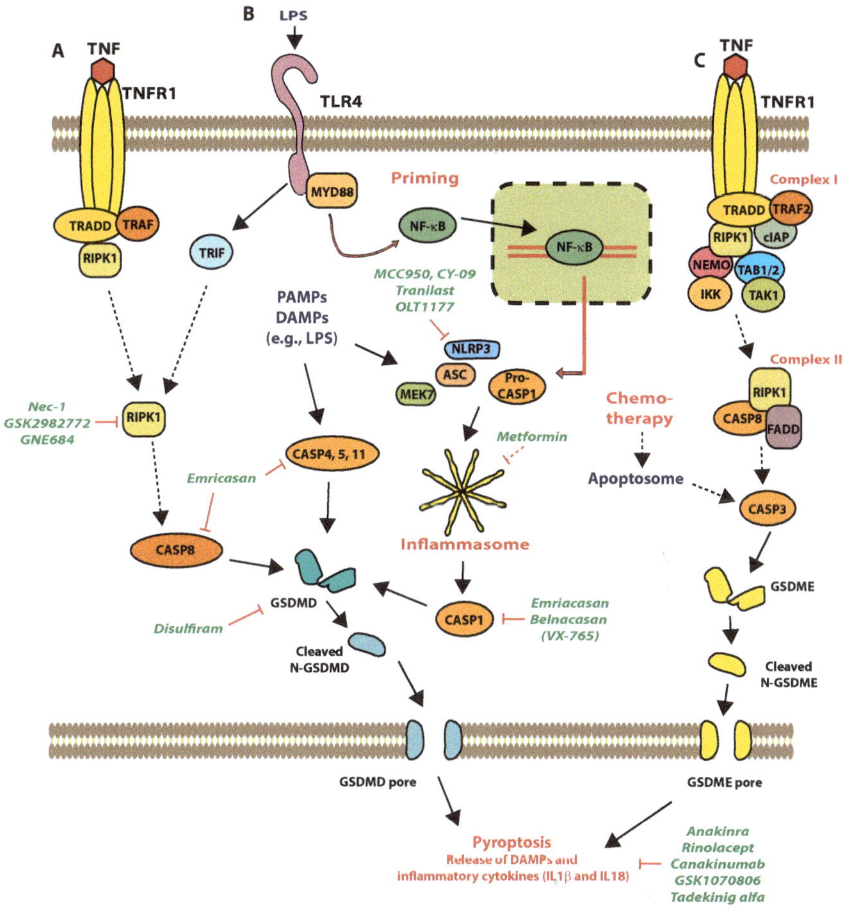

Figure 1.3 ***Mechanisms of pyroptosis, major drug targets, and modulators based on small molecules and biologics.*** (A) TNF/TNFR engagement and active TLR4 signaling activate RIPK1, which leads to the formation of RIPK1-FADD-caspase 8 complex assembled with the Rag-Ragulator complex on lysosomal membrane (not depicted here), thereby activating caspase 8, which cleaves GSDMD to form a pore to induce pyroptosis. (B) Signaling from TLR4, stimulated by, for example, lipopolysaccharides (LPS), activates the NF-κB pathway, which upregulates transcriptionally key components required for inflammasome assembly and activation. This is known as the priming step of the inflammasome activation pathway. In the canonical pathway, NLRP3, ASC, procaspase 1 and NEK7 assemble to form and activate the NLRP3 inflammasome, resulting in caspase 1 autoactivation, GSDMD cleavage, and ensuing GSDMD pore formation and pyroptosis. In the noncanonical pathway, PAMPs and DAMPs bind to and activate human caspases 4 and 5 or mouse caspase 11, which cleave GSDMD to promote pyroptosis. (C) Activation of caspase 3 through TNF signaling, presumably via complex II, and the intrinsic apoptosis pathway induced by chemotherapy, leads to GSDME cleavage (also known as DFNA5) and pyroptosis. Small molecule inhibitors and biologics targeting various proteins in the pyroptotic pathways are depicted in green.

family NLRP1, NLRP3, and NLRC4, as well as several other proteins such as absent in melanoma 2 (AIM2). The NLR family members feature the nucleotide-binding oligomerization domain (NOD) and leucine-rich repeat (LRR). NOD in the NLR proteins is also known as NACHT for domain present in NAIP, CIITA, HET-E, and TP1, which harbors a nucleotide-binding subdomain (NBD).

4.1 NLRP1 inflammasomes

The first described inflammasome comprises NLRP1, caspase 1, and the adaptor protein ASC (apoptosis-associated speck-like protein containing a CARD, encoded by the PYCARD gene), which activates caspase 1 to covert prointerleukin-1β (pro-IL-1β) (encoded by the IL1B gene) to mature IL-1β [211,212]. NLRP1 is a protein with modular structural domains including the N-terminal PYD, followed by NOD, LRR, a function-to-find domain (FIIND), and the C-terminal CARD, akin to the apoptosome scaffold protein APAF1 and the PIDDosome scaffold PIDD (p53-induced protein with a death domain) [213–215]. PYD and CARD are known as death-fold domains that facilitate the oligomerization of these scaffold proteins via homotypic (CARD–CARD or PYD–PYD) interactions [216]. Analogous to apoptosome, inflammasome serves to increase local concentration of caspase 1 for autocleavage and activation. PAMPs such as viral dsRNAs can interact directly with NLRP1 to trigger a conformational change that leads to the release of a C-terminal fragment (due to autoproteolysis within FIIND). The C-terminal fragment containing the UPA (conserved in UNC5, PIDD, and ankyrins) and CARD self-oligomerizes to form a platform for recruiting the adaptor protein ASC and caspase 1, resulting in caspase 1 activation and the release of inflammatory cytokines such as IL-1β and IL-18 [217,218].

4.2 NLRP3 inflammasomes

The NLRP3 inflammasome was discovered in 2004 [219] and has since been extensively studied [220–224]. The NLRP3 protein contains an N-terminal PYD domain, a central NOD, helical domains 1 and 2 (HD1 and HD2) linked with the winged helical domain (WHD) in between, and a C-terminal LRR domain. When stimulated by PAMPs or DAMPs, NLRP3 initially oligomerizes through a homotypic interaction of the NOD/NACHT domain. NLRP3 then interacts with ASC via the homotypic interactions between their PYDs. This nucleates the polymerization of ASC to form prion-like filaments [225,226]. Caspase 1 is recruited to the NLRP3-ASC oligomer via the CARD homotypic interactions between caspase 1 and ASC. Procaspase 1 concentrated in the inflammasome undergoes self-cleavage at the residue D316 in the linker between the large catalytic subunit (p20) and the small catalytic subunit (p10) in the C-terminal region. The cleaved fragments of 33-kDa (containing the N-terminal CARD and the p20 domain) and p10 form a catalytically active complex that remains associated with ASC in the inflammasome. The requirement for linker cleavage for caspase 1 activation is analogous to the activation of executioner caspases 3, 6, and 7 in apoptosis. By contrast, the cleavage of the linker is optional for the activation of the apoptotic initiator caspases such as caspase 9 [50]. For the activated caspase 1, the 33-kDa fragment undergoes further autocleavage at the residue D119, which results in the release of the p20 and p10 heterotetrameric complex and their subsequent degradation. With this, the caspase 1 activity is abolished and the inflammasome activity is terminated [227]. Recent studies show that NIMA (never in mitosis gene a)-related kinase 7 (NEK7) is a component of NLRP3 inflammasome and critical to its activation [222,228–231]. In contrast, NEK7 does not bind to NLRC4 or AIM2. Although its kinase activity is dispensable, the kinase domain directly binds to the LRR domain and the HD2 region of NLRP3 and these interactions are required for NLRP3 inflammasome activation [228,229,231].

Typically, the expression of NLRP3 and other important proinflammatory genes is upregulated in response to PAMPs or DAMPs by TLR-mediated signaling cascade through NF-κB activation (Fig. 1.3B). This process is often referred to as the priming step in NLRP3 inflammasome activation [222,232–234]. Additionally, TRAF6 was shown to mediate nontranscriptional priming of NLRP3 inflammasome formation when TLR/IL-1R senses PAMPs/DAMPs [235]. Priming leads to increased amounts of key components for the assembly of inflammasome. As discussed above, NLRP3 requires a conformational change for oligomerization, which can be triggered by multiple PAMPs/DAMPs in the canonical inflammasome pathway. Other triggers of NLRP3 inflammasome activation include efflux of potassium (K^+) and chloride (Cl^-) ions, intracellular calcium ion (Ca^{2+}) flux, ROS accumulation, the cytosolic presence of mitochondrial DNAs, and lysosomal rupture that releases the protease cathepsin and Ca^{2+} [222,223,236]. Growing evidence shows that lipid metabolism plays important roles in regulating inflammatory response and inflammasome activation [237–240]. For example, cholesterol trafficking to the ER membranes is critical to NLRP3 inflammasome assembly and activation [241].

Although the requirement of caspase 1 activation for pyroptosis was observed in the 1990s [203–207], the final effectors or executors of pyroptotic cell death were unclear until 2015 when the cleavage of GSDMD by human and mouse caspase 1, the mouse caspase 11, and its human orthologs caspases 4 and 5 was shown to be essential to pyroptosis [242–246]. The caspase-cleaved N-terminal fragment of GSDMD oligomerizes and binds to certain acidic lipids such as phosphatidylinositols and cardiolipins in the cytosolic leaflets of the lipid bilayers, resulting in the rapture of plasma membranes [246–250]. A recent study demonstrates that ROS promotes GSDMD oligomerization, pore formation, and pyroptosis [251]. In addition to the classical inflammatory caspases, other caspases including the apoptotic executioner caspase 3 and the initiator caspase 8 can cleave GSDME (DFNA5) and/or GADMD [252–256] (Fig. 1.3).

4.3 NLRC4 inflammasomes

NLRC4 was initially identified as a caspase 1-activating protein related to the apoptosome scaffold APAF1 [257]. NLRC4 comprises an N-terminal CARD domain, the NOD/NACHT domain in the central region, and a C-terminal LRR, thus sharing structural features with NLRP1 and NLRP3. Transient expression of a truncated NLRC4 lacking the C-terminal LRR but not the full-length protein could activate caspase 1 [257], providing the initial evidence that the full-length NLRC4 might exist in an inactive state and thus is unable to activate caspase 1. In 2006, two independent reports show that NLRC4 is activated in response to the presence of the bacterial protein flagellin in the cytoplasm [258,259]. It appears that NLRC4 responds highly specifically to bacterial flagellin and type 3 secretion system needle and rod subunits [208,260–262], in contrast to NLRP1 and NLRP3, which can be activated by diverse PAMPs and DAMPs as discussed above. A structural study indicates that the interactions between the NBD and the WHD in NACHT of NLRC4 maintain NLRC4 in a closed conformation and thus in an autoinhibitory state that is refractory to oligomerization [263]. Upon the binding of the aforementioned bacterial proteins to the sensor protein NAIP (baculoviral IAP repeat-containing protein 1, also known as BIRC1, a member of the IAP family), the ligand-bound NAIP interacts with NLRC4. Once activated, NLRC4 interacts with and activates another quiescent NLRC4 protein in a self-propagating manner to ultimately form an oligomer with a disk-like architecture [262,264–267]. Through homotypic CARD–CARD interactions, the adaptor ASC is recruited to NLRC4, with subsequent binding of caspase 1 to ASC. In this way, caspase 1 is activated, ultimately leading to pyroptosis.

4.4 Noncanonical inflammasomes

The human pyrin protein (encoded by the MEFV gene) consists of an N-terminal PYD domain and a C-terminal B30.2/SPRY domain. Pyrin is a member of the tripartite motif receptors. Certain cytosolic PAMPs such as *Clostridioides difficile* toxin B can inactivate RHOA (Rho GTPase A) due to posttranslational modifications (e.g., glucosylation and ADP ribosylation) [268]. Active RHOA induces the activation of PKN1/2 that phosphorylates pyrin, which binds to 14-3-3 to maintain pyrin in an inactive state. RHOA inactivation by PAMPs prevents pyrin phosphorylation, which enables pyrin to interact with the adaptor protein ASC [268,269] to form the pyrin inflammasome, leading to the recruitment and activation of caspase 1 [268,270].

AIM2 (absence in melanoma 2) and IFI16 (gamma-interferon—inducible protein 16) detect cytosolic double-stranded DNAs (dsDNAs) of microbial and host origins through their hematopoietic interferon-inducible nuclear (HIN, also known as HIN-200) domain (IFI16 contains two HIN domains). The negatively charged dsDNA molecules interact electrostatically with the positively charged HIN, thereby activating AIM2/IFI16 from their autoinhibitory conformation to oligomerize [271,272]. Both AIM2 and IFI16 have an N-terminal PYD domain. The adaptor ASC is then assembled into the oligomers through the PYD:PYD homotypic interaction [226,273—275]. The CARD of ASC in the inflammasomes recruits caspase 1 for its activation and subsequent pyroptosis [267].

4.5 Noncanonical pyroptotic pathway

In 2011, Kayagaki et al. reported that the mouse caspase 11 (the ortholog of human caspases 4 and 5) is critical for caspase 1 activation and mature IL-1β production in macrophages infected with several gram-negative bacterial species such as *Escherichia coli*, *Citrobacter rodentium*, or *Vibrio cholerae*. Caspase 11 does not directly process pro-IL-1β, while caspase 1 does not cleave caspase 11. Thus, caspase 11 likely acts upstream of caspase 1 activation. The authors termed this as a new form, noncanonical inflammasome pathway [276]. They demonstrated that the canonical inflammasome pathway can be activated upon the activation of the noncanonical inflammasome pathway.

The presence of LPS in cytoplasm rather than the LPS receptor TLR4 signaling is required for the activation of this noncanonical pathway [277,278]. A later study shows that the direct binding of cytosolic LPS to the CARD domain of the mouse caspase 11 or human caspases 4 and 5 triggers the oligomerization and activation of these caspases [279]. The activated caspases 11/4/5 then cleave GSDMD to induce pyroptotic cell death [242–244]. Active caspase 11 induces noncanonical activation of NLRP3 possibly by processing pannexin 1 that leads to potassium efflux. Assembly of the NLRP3 inflammasome results in caspase 1 activation and the processing of pro-IL-1β and pro-IL-18, which are then released by GSDMD-induced pyroptosis.

4.6 Inflammasome dysregulation in diseases

A variety of inflammatory diseases are associated with inflammasome dysregulation [221–224,280,281]. Certain germline gain-of-function NLRP1 mutations in humans are associated with several inflammatory diseases such as multiple self-healing palmoplantar carcinoma, familial keratosis lichenoides chronica [282], vitiligo [283,284], autoinflammation with arthritis and dyskeratosis [285,286], and juvenile-onset recurrent respiratory papillomatosis [287]. These mutations cause constitutive NLRP1 inflammasome activation and ensuing pyroptosis [283,285], resulting in persistent inflammation. Mutations in MEFV encoding pyrin are associated with a human autoinflammatory disease known as familial Mediterranean fever (FMF) [288,289]. Genetic variants in the human CASP1 gene that encodes caspase 1 with reduced protease activity and IL-1β production are also associated with the development of autoinflammatory disease [290–293].

Aberrant NLRP3 inflammasome activation is linked to diverse diseases including neurological disorders such as AD [294–298], metabolic disorders such as type 2 diabetes [299–301], autoimmune diseases [302], and cardiovascular maladies [303,304]. Gain-of-function NLRP3 mutations are also associated with autoinflammatory conditions, such as familial cold autoinflammatory syndrome, Muckle–Wells syndrome (MWS), and chronic infantile neurological cutaneous and articular syndrome (also known as neonatal-onset multisystem inflammatory disorder, NOMID) [305–308]. Consistent with a key role for NEK7 in NLRP3 inflammasome activation, it is required for caspase 1 activation by the R258W NLRP3 mutant. R258W in NLRP3 is a gain-of-function mutation that causes MWS [228]. The group of diseases associated with NLRP3 gain-of-function mutations are collectively known as cryopyrin-associated periodic syndrome (CAPS). Additionally, the TNFα-caspase 3-GSDME pyroptotic pathway is linked to the initiation of ureteral obstruction–induced renal tubule injury, which subsequentially leads to the late-stage progression of hydronephrosis, inflammation, and fibrosis in obstructive nephropathy [309–312].

Inflammasome activation represents an important mechanism of host antimicrobial response [313–315]. Active GSDMD-mediated pyroptosis confers host antimicrobial defense by trapping intracellular pathogens in pyroptotic cell debris and subsequent clearance by phagocytes and neutrophil extracellular traps (NETs) consisting of web-like chromatin structures and concomitant cell death known as NETosis [316–318]. The cleaved IL-1β, and other potent proinflammatory cytokines, can be released from lysed pyroptotic cells or through GSDMD pores from live macrophages [319–321], which have further effects in controlling microbial infection and immune regulation [322,323]. A recent study links RIPK1 kinase to caspase 8-mediated GSDME cleavage, neutrophil pyroptosis, and IL-1β release, which can limit Yersinia (gram-negative coccobacilli bacteria) infection [324]. Both canonical and noncanonical inflammasome pathways can be activated by viral infection. As noted above in the discussion of necroptosis, the presence of a high level of proinflammatory cytokines in the bloodstream can cause excessive inflammatory responses called cytokine storm in severe COVID-19 cases [325]. The activation of NLRP3 inflammasome pathway appears to play an important role in inflammation induced by SARS-CoV-2 infection [223,326–331]. In severe COVID-19 patients, NLRP3 inflammasome is activated in peripheral blood mononuclear cells and tissues of postmortem patients upon autopsy. Molecules produced due to inflammasome activation, such as caspase 1 p20 and mature IL-18, are found in patient sera along with certain markers of COVID-19 severity (e.g., IL-6 and LDH). Concordantly, higher levels of IL-18 and caspase 1 p20 correlate with COVID-19 disease severity and poor clinical outcome [328].

Inflammasome activation plays both tumor promoting and suppressive roles [4,16,281,332–340]. Indeed, it has been shown that inflammasome activation in the TMEs has both pro- or anticancer effects. It is reported that NLRP3 inflammasome activation in cancer-associated fibroblasts (CAFs) can promote breast tumor growth and metastatic progression. IL-1β derived from CAFs in the TME facilitates the recruitment of myeloid-derived suppressor cells (MDSCs) [341]. Consistently, chemotherapy-induced NLRP3 inflammasome activation in MDSCs and the subsequent secretion of IL-1β promote an immunosuppressive TME, thereby blunting the therapeutic efficacy of gemcitabine and 5-fluorouracil in a mouse model—based study [342]. IL-1β blockade reverses immunosuppression and enhances anti-PD-1 immunotherapy in mouse breast cancer models [343]. The NLRP3 inflammasome was shown to promote metastasis via the lymphatics downstream of the sphingolipid sphingosine-1-phosphate receptor signaling in tumor-associated macrophages (TAMs) in breast cancer. In human breast cancers, stromal TAMs with NLRP3 expression correlate with disease aggressiveness, metastasis, and poor survival [344]. In multiple myelomas, NLRP3 inflammasome activation in TAMs is associated with cancer progression [345]. On the other hand, NLRP3 inflammasome activation in Kupffer cells promotes the release of IL-18, which enhances maturation of hepatic natural killer cells (NKs), and the induction of FAS-mediated extrinsic apoptotic cell death of tumor cells in a mouse model of liver metastasis of colorectal cancer [346]. It has been shown that the tumor suppressor PTEN in myeloid cells directly interacts with and dephosphorylates NLRP3, thus promoting NLRP3 inflammasome assembly and activation. The secreted IL-1β enhances antitumor immunity and chemotherapy sensitivity. Therefore, PTEN in the stromal myeloid cells appears to determine treatment response to chemotherapy by promoting NLRP3-dependent antitumor immunity [347]. The antitumor effects of IL-1β have been observed in other studies [348—350]. Caspase 3 activation induced by chemotherapy cleaves GSDME [252] (Fig. 1.3C). Notably, reduced GSDME levels correlate with poor survival in patients with breast cancer [351—353]. Cancer-associated GSDME mutations compromise GSDME function [354,355]. Importantly, GSDME expression in tumor cells promotes antitumor immunity by increasing the phagocytosis of tumor cells by TAMs and the number and functions of NKs and CD8(+) T cells [354]. Another study shows that induction of pyroptosis promotes T cells-mediated tumor regression in a syngeneic model of mouse 4T1 breast cancer [356]. The concept of activation of the NLRP3 inflammasome for cancer therapy has been translated into the clinic. Currently, the novel small molecule NLRP3 agonist BMS-986299 is being evaluated in a phase I clinical trial as monotherapy and in combination with the immuno-checkpoint blockers nivolumab and ipilimumab for treating advanced solid tumors (ClinicalTrials.gov Identifier: NCT03444753).

4.7 Targeting inflammasome for treating diseases

Because of the association of inflammasome activation with a variety of serious human diseases ranging from severe systemic infections, and chronic inflammatory disorders to metabolic, cardiovascular, and neurodegenerative diseases, there have been markedly increased efforts in developing therapeutics to modulate inflammasome activation and limit deleterious effects of released DAMPs in inflammatory responses in recent years [221,222,357–360].

A focal point for developing therapeutics targeting the inflammasomes is the ATP-binding pockets of NLRP3 and related inflammasome organizers [359,361]. Several diarylsulfonylurea-containing compounds were discovered as new inhibitors of IL-1β processing in 2001 [362]. In 2009, Lamkani et al. identified the sulfonylureas compound glyburide that has been used for treating type 2 diabetes as a selective inhibitor of NLRP3 inflammasome activation [363]. Glyburide is structurally related to diarylsulfonylureas. In 2015, one of the diarylsulfonylurea compounds MCC950 (also known as CRID3) was identified as the most potent inhibitors of the NLRP3 inflammasome with an IC_{50} value of ~20 nM [364]. MCC950 specifically inhibits NLRP3 activation without affecting NLRP1, NLRC4, and AIM2 inflammasome activation. MCC950 reduces IL-1β levels in vivo and disease severity of autoimmune symptom in a mouse model of multiple sclerosis. It also prevents neonatal lethality in a mouse model of CAPS. In a recent study, MCC950 was shown to directly target the NLRP3 ATP hydrolysis motif to block inflammasome activation [365].

In a screen for inhibitors of NLRP3 inflammasome, the cystic fibrosis transmembrane conductance regulator (CFTR) channel inhibitor C172 was identified [366]. A C172 analog CY-09 that does not inhibit CTFR activity directly binds to the Walker A motif within the central NLRP3 NOD/NACHT, thus interfering with the ATPase activity of NLRP3. CY-09 does not inhibit the priming phase of NLRP3 inflammasome activation. CY-09 attenuates IL-1β release and exhibits therapeutic effects in mouse models of CAPS and type 2 diabetes [366]. CY-09 shows IC_{50} values in the micromolar concentrations and further optimization to increase inhibitory potency against NLRP3 is anticipated.

Tranilast (TR), a tryptophan derivative, has proven antiinflammatory effects and been used as an antiallergic drug. It binds directly to the NOD/NACHT domain of NLRP3, inhibits inflammasome activation, and exhibits beneficial effects for NLRP3-driven diseases [222,359,361,367].

Because of its proven safety profile and clinical antiinflammatory efficacy as well as the identification of NLRP3 as its target, TR has undergone clinical trials for treating patients with various inflammation-related diseases including CAPS.

Other inhibitors of NLRP3 inflammasome such as parthenolide, Bay 11−7082, 3,4-methylenedioxy-β-nitrostyrene (MNS), oridonin, and OLT1177 have also been identified [358,368−372], although these compounds appear to have additional targets in the inflammatory pathways [221,359]. Metformin, a major medication for type 2 diabetes, has recently been shown to inhibit ATP and DNA synthesis in mitochondria and indirectly blocks NLRP3 inflammasome activation and pulmonary inflammation [329] (Fig. 1.3).

Inflammatory caspases are important targets for developing therapeutics for treating inflammatory diseases [359,373]. The caspase 1 inhibitor VX-765 (belnacasan) prevents glial inflammasome activation, and pyroptosis in models of multiple sclerosis as well as other models of neuroinflammation and neurodegenerative diseases [374−377]. For example, caspase 1 inhibition has been shown to improve cognitive impairment and neuropathology in a mouse model of AD [378]. Nonetheless, previous clinical trials of caspase 1 inhibitors have been disappointing [359]. The pan-caspase inhibitor emricasan (IDN-6556/PF 03491390) [98,99] inhibits caspase 3-dependent apoptosis [379] and is being tested in clinical trials for treating liver diseases such as cirrhosis and nonalcoholic steatohepatitis (NASH) [100−103].

Interestingly, a recent study shows that the FDA-approved drug disulfiram covalently binds to Cys191 (mouse Cys192) in GSDMD and specifically blocks GSDMD pore formation at a nanomolar concentration. Importantly, disulfiram does not inhibit IL-1β and GSDMD processing by caspases [380], suggesting disulfiram is a highly potent and specific inhibitor of GSDMD to block IL-1β release and pyroptosis. Notably, the Cys-reactive covalent MLKL inhibitor NSA discussed above [116] (Fig. 1.2) also binds directly to full-length GSDMD via Cys191 and inhibits pyroptosis [381]. However, NSA and Bay 11-7802, another inhibitor of NLRP3 inflammasome, were shown to inhibit proteolytic processing of caspase 1 and pro-IL-1β [380].

As discussed above, in the priming step of NLRP3 inflammasome pathway, NF-κB is activated to upregulate NLRP3 and other genes in the inflammatory response pathways (Fig. 1.3B). There are many upstream signaling pathways such as TLRs, TNFR, and cytokine receptors that can lead to NF-κB activation. Several compounds described above such as Bay 11-7082 and parthenolide also inhibit the priming step [359]. The biology NF-κB pathway and agents that target it have been reviewed extensively [382–387].

In addition to small molecule inhibitors targeting key components of the inflammasome activation pathways, blocking or neutralizing biologics targeting IL-1β have been widely explored to treat inflammatory disorders including rheumatic diseases (gout, pseudogout, systemic juvenile idiopathic arthritis, and Still's disease) and hereditary autoinflammatory conditions such as CAPS and FMF as well as in oncology [212,388–390]. Pro-IL-1β is transcriptionally upregulated in immune cells of myeloid lineage upon stimulation by inflammatory cues. As discussed above, pro-IL-1β is cleaved by the activated caspase 1 in the inflammasomes and the active IL-1β is then secreted and binds to IL-1 receptor 1 (IL-1R1) on target cells to trigger heterodimerization of IL-1R1 with the IL-1R coreceptor IL1RAP (also known as IL-1RAcP), thereby activating a signal transduction pathway leading to the activation of the NF-κB and MAPK pathways to propagate inflammatory activation. There are two intrinsic mechanisms to prevent unnecessary IL-1-mediated IL-1R activation: (a) cells also express IL-1 receptor antagonist IL-1Ra (encoded by IL1RN), which binds IL-1R1 and prevents it from dimerizing with IL1RAP; (b) the IL-1R1 decoy receptor known as IL-1R2 that lacks the cytoplasmic TIR (the Toll/interleukin-1 receptor) domain is also expressed. Although IL-1α and IL-1β can bind to IL-1R2, this engagement cannot lead to downstream inflammatory signaling, as TIR is required for assembling the cytoplasmic signaling complex. There are several FDA-approved IL-1β blocking biologics, including a recombinant IL-1Ra (anakinra), the monoclonal antibody against IL-1β (canakinumab), and a recombinant fusion protein consisting of the ligand-binding regions of IL-1R1 and IL1ARP linked to the Fc region of human IgG1 (rilonacept). Anakinra was initially approved by the FDA in 2001 for the treatment of moderate-to-severe active rheumatoid arthritis, and then for NOMID in 2013. In 2000, anakinra was approved for treating patients with IL-1Ra deficiency due to IL1RN mutation, an ultrarare autoinflammatory disease. Canakinumab is an FDA-approved medication for treating several inflammatory diseases including CAPS, TNFR-associated periodic syndrome, hyperimmunoglobulin D syndrome/mevalonate kinase deficiency, and FMF, and active Still's disease. Rilonacept has also been approved by the FDA for treating CAPS [391] and recurrent pericarditis, a condition characterized by inflammation of the pericardium [392]. Antiinflammatory therapy with canakinumab targeting IL-1β significantly reduces the rate of recurrent cardiovascular events, independent of lipid-level lowering effects [212]. Canakinumab use in patients with atherosclerosis also appears to significantly reduce incident lung cancer and lung cancer—related mortality [393], suggesting that blocking IL-1β-mediated inflammation may prevent lung cancer development. However, IL-1β blockade therapy is associated with a higher incidence of fatal infection and sepsis.

In the COVID-19 pandemic, the IL-1 targeting biologics have been tested for improving survival of critically ill COVID-19 patients. Anakinra appears effective in reducing clinical signs of hyperinflammation in COVID-19 patients with critical symptoms [394,395]. However, in a randomized clinical trial, canakinumab does not appear to improve survival of patients with severe COVID-19 [396].

GSK1070806 is a humanized monoclonal antibody against IL-18, whereas Tadekinig alfa is a soluble recombinant IL-18 binding protein (Fig. 1.3). GSK1070806 has been tested in clinical trials for treating type 2 diabetes, and delayed graft function in patients after kidney transplantation [397,398] as well as in patients with moderate-to-severe Crohn's disease (ClinicalTrials.gov Identifier: NCT03681067). Tadekinig alfa has been tested in clinical studies for treating conditions with elevated IL-18 levels such as adult-onset Still's disease [399] and patients with NLRC4-macrophage activation syndrome mutation or X-linked inhibitor of apoptosis (XIAP) deficiency [400] (ClinicalTrials.gov Identifier: NCT03113760).

5. Concluding remarks

As our understanding of the molecular mechanisms underlying apoptosis, necroptosis, and pyroptosis has become increasingly sophisticated, more potent and specific agents modulating these forms of RCD are anticipated in the near future. Drug discovery efforts on targeting the apoptosis pathways have mainly been focused on BCL-2 family proteins, IAPs, and PARAs thus far. Other emerging targets such as FLIP and VDAC will attract more attention. The complexity of pyroptosis pathway and its extensive connection to many human diseases provide ample opportunities for drug discovery and development. While NLRP3 inhibitors/agonists have been discovered and some of them have entered clinical trials, medicinal chemistry optimization of known compounds and discovery of new compounds will continue to drive research in this area. Few modulators of other inflammasome pathways have been described. These less explored targets will undoubtedly become subjects of active investigation. The immunogenic nature of necroptosis and pyroptosis has been increasingly explored for enhancing current cancer immunotherapy [16].

There is an increasing recognition that different modes of RCD are interconnected [2,3,401]. Activated caspases including the initiator caspase 8 and the executioner caspase 3 in apoptosis have been shown to cleave GSDMD and GSDME, respectively, thus promoting pyroptosis [252–256]. A recent study shows that TNF/TNFR signaling promotes the assembly of the RIPK1-FADD-caspase 8 complex along with the Rag-Ragulator complex on lysosomal membrane for caspase 8 activation and GSDMD cleavage to induce pyroptosis [256] (Fig. 1.3). Likewise, activated caspase 1 during pyroptosis can cleave and activate caspases 3 and 7 and convert BID to tBID to promote apoptosis [2]. While necroptosis and pyroptosis are better known to control pathogen infections [78,313], apoptosis also plays an important role in limiting pathogen infections [2,402]. The intertwined nature of apoptosis, necroptosis, and pyroptosis will be increasingly explored for therapeutic development and applications.

References

[1] D. Tang, et al., The molecular machinery of regulated cell death, Cell Res. 29 (5) (2019) 347–364.

[2] S. Bedoui, M.J. Herold, A. Strasser, Emerging connectivity of programmed cell death pathways and its physiological implications, Nat. Rev. Mol. Cell Biol. 21 (11) (2020) 678–695.

[3] D. Bertheloot, E. Latz, B.S. Franklin, Necroptosis, pyroptosis and apoptosis: an intricate game of cell death, Cell. Mol. Immunol. 18 (5) (2021) 1106–1121.

[4] E. Koren, Y. Fuchs, Modes of regulated cell death in cancer, Cancer Discov. 11 (2) (2021) 245–265.

[5] J.F. Kerr, Shrinkage necrosis: a distinct mode of cellular death, J. Pathol. 105 (1) (1971) 13–20.

[6] J.F. Kerr, A.H. Wyllie, A.R. Currie, Apoptosis: a basic biological phenomenon with wide-ranging implications in tissue kinetics, Br. J. Cancer 26 (4) (1972) 239–257.

[7] D.R. Green, L. Galluzzi, G. Kroemer, Cell biology. Metabolic control of cell death, Science 345 (6203) (2014) 1250256.

[8] C.B. Medina, K.S. Ravichandran, Do not let death do us part: 'find-me' signals in communication between dying cells and the phagocytes, Cell Death Differ. 23 (6) (2016) 979–989.

[9] P.M. Henson, Cell removal: efferocytosis, Annu. Rev. Cell Dev. Biol. 33 (2017) 127–144.

[10] D.R. Green, The coming decade of cell death research: five riddles, Cell 177 (5) (2019) 1094–1107.
[11] C.B. Medina, et al., Metabolites released from apoptotic cells act as tissue messengers, Nature 580 (7801) (2020) 130–135.
[12] E. Boada-Romero, et al., The clearance of dead cells by efferocytosis, Nat. Rev. Mol. Cell Biol. 21 (7) (2020) 398–414.
[13] A.D. Garg, P. Agostinis, Cell death and immunity in cancer: from danger signals to mimicry of pathogen defense responses, Immunol. Rev. 280 (1) (2017) 126–148.
[14] L. Galluzzi, et al., Immunogenic cell death in cancer and infectious disease, Nat. Rev. Immunol. 17 (2) (2017) 97–111.
[15] H. Anderton, I.P. Wicks, J. Silke, Cell death in chronic inflammation: breaking the cycle to treat rheumatic disease, Nat. Rev. Rheumatol. 16 (9) (2020) 496–513.
[16] S.R. Rosenbaum, N.A. Wilski, A.E. Aplin, Fueling the fire: inflammatory forms of cell death and implications for cancer immunotherapy, Cancer Discov. 11 (2) (2021) 266–281.
[17] L. Galluzzi, et al., Molecular mechanisms of cell death: recommendations of the nomenclature committee on cell death 2018, Cell Death Differ. 25 (3) (2018) 486–541.
[18] X.M. Hu, et al., Guidelines for regulated cell death assays: a systematic summary, a categorical comparison, a prospective, Front. Cell Dev. Biol. 9 (2021) 634690.
[19] Y. Fuchs, H. Steller, Programmed cell death in animal development and disease, Cell 147 (4) (2011) 742–758.
[20] F.F.S. Ke, et al., Embryogenesis and adult life in the absence of intrinsic apoptosis effectors BAX, BAK, and BOK, Cell 173 (5) (2018) 1217–1230 e17.
[21] T.T.M. Nguyen, G. Gillet, N. Popgeorgiev, Caspases in the developing central nervous system: apoptosis and beyond, Front. Cell Dev. Biol. 9 (2021) 702404.
[22] C.V. Rothlin, et al., TAM receptor signaling in immune homeostasis, Annu. Rev. Immunol. 33 (2015) 355–391.
[23] S.J. Riedl, Y. Shi, Molecular mechanisms of caspase regulation during apoptosis, Nat. Rev. Mol. Cell Biol. 5 (11) (2004) 897–907.
[24] R.C. Taylor, S.P. Cullen, S.J. Martin, Apoptosis: controlled demolition at the cellular level, Nat. Rev. Mol. Cell Biol. 9 (3) (2008) 231–241.
[25] D. Hanahan, R.A. Weinberg, Hallmarks of cancer: the next generation, Cell 144 (5) (2011) 646–674.
[26] J. Chai, et al., Structural and biochemical basis of apoptotic activation by Smac/DIABLO, Nature 406 (6798) (2000) 855–862.
[27] L. Li, et al., A small molecule Smac mimic potentiates TRAIL- and TNFalpha-mediated cell death, Science 305 (5689) (2004) 1471–1474.
[28] T. Oltersdorf, et al., An inhibitor of Bcl-2 family proteins induces regression of solid tumours, Nature 435 (7042) (2005) 677–681.
[29] C. Billard, BH3 mimetics: status of the field and new developments, Mol. Cancer Therapeut. 12 (9) (2013) 1691–1700.
[30] A.R. Delbridge, et al., Thirty years of BCL-2: translating cell death discoveries into novel cancer therapies, Nat. Rev. Cancer 16 (2) (2016) 99–109.
[31] A. Ashkenazi, et al., From basic apoptosis discoveries to advanced selective BCL-2 family inhibitors, Nat. Rev. Drug Discov. 16 (4) (2017) 273–284.
[32] B.L. Lampson, M.S. Davids, The development and current use of BCL-2 inhibitors for the treatment of chronic lymphocytic leukemia, Curr. Hematol. Malig. Rep. 12 (1) (2017) 11–19.
[33] R.W. Birkinshaw, et al., Structures of BCL-2 in complex with venetoclax reveal the molecular basis of resistance mutations, Nat. Commun. 10 (1) (2019) 2385.

[34] A.D. Cowan, et al., BAK core dimers bind lipids and can be bridged by them, Nat. Struct. Mol. Biol. 27 (11) (2020) 1024–1031.
[35] E. Morrish, G. Brumatti, J. Silke, Future therapeutic directions for Smac-mimetics, Cells 9 (2) (2020).
[36] X.Y. Zhao, et al., Potency and selectivity of SMAC/DIABLO mimetics in solid tumor therapy, Cells 9 (4) (2020).
[37] R. Abbas, S. Larisch, Targeting XIAP for promoting cancer cell death-the story of ARTS and SMAC, Cells 9 (3) (2020).
[38] C. Hamilton, et al., Therapeutics targeting the core apoptotic machinery, Cancers 13 (11) (2021).
[39] A. Ashkenazi, Directing cancer cells to self-destruct with pro-apoptotic receptor agonists, Nat. Rev. Drug Discov. 7 (12) (2008) 1001–1012.
[40] A. Ashkenazi, Targeting the extrinsic apoptotic pathway in cancer: lessons learned and future directions, J. Clin. Invest. 125 (2) (2015) 487–489.
[41] Y.P. Ow, et al., Cytochrome c: functions beyond respiration, Nat. Rev. Mol. Cell Biol. 9 (7) (2008) 532–542.
[42] L. Galluzzi, et al., Molecular definitions of cell death subroutines: recommendations of the nomenclature committee on cell death 2012, Cell Death Differ. 19 (1) (2012) 107–120.
[43] F. Llambi, et al., BOK is a non-canonical BCL-2 family effector of apoptosis regulated by ER-associated degradation, Cell 165 (2) (2016) 421–433.
[44] S.W. Tait, D.R. Green, Mitochondria and cell death: outer membrane permeabilization and beyond, Nat. Rev. Mol. Cell Biol. 11 (9) (2010) 621–632.
[45] P.E. Czabotar, et al., Control of apoptosis by the BCL-2 protein family: implications for physiology and therapy, Nat. Rev. Mol. Cell Biol. 15 (1) (2014) 49–63.
[46] P.D. Bhola, A. Letai, Mitochondria-judges and executioners of cell death sentences, Mol. Cell 61 (5) (2016) 695–704.
[47] R. Singh, A. Letai, K. Sarosiek, Regulation of apoptosis in health and disease: the balancing act of BCL-2 family proteins, Nat. Rev. Mol. Cell Biol. 20 (3) (2019) 175–193.
[48] X. Wang, The expanding role of mitochondria in apoptosis, Genes Dev. 15 (22) (2001) 2922–2933.
[49] D. Acehan, et al., Three-dimensional structure of the apoptosome: implications for assembly, procaspase-9 binding, and activation, Mol. Cell 9 (2) (2002) 423–432.
[50] S.J. Riedl, G.S. Salvesen, The apoptosome: signalling platform of cell death, Nat. Rev. Mol. Cell Biol. 8 (5) (2007) 405–413.
[51] L. Bouchier-Hayes, D.R. Green, Caspase-2: the orphan caspase, Cell Death Differ. 19 (1) (2012) 51–57.
[52] Y. Lim, L. Dorstyn, S. Kumar, The p53-caspase-2 axis in the cell cycle and DNA damage response, Exp. Mol. Med. 53 (4) (2021) 517–527.
[53] A.J. Souers, et al., ABT-199, a potent and selective BCL-2 inhibitor, achieves antitumor activity while sparing platelets, Nat. Med. 19 (2) (2013) 202–208.
[54] D.R. Green, A BH3 mimetic for killing cancer cells, Cell 165 (7) (2016) 1560.
[55] S. Khan, et al., A selective BCL-XL PROTAC degrader achieves safe and potent antitumor activity, Nat. Med. (2019).
[56] R. Kolb, et al., Proteolysis-targeting chimera against BCL-XL destroys tumor-infiltrating regulatory T cells, Nat. Commun. 12 (1) (2021) 1281.
[57] A. Mullard, Targeted protein degraders crowd into the clinic, Nat. Rev. Drug Discov. 20 (4) (2021) 247–250.

[58] Y. Suzuki, Y. Nakabayashi, R. Takahashi, Ubiquitin-protein ligase activity of X-linked inhibitor of apoptosis protein promotes proteasomal degradation of caspase-3 and enhances its anti-apoptotic effect in Fas-induced cell death, Proc. Natl. Acad. Sci. USA 98 (15) (2001) 8662–8667.
[59] A.J. Schile, M. Garcia-Fernandez, H. Steller, Regulation of apoptosis by XIAP ubiquitin-ligase activity, Genes Dev. 22 (16) (2008) 2256–2266.
[60] B. Dumetier, A. Zadoroznyj, L. Dubrez, IAP-mediated protein ubiquitination in regulating cell signaling, Cells 9 (5) (2020).
[61] M. Gyrd-Hansen, et al., IAPs contain an evolutionarily conserved ubiquitin-binding domain that regulates NF-kappaB as well as cell survival and oncogenesis, Nat. Cell Biol. 10 (11) (2008) 1309–1317.
[62] E.C. Dueber, et al., Antagonists induce a conformational change in cIAP1 that promotes autoubiquitination, Science 334 (6054) (2011) 376–380.
[63] J. Lopez, et al., CARD-mediated autoinhibition of cIAP1's E3 ligase activity suppresses cell proliferation and migration, Mol. Cell 42 (5) (2011) 569–583.
[64] R. Budhidarmo, C.L. Day, IAPs: modular regulators of cell signalling, Semin. Cell Dev. Biol. 39 (2015) 80–90.
[65] R. Jia, J.S. Bonifacino, Negative regulation of autophagy by UBA6-BIRC6-mediated ubiquitination of LC3, Elife 8 (2019) e50034.
[66] S. Fulda, D. Vucic, Targeting IAP proteins for therapeutic intervention in cancer, Nat. Rev. Drug Discov. 11 (2) (2012) 109–124.
[67] S. Fulda, Smac mimetics as IAP antagonists, Semin. Cell Dev. Biol. 39 (2015) 132–138.
[68] Y. Huang, et al., Structural basis of caspase inhibition by XIAP: differential roles of the linker versus the BIR domain, Cell 104 (5) (2001) 781–790.
[69] S. Wang, Design of small-molecule Smac mimetics as IAP antagonists, Curr. Top. Microbiol. Immunol. 348 (2011) 89–113.
[70] F. Cossu, et al., Targeting the BIR domains of inhibitor of apoptosis (IAP) proteins in cancer treatment, Comput. Struct. Biotechnol. J. 17 (2019) 142–150.
[71] H. Zhu, et al., Bivalent SMAC mimetics for treating cancer by antagonizing inhibitor of apoptosis proteins, ChemMedChem 14 (23) (2019) 1951–1962.
[72] A.M. Verhagen, et al., Identification of DIABLO, a mammalian protein that promotes apoptosis by binding to and antagonizing IAP proteins, Cell 102 (1) (2000) 43–53.
[73] A.M. Verhagen, et al., Identification of mammalian mitochondrial proteins that interact with IAPs via N-terminal IAP binding motifs, Cell Death Differ. 14 (2) (2007) 348–357.
[74] H. Sun, et al., Structure-based design of potent, conformationally constrained Smac mimetics, J. Am. Chem. Soc. 126 (51) (2004) 16686–16687.
[75] H. Sun, et al., Design of small-molecule peptidic and nonpeptidic Smac mimetics, Acc. Chem. Res. 41 (10) (2008) 1264–1277.
[76] S.M. Condon, et al., Birinapant, a Smac-mimetic with improved tolerability for the treatment of solid tumors and hematological malignancies, J. Med. Chem. 57 (9) (2014) 3666–3677.
[77] D.E. Christofferson, Y. Li, J. Yuan, Control of life-or-death decisions by RIP1 kinase, Annu. Rev. Physiol. 76 (2014) 129–150.
[78] F.K. Chan, N.F. Luz, K. Moriwaki, Programmed necrosis in the cross talk of cell death and inflammation, Annu. Rev. Immunol. 33 (2015) 79–106.
[79] B. Shan, et al., Necroptosis in development and diseases, Genes Dev. 32 (5–6) (2018) 327–340.
[80] S. Fulda, Promises and challenges of Smac mimetics as cancer therapeutics, Clin. Cancer Res. 21 (22) (2015) 5030–5036.

[81] G.A. Ward, et al., ASTX660, a novel non-peptidomimetic antagonist of cIAP1/2 and XIAP, potently induces TNFalpha-dependent apoptosis in cancer cell lines and inhibits tumor growth, Mol. Cancer Therapeut. 17 (7) (2018) 1381–1391.
[82] A. Ashkenazi, et al., Safety and antitumor activity of recombinant soluble Apo2 ligand, J. Clin. Invest. 104 (2) (1999) 155–162.
[83] F. Gonzalvez, A. Ashkenazi, New insights into apoptosis signaling by Apo2L/TRAIL, Oncogene 29 (34) (2010) 4752–4765.
[84] D. de Miguel, et al., Onto better TRAILs for cancer treatment, Cell Death Differ. 23 (5) (2016) 733–747.
[85] S. von Karstedt, A. Montinaro, H. Walczak, Exploring the TRAILs less travelled: TRAIL in cancer biology and therapy, Nat. Rev. Cancer 17 (6) (2017) 352–366.
[86] L. Pukac, et al., HGS-ETR1, a fully human TRAIL-receptor 1 monoclonal antibody, induces cell death in multiple tumour types in vitro and in vivo, Br. J. Cancer 92 (8) (2005) 1430–1441.
[87] C. Adams, et al., Structural and functional analysis of the interaction between the agonistic monoclonal antibody Apomab and the proapoptotic receptor DR5, Cell Death Differ. 15 (4) (2008) 751–761.
[88] H.M. Rosevear, A.J. Lightfoot, T.S. Griffith, Conatumumab, a fully human mAb against death receptor 5 for the treatment of cancer, Curr. Opin. Invest. Drugs 11 (6) (2010) 688–698.
[89] G.V. Georgakis, et al., Activity of selective fully human agonistic antibodies to the TRAIL death receptors TRAIL-R1 and TRAIL-R2 in primary and cultured lymphoma cells: induction of apoptosis and enhancement of doxorubicin- and bortezomib-induced cell death, Br. J. Haematol. 130 (4) (2005) 501–510.
[90] A. Yada, et al., A novel humanized anti-human death receptor 5 antibody CS-1008 induces apoptosis in tumor cells without toxicity in hepatocytes, Ann. Oncol. 19 (6) (2008) 1060–1067.
[91] D.C. Phillips, et al., Hexavalent TRAIL fusion protein Eftozanermin alfa optimally clusters apoptosis-inducing TRAIL receptors to induce on-target antitumor activity in solid tumors, Cancer Res. 81 (12) (2021) 3402–3414.
[92] M.J. Ratain, et al., Phase 1, first-in-human study of TRAIL receptor agonist fusion protein ABBV-621, J. Clin. Oncol. 37 (15) (2019) 3013.
[93] M.J.A. de Jonge, et al., First-in-Human study of Abbv-621, a TRAIL receptor agonist fusion protein, in patients (Pts) with relapsed/refractory (RR) acute myeloid leukemia (AML) and diffuse large B-cell lymphoma (DLBCL), Blood 134 (Suppl. 1) (2019) 3924.
[94] Y. Tsuchiya, O. Nakabayashi, H. Nakano, FLIP the switch: regulation of apoptosis and necroptosis by cFLIP, Int. J. Mol. Sci. 16 (12) (2015) 30321–30341.
[95] L. Humphreys, M. Espona-Fiedler, D.B. Longley, FLIP as a therapeutic target in cancer, FEBS J. 285 (22) (2018) 4104–4123.
[96] M.A. Hughes, et al., Co-operative and hierarchical binding of c-FLIP and caspase-8: a unified model defines how c-FLIP isoforms differentially control cell fate, Mol. Cell 61 (6) (2016) 834–849.
[97] H.I. Muendlein, et al., cFLIPL protects macrophages from LPS-induced pyroptosis via inhibition of complex II formation, Science 367 (6484) (2020) 1379–1384.
[98] S.D. Linton, et al., First-in-class pan caspase inhibitor developed for the treatment of liver disease, J. Med. Chem. 48 (22) (2005) 6779–6782.
[99] S.D. Linton, Caspase inhibitors: a pharmaceutical industry perspective, Curr. Top. Med. Chem. 5 (16) (2005) 1697–1717.
[100] C.T. Frenette, et al., Emricasan improves liver function in patients with cirrhosis and high model for end-stage liver disease scores compared with placebo, Clin. Gastroenterol. Hepatol. 17 (4) (2019) 774–783 e4.

[101] S.A. Harrison, et al., A randomized, placebo-controlled trial of emricasan in patients with NASH and F1-F3 fibrosis, J. Hepatol. 72 (5) (2020) 816–827.
[102] G. Garcia-Tsao, et al., Randomized placebo-controlled trial of emricasan for non-alcoholic steatohepatitis-related cirrhosis with severe portal hypertension, J. Hepatol. 72 (5) (2020) 885–895.
[103] K. Li, M.F. van Delft, G. Dewson, Too much death can kill you: inhibiting intrinsic apoptosis to treat disease, EMBO J. 40 (14) (2021) e107341.
[104] T.K. Rostovtseva, et al., Targeting the multiple physiologic roles of VDAC with steroids and hydrophobic drugs, Front. Physiol. 11 (2020) 446.
[105] M.F. van Delft, et al., A small molecule interacts with VDAC2 to block mouse BAK-driven apoptosis, Nat. Chem. Biol. 15 (11) (2019) 1057–1066.
[106] D. Mooser, et al., The menaquinol-oxidizing cytochrome bc complex from Thermus thermophilus: protein domains and subunits, Biochim. Biophys. Acta 1757 (9–10) (2006) 1084–1095.
[107] A. Linkermann, D.R. Green, Necroptosis, N. Engl. J. Med. 370 (5) (2014) 455–465.
[108] M. Pasparakis, P. Vandenabeele, Necroptosis and its role in inflammation, Nature 517 (7534) (2015) 311–320.
[109] R. Weinlich, et al., Necroptosis in development, inflammation and disease, Nat. Rev. Mol. Cell Biol. 18 (2) (2017) 127–136.
[110] W. Zhou, J. Yuan, Necroptosis in health and diseases, Semin. Cell Dev. Biol. 35 (2014) 14–23.
[111] L. Galluzzi, et al., Necroptosis: mechanisms and relevance to disease, Annu. Rev. Pathol. 12 (2017) 103–130.
[112] T.W. Day, S. Huang, A.R. Safa, c-FLIP knockdown induces ligand-independent DR5-, FADD-, caspase-8-, and caspase-9-dependent apoptosis in breast cancer cells, Biochem. Pharmacol. 76 (12) (2008) 1694–1704.
[113] H.H. Cheung, et al., Down-regulation of c-FLIP Enhances death of cancer cells by Smac mimetic compound, Cancer Res. 69 (19) (2009) 7729–7738.
[114] P. Vandenabeele, et al., The role of the kinases RIP1 and RIP3 in TNF-induced necrosis, Sci. Signal. 3 (115) (2010) re4.
[115] J. Li, et al., The RIP1/RIP3 necrosome forms a functional amyloid signaling complex required for programmed necrosis, Cell 150 (2) (2012) 339–350.
[116] L. Sun, et al., Mixed lineage kinase domain-like protein mediates necrosis signaling downstream of RIP3 kinase, Cell 148 (1–2) (2012) 213–227.
[117] P. Mandal, et al., RIP3 induces apoptosis independent of pronecrotic kinase activity, Mol. Cell 56 (4) (2014) 481–495.
[118] A. Takaoka, et al., DAI (DLM-1/ZBP1) is a cytosolic DNA sensor and an activator of innate immune response, Nature 448 (7152) (2007) 501–505.
[119] L. Sun, X. Wang, A new kind of cell suicide: mechanisms and functions of programmed necrosis, Trends Biochem. Sci. 39 (12) (2014) 587–593.
[120] J. Zhao, et al., Mixed lineage kinase domain-like is a key receptor interacting protein 3 downstream component of TNF-induced necrosis, Proc. Natl. Acad. Sci. USA 109 (14) (2012) 5322–5327.
[121] G. Quarato, et al., Sequential engagement of distinct MLKL phosphatidylinositol-binding sites executes necroptosis, Mol. Cell 61 (4) (2016) 589–601.
[122] S. Martens, et al., MLKL in cancer: more than a necroptosis regulator, Cell Death Differ. 28 (6) (2021) 1757–1772.
[123] A. Degterev, et al., Identification of RIP1 kinase as a specific cellular target of necrostatins, Nat. Chem. Biol. 4 (5) (2008) 313–321.
[124] A. Degterev, et al., Chemical inhibitor of nonapoptotic cell death with therapeutic potential for ischemic brain injury, Nat. Chem. Biol. 1 (2) (2005) 112–119.

[125] S. He, et al., Receptor interacting protein kinase-3 determines cellular necrotic response to TNF-alpha, Cell 137 (6) (2009) 1100–1111.

[126] Y.S. Cho, et al., Phosphorylation-driven assembly of the RIP1-RIP3 complex regulates programmed necrosis and virus-induced inflammation, Cell 137 (6) (2009) 1112–1123.

[127] D.W. Zhang, et al., RIP3, an energy metabolism regulator that switches TNF-induced cell death from apoptosis to necrosis, Science 325 (5938) (2009) 332–336.

[128] K. Newton, et al., Activity of protein kinase RIPK3 determines whether cells die by necroptosis or apoptosis, Science 343 (6177) (2014) 1357–1360.

[129] N. Robinson, et al., Type I interferon induces necroptosis in macrophages during infection with *Salmonella enterica* serovar Typhimurium, Nat. Immunol. 13 (10) (2012) 954–962.

[130] R.J. Thapa, et al., Interferon-induced RIP1/RIP3-mediated necrosis requires PKR and is licensed by FADD and caspases, Proc. Natl. Acad. Sci. USA 110 (33) (2013) E3109–E3118.

[131] S. McComb, et al., Type-I interferon signaling through ISGF3 complex is required for sustained Rip3 activation and necroptosis in macrophages, Proc. Natl. Acad. Sci. USA 111 (31) (2014) E3206–E3213.

[132] T. Frank, et al., Cell cycle arrest in mitosis promotes interferon-induced necroptosis, Cell Death Differ. 26 (10) (2019) 2046–2060.

[133] J. Sarhan, et al., Constitutive interferon signaling maintains critical threshold of MLKL expression to license necroptosis, Cell Death Differ. 26 (2) (2019) 332–347.

[134] T. Molnar, et al., Current translational potential and underlying molecular mechanisms of necroptosis, Cell Death Dis. 10 (11) (2019) 860.

[135] K. Newton, X. Sun, V.M. Dixit, Kinase RIP3 is dispensable for normal NF-kappa Bs, signaling by the B-cell and T-cell receptors, tumor necrosis factor receptor 1, and Toll-like receptors 2 and 4, Mol. Cell Biol. 24 (4) (2004) 1464–1469.

[136] J. Wu, et al., Mlkl knockout mice demonstrate the indispensable role of Mlkl in necroptosis, Cell Res. 23 (8) (2013) 994–1006.

[137] E.E. Varfolomeev, et al., Targeted disruption of the mouse Caspase 8 gene ablates cell death induction by the TNF receptors, Fas/Apo1, and DR3 and is lethal prenatally, Immunity 9 (2) (1998) 267–276.

[138] W.J. Kaiser, et al., RIP3 mediates the embryonic lethality of caspase-8-deficient mice, Nature 471 (7338) (2011) 368–372.

[139] A. Oberst, et al., Catalytic activity of the caspase-8-FLIP(L) complex inhibits RIPK3-dependent necrosis, Nature 471 (7338) (2011) 363–367.

[140] N. Yatim, et al., RIPK1 and NF-kappaB signaling in dying cells determines cross-priming of CD8(+) T cells, Science 350 (6258) (2015) 328–334.

[141] J. Ren, et al., The RIP3-RIP1-NF-kappaB signaling axis is dispensable for necroptotic cells to elicit cross-priming of CD8(+) T cells, Cell. Mol. Immunol. 14 (7) (2017) 639–642.

[142] T.L. Aaes, P. Vandenabeele, The intrinsic immunogenic properties of cancer cell lines, immunogenic cell death, and how these influence host antitumor immune responses, Cell Death Differ. 28 (3) (2021) 843–860.

[143] A. Polykratis, et al., Cutting edge: RIPK1 Kinase inactive mice are viable and protected from TNF-induced necroptosis in vivo, J. Immunol. 193 (4) (2014) 1539–1543.

[144] X. Wang, et al., Direct activation of RIP3/MLKL-dependent necrosis by herpes simplex virus 1 (HSV-1) protein ICP6 triggers host antiviral defense, Proc. Natl. Acad. Sci. USA 111 (43) (2014) 15438–15443.

[145] Z. Huang, et al., RIP1/RIP3 binding to HSV-1 ICP6 initiates necroptosis to restrict virus propagation in mice, Cell Host Microbe 17 (2) (2015) 229–242.

[146] R.K. Lane, et al., Necroptosis-based CRISPR knockout screen reveals Neuropilin-1 as a critical host factor for early stages of murine cytomegalovirus infection, Proc. Natl. Acad. Sci. USA 117 (33) (2020) 20109–20116.
[147] C. Wen, et al., RIPK3-Dependent necroptosis is induced and restricts viral replication in human astrocytes infected with Zika virus, Front. Cell. Infect. Microbiol. 11 (2021) 637710.
[148] A.G. Snyder, A. Oberst, The antisocial network: cross talk between cell death programs in host defense, Annu. Rev. Immunol. 39 (2021) 77–101.
[149] T. Zhang, et al., CaMKII is a RIP3 substrate mediating ischemia- and oxidative stress-induced myocardial necroptosis, Nat. Med. 22 (2) (2016) 175–182.
[150] Z. Yang, et al., Melatonin attenuates chronic pain related myocardial ischemic susceptibility through inhibiting RIP3-MLKL/CaMKII dependent necroptosis, J. Mol. Cell. Cardiol. 125 (2018) 185–194.
[151] A. Szobi, et al., Cardioprotection of ischaemic preconditioning is associated with inhibition of translocation of MLKL within the plasma membrane, J. Cell Mol. Med. 22 (9) (2018) 4183–4196.
[152] Y. Zhou, et al., Insight into crosstalk between ferroptosis and necroptosis: novel therapeutics in ischemic stroke, Oxid. Med. Cell. Longev. 2021 (2021) 9991001.
[153] D. Rosentreter, et al., RIP1-Dependent programmed necrosis is negatively regulated by caspases during hepatic ischemia-reperfusion, Shock 44 (1) (2015) 72–76.
[154] S. Shi, et al., Necroptotic cell death in liver transplantation and underlying diseases: mechanisms and clinical perspective, Liver Transplant. 25 (7) (2019) 1091–1104.
[155] H.M. Ni, et al., Receptor-interacting serine/threonine-protein kinase 3 (RIPK3)-Mixed lineage kinase domain-like protein (MLKL)-Mediated necroptosis contributes to ischemia-reperfusion injury of steatotic livers, Am. J. Pathol. 189 (7) (2019) 1363–1374.
[156] S. Liao, et al., The possible roles of necroptosis during cerebral ischemia and ischemia/reperfusion injury, Arch. Biochem. Biophys. 695 (2020) 108629.
[157] R. Baidya, et al., Necroptosis in hepatosteatotic ischaemia-reperfusion injury, Int. J. Mol. Sci. 21 (16) (2020) 5931.
[158] L. Jin, et al., RIPK1: a rising star in inflammatory and neoplastic skin diseases, J. Dermatol. Sci. 99 (3) (2020) 146–151.
[159] H. Lu, et al., RIPK1 inhibitor ameliorates colitis by directly maintaining intestinal barrier homeostasis and regulating following IECs-immuno crosstalk, Biochem. Pharmacol. 172 (2020) 113751.
[160] N. Oikonomou, et al., Airway epithelial cell necroptosis contributes to asthma exacerbation in a mouse model of house dust mite-induced allergic inflammation, Mucosal Immunol. 14 (5) (2021) 1160–1171.
[161] M. Speir, et al., Targeting RIP kinases in chronic inflammatory disease, Biomolecules 11 (5) (2021) 646.
[162] S. Li, et al., SARS-CoV-2 triggers inflammatory responses and cell death through caspase-8 activation, Signal Transduct. Target Ther. 5 (1) (2020) 235.
[163] R. Karki, et al., Synergism of TNF-alpha and IFN-gamma triggers inflammatory cell death, tissue damage, and mortality in SARS-CoV-2 infection and cytokine shock syndromes, Cell 184 (1) (2021) 149–168 e17.
[164] J. Taft, et al., Human TBK1 deficiency leads to autoinflammation driven by TNF-induced cell death, Cell 184 (17) (2021) 4447–4463.e20.
[165] N. Werthmoller, et al., Modulation of radiochemoimmunotherapy-induced B16 melanoma cell death by the pan-caspase inhibitor zVAD-fmk induces anti-tumor immunity in a HMGB1-, nucleotide- and T-cell-dependent manner, Cell Death Dis. 6 (5) (2015) e1761.

[166] H. Yang, et al., Contribution of RIP3 and MLKL to immunogenic cell death signaling in cancer chemotherapy, OncoImmunology 5 (6) (2016) e1149673.
[167] S.T. Workenhe, J. Pol, G. Kroemer, Tumor-intrinsic determinants of immunogenic cell death modalities, OncoImmunology 10 (1) (2021) 1893466.
[168] G.B. Koo, et al., Methylation-dependent loss of RIP3 expression in cancer represses programmed necrosis in response to chemotherapeutics, Cell Res. 25 (6) (2015) 707–725.
[169] C. Yang, et al., Regulation of RIP3 by the transcription factor Sp1 and the epigenetic regulator UHRF1 modulates cancer cell necroptosis, Cell Death Dis. 8 (10) (2017) e3084.
[170] Y. Gong, et al., The role of necroptosis in cancer biology and therapy, Mol. Cancer 18 (1) (2019) 100.
[171] W. Chen, et al., Ppm1b negatively regulates necroptosis through dephosphorylating Rip3, Nat. Cell Biol. 17 (4) (2015) 434–444.
[172] Y. Wu, G. Dong, C. Sheng, Targeting necroptosis in anticancer therapy: mechanisms and modulators, Acta Pharm. Sin. B 10 (9) (2020) 1601–1618.
[173] S. Park, et al., The receptor interacting protein 1 inhibits p53 induction through NF-kappaB activation and confers a worse prognosis in glioblastoma, Cancer Res. 69 (7) (2009) 2809–2816.
[174] K.D. McCormick, et al., Innate immune signaling through differential RIPK1 expression promote tumor progression in head and neck squamous cell carcinoma, Carcinogenesis 37 (5) (2016) 522–529.
[175] B. Strilic, et al., Tumour-cell-induced endothelial cell necroptosis via death receptor 6 promotes metastasis, Nature 536 (7615) (2016) 215–218.
[176] L. Seifert, et al., The necrosome promotes pancreatic oncogenesis via CXCL1 and Mincle-induced immune suppression, Nature 532 (7598) (2016) 245–249.
[177] S. Martens, et al., Inhibitors targeting RIPK1/RIPK3: old and new drugs, Trends Pharmacol. Sci. 41 (3) (2020) 209–224.
[178] A. Degterev, J.L. Maki, J. Yuan, Activity and specificity of necrostatin-1, small-molecule inhibitor of RIP1 kinase, Cell Death Differ. 20 (2) (2013) 366.
[179] T. Xie, et al., Structural basis of RIP1 inhibition by necrostatins, Structure 21 (3) (2013) 493–499.
[180] S. Patel, et al., RIP1 inhibition blocks inflammatory diseases but not tumor growth or metastases, Cell Death Differ. 27 (1) (2020) 161–175.
[181] A.A. Zarrin, et al., Kinase inhibition in autoimmunity and inflammation, Nat. Rev. Drug Discov. 20 (1) (2021) 39–63.
[182] S.B. Berger, et al., Characterization of GSK'963: a structurally distinct, potent and selective inhibitor of RIP1 kinase, Cell Death Discov. 1 (2015) 15009.
[183] P.A. Harris, et al., Discovery of a first-in-class receptor interacting protein 1 (RIP1) kinase specific clinical candidate (GSK2982772) for the treatment of inflammatory diseases, J. Med. Chem. 60 (4) (2017) 1247–1261.
[184] L. Mifflin, D. Ofengeim, J. Yuan, Receptor-interacting protein kinase 1 (RIPK1) as a therapeutic target, Nat. Rev. Drug Discov. 19 (8) (2020) 553–571.
[185] K. Weisel, et al., Randomized clinical study of safety, pharmacokinetics, and pharmacodynamics of RIPK1 inhibitor GSK2982772 in healthy volunteers, Pharmacol. Res. Perspect. 5 (6) (2017) e00365.
[186] K. Weisel, et al., Response to inhibition of receptor-interacting protein kinase 1 (RIPK1) in active plaque psoriasis: a randomized placebo-controlled study, Clin. Pharmacol. Ther. 108 (4) (2020) 808–816.

[187] K. Weisel, et al., A randomized, placebo-controlled experimental medicine study of RIPK1 inhibitor GSK2982772 in patients with moderate to severe rheumatoid arthritis, Arthritis Res. Ther. 23 (1) (2021) 85.
[188] P.A. Harris, et al., Identification of a RIP1 kinase inhibitor clinical candidate (GSK3145095) for the treatment of pancreatic cancer, ACS Med. Chem. Lett. 10 (6) (2019) 857–862.
[189] K. Xia, et al., Discovery of a potent RIPK3 inhibitor for the amelioration of necroptosis-associated inflammatory injury, Front. Cell Dev. Biol. 8 (2020) 606119.
[190] A.C. Hart, et al., Identification of RIPK3 type II inhibitors using high-throughput mechanistic studies in hit triage, ACS Med. Chem. Lett. 11 (3) (2020) 266–271.
[191] B. Yan, et al., Discovery of a new class of highly potent necroptosis inhibitors targeting the mixed lineage kinase domain-like protein, Chem. Commun. 53 (26) (2017) 3637–3640.
[192] M. Rübbelke, et al., Locking mixed-lineage kinase domain-like protein in its auto-inhibited state prevents necroptosis, Proc. Natl. Acad. Sci. USA 117 (52) (2020) 33272–33281.
[193] J.M. Hildebrand, et al., Activation of the pseudokinase MLKL unleashes the four-helix bundle domain to induce membrane localization and necroptotic cell death, Proc. Natl. Acad. Sci. USA 111 (42) (2014) 15072–15077.
[194] C. Zhan, et al., MLKL: functions beyond serving as the executioner of necroptosis, Theranostics 11 (10) (2021) 4759–4769.
[195] G. Rogler, Resolution of inflammation in inflammatory bowel disease, Lancet Gastroenterol. Hepatol. 2 (7) (2017) 521–530.
[196] J.V. Patankar, C. Becker, Cell death in the gut epithelium and implications for chronic inflammation, Nat. Rev. Gastroenterol. Hepatol. 17 (9) (2020) 543–556.
[197] J.U. Scher, M.H. Pillinger, The anti-inflammatory effects of prostaglandins, J. Invest. Med. 57 (6) (2009) 703–708.
[198] C.A. Loynes, et al., PGE2 production at sites of tissue injury promotes an anti-inflammatory neutrophil phenotype and determines the outcome of inflammation resolution in vivo, Sci. Adv. 4 (9) (2018) eaar8320.
[199] J.V. Patankar, et al., E-type prostanoid receptor 4 drives resolution of intestinal inflammation by blocking epithelial necroptosis, Nat. Cell Biol. 23 (7) (2021) 796–807.
[200] Y. Murahashi, et al., Oral administration of EP4-selective agonist KAG-308 suppresses mouse knee osteoarthritis development through reduction of chondrocyte hypertrophy and TNF secretion, Sci. Rep. 9 (1) (2019) 20329.
[201] I. Aringer, et al., Agonism of prostaglandin E2 receptor 4 ameliorates tubulointerstitial injury in nephrotoxic serum nephritis in mice, J. Clin. Med. 10 (4) (2021) 832.
[202] A. Zychlinsky, M.C. Prevost, P.J. Sansonetti, *Shigella flexneri* induces apoptosis in infected macrophages, Nature 358 (6382) (1992) 167–169.
[203] H. Hilbi, et al., The interleukin 1beta-converting enzyme, caspase 1, is activated during *Shigella flexneri*-induced apoptosis in human monocyte-derived macrophages, Infect. Immun. 65 (12) (1997) 5165–5170.
[204] B.T. Cookson, M.A. Brennan, Pro-inflammatory programmed cell death, Trends Microbiol. 9 (3) (2001) 113–114.
[205] S.L. Fink, B.T. Cookson, Apoptosis, pyroptosis, and necrosis: mechanistic description of dead and dying eukaryotic cells, Infect. Immun. 73 (4) (2005) 1907–1916.
[206] S.L. Fink, B.T. Cookson, Caspase-1-dependent pore formation during pyroptosis leads to osmotic lysis of infected host macrophages, Cell Microbiol. 8 (11) (2006) 1812–1825.

[207] M.K. Stewart, B.T. Cookson, Evasion and interference: intracellular pathogens modulate caspase-dependent inflammatory responses, Nat. Rev. Microbiol. 14 (6) (2016) 346–359.
[208] P. Broz, V.M. Dixit, Inflammasomes: mechanism of assembly, regulation and signalling, Nat. Rev. Immunol. 16 (7) (2016) 407–420.
[209] M. Lamkanfi, et al., Inflammasome-dependent release of the alarmin HMGB1 in endotoxemia, J. Immunol. 185 (7) (2010) 4385–4392.
[210] O. Gross, et al., Inflammasome activators induce interleukin-1alpha secretion via distinct pathways with differential requirement for the protease function of caspase-1, Immunity 36 (3) (2012) 388–400.
[211] F. Martinon, K. Burns, J. Tschopp, The inflammasome: a molecular platform triggering activation of inflammatory caspases and processing of proIL-beta, Mol. Cell 10 (2) (2002) 417–426.
[212] P.M. Ridker, et al., Antiinflammatory therapy with canakinumab for atherosclerotic disease, N. Engl. J. Med. 377 (12) (2017) 1119–1131.
[213] A. Degterev, J. Yuan, Expansion and evolution of cell death programmes, Nat. Rev. Mol. Cell Biol. 9 (5) (2008) 378–390.
[214] S. Janssens, A. Tinel, The PIDDosome, DNA-damage-induced apoptosis and beyond, Cell Death Differ. 19 (1) (2012) 13–20.
[215] V.C. Sladky, A. Villunger, Uncovering the PIDDosome and caspase-2 as regulators of organogenesis and cellular differentiation, Cell Death Differ. 27 (7) (2020) 2037–2047.
[216] L. Robert Hollingsworth, et al., Mechanism of filament formation in UPA-promoted CARD8 and NLRP1 inflammasomes, Nat. Commun. 12 (1) (2021) 189.
[217] S. Bauernfried, et al., Human NLRP1 is a sensor for double-stranded RNA, Science 371 (6528) (2021), eabd0811.
[218] D.A. Bachovchin, NLRP1: a jack of all trades, or a master of one? Mol. Cell 81 (3) (2021) 423–425.
[219] L. Agostini, et al., NALP3 forms an IL-1beta-processing inflammasome with increased activity in Muckle-Wells autoinflammatory disorder, Immunity 20 (3) (2004) 319–325.
[220] H. Guo, J.B. Callaway, J.P. Ting, Inflammasomes: mechanism of action, role in disease, and therapeutics, Nat. Med. 21 (7) (2015) 677–687.
[221] M.S.J. Mangan, et al., Targeting the NLRP3 inflammasome in inflammatory diseases, Nat. Rev. Drug Discov. 17 (8) (2018) 588–606.
[222] K.V. Swanson, M. Deng, J.P. Ting, The NLRP3 inflammasome: molecular activation and regulation to therapeutics, Nat. Rev. Immunol. 19 (8) (2019) 477–489.
[223] N. Zhao, B. Di, L.L. Xu, The NLRP3 inflammasome and COVID-19: activation, pathogenesis and therapeutic strategies, Cytokine Growth Factor Rev. (2021), https://doi.org/10.1016/j.cytogfr.2021.06.002.
[224] N. Zhao, et al., Recent advances in the NEK7-licensed NLRP3 inflammasome activation: mechanisms, role in diseases and related inhibitors, J. Autoimmun. 113 (2020) 102515.
[225] X. Cai, et al., Prion-like polymerization underlies signal transduction in antiviral immune defense and inflammasome activation, Cell 156 (6) (2014) 1207–1222.
[226] A. Lu, et al., Unified polymerization mechanism for the assembly of ASC-dependent inflammasomes, Cell 156 (6) (2014) 1193–1206.
[227] D. Boucher, et al., Caspase-1 self-cleavage is an intrinsic mechanism to terminate inflammasome activity, J. Exp. Med. 215 (3) (2018) 827–840.

[228] Y. He, et al., NEK7 is an essential mediator of NLRP3 activation downstream of potassium efflux, Nature 530 (7590) (2016) 354–357.
[229] H. Shi, et al., NLRP3 activation and mitosis are mutually exclusive events coordinated by NEK7, a new inflammasome component, Nat. Immunol. 17 (3) (2016) 250–258.
[230] J.L. Schmid-Burgk, et al., A genome-wide CRISPR (Clustered Regularly Interspaced short palindromic repeats) screen identifies NEK7 as an essential component of NLRP3 inflammasome activation, J. Biol. Chem. 291 (1) (2016) 103–109.
[231] H. Sharif, et al., Structural mechanism for NEK7-licensed activation of NLRP3 inflammasome, Nature 570 (7761) (2019) 338–343.
[232] F.G. Bauernfeind, et al., Cutting edge: NF-kappaB activating pattern recognition and cytokine receptors license NLRP3 inflammasome activation by regulating NLRP3 expression, J. Immunol. 183 (2) (2009) 787–791.
[233] L. Franchi, T. Eigenbrod, G. Nunez, Cutting edge: TNF-alpha mediates sensitization to ATP and silica via the NLRP3 inflammasome in the absence of microbial stimulation, J. Immunol. 183 (2) (2009) 792–796.
[234] Y. Huang, W. Xu, R. Zhou, NLRP3 inflammasome activation and cell death, Cell. Mol. Immunol. (2021), https://doi.org/10.1038/s41423-021-00740-6.
[235] Y. Xing, et al., Cutting edge: TRAF6 mediates TLR/IL-1R signaling-induced nontranscriptional priming of the NLRP3 inflammasome, J. Immunol. 199 (5) (2017) 1561–1566.
[236] M.Y. Chen, et al., The signaling pathways regulating NLRP3 inflammasome activation, Inflammation 44 (4) (2021) 1229–1245.
[237] S.S. Im, et al., Linking lipid metabolism to the innate immune response in macrophages through sterol regulatory element binding protein-1a, Cell Metabol. 13 (5) (2011) 540–549.
[238] J. Hawiger, J. Zienkiewicz, Decoding inflammation, its causes, genomic responses, and emerging countermeasures, Scand. J. Immunol. 90 (6) (2019) e12812.
[239] J.H. Lee, et al., SREBP-1a-stimulated lipid synthesis is required for macrophage phagocytosis downstream of TLR4-directed mTORC1, Proc. Natl. Acad. Sci. USA 115 (52) (2018) E12228–E12234.
[240] S.R. Liebergall, et al., Inflammation triggers liver X receptor-dependent lipogenesis, Mol. Cell Biol. 40 (2) (2020) 19, e00364-19.
[241] M. de la Roche, et al., Trafficking of cholesterol to the ER is required for NLRP3 inflammasome activation, J. Cell Biol. 217 (10) (2018) 3560–3576.
[242] J. Shi, et al., Cleavage of GSDMD by inflammatory caspases determines pyroptotic cell death, Nature 526 (7575) (2015) 660–665.
[243] N. Kayagaki, et al., Caspase-11 cleaves gasdermin D for non-canonical inflammasome signalling, Nature 526 (7575) (2015) 666–671.
[244] W.T. He, et al., Gasdermin D is an executor of pyroptosis and required for interleukin-1beta secretion, Cell Res. 25 (12) (2015) 1285–1298.
[245] S.B. Kovacs, E.A. Miao, Gasdermins: effectors of pyroptosis, Trends Cell Biol. 27 (9) (2017) 673–684.
[246] P. Broz, P. Pelegrin, F. Shao, The gasdermins, a protein family executing cell death and inflammation, Nat. Rev. Immunol. 20 (3) (2020) 143–157.
[247] R.A. Aglietti, E.C. Dueber, Recent insights into the molecular mechanisms underlying pyroptosis and gasdermin family functions, Trends Immunol. 38 (4) (2017) 261–271.
[248] I. Jorgensen, M. Rayamajhi, E.A. Miao, Programmed cell death as a defence against infection, Nat. Rev. Immunol. 17 (3) (2017) 151–164.
[249] J. Shi, W. Gao, F. Shao, Pyroptosis: gasdermin-mediated programmed necrotic cell death, Trends Biochem. Sci. 42 (4) (2017) 245–254.

[250] J. Ruan, Structural insight of gasdermin family driving pyroptotic cell death, Adv. Exp. Med. Biol. 1172 (2019) 189–205.
[251] C.L. Evavold, et al., Control of gasdermin D oligomerization and pyroptosis by the Ragulator-Rag-mTORC1 pathway, Cell 184 (17) (2021) 4495–4511.
[252] Y. Wang, et al., Chemotherapy drugs induce pyroptosis through caspase-3 cleavage of a gasdermin, Nature 547 (7661) (2017) 99–103.
[253] C. Rogers, et al., Cleavage of DFNA5 by caspase-3 during apoptosis mediates progression to secondary necrotic/pyroptotic cell death, Nat. Commun. 8 (2017) 14128.
[254] P. Orning, et al., Pathogen blockade of TAK1 triggers caspase-8-dependent cleavage of gasdermin D and cell death, Science 362 (6418) (2018) 1064–1069.
[255] J. Sarhan, et al., Caspase-8 induces cleavage of gasdermin D to elicit pyroptosis during *Yersinia* infection, Proc. Natl. Acad. Sci. USA 115 (46) (2018) E10888–E10897.
[256] Z.Z. Zheng, et al., The lysosomal Rag-Ragulator complex licenses RIPK1-and caspase-8-mediated pyroptosis by *Yersinia*, Science 372 (6549) (2021) 1412, https://doi.org/10.1126/science.abg0269.
[257] J.L. Poyet, et al., Identification of Ipaf, a human caspase-1-activating protein related to Apaf-1, J. Biol. Chem. 276 (30) (2001) 28309–28313.
[258] E.A. Miao, et al., Cytoplasmic flagellin activates caspase-1 and secretion of interleukin 1beta via Ipaf, Nat. Immunol. 7 (6) (2006) 569–575.
[259] L. Franchi, et al., Cytosolic flagellin requires Ipaf for activation of caspase-1 and interleukin 1beta in salmonella-infected macrophages, Nat. Immunol. 7 (6) (2006) 576–582.
[260] Y. Zhao, et al., The NLRC4 inflammasome receptors for bacterial flagellin and type III secretion apparatus, Nature 477 (7366) (2011) 596–600.
[261] E.M. Kofoed, R.E. Vance, Innate immune recognition of bacterial ligands by NAIPs determines inflammasome specificity, Nature 477 (7366) (2011) 592–595.
[262] J.A. Duncan, S.W. Canna, The NLRC4 inflammasome, Immunol. Rev. 281 (1) (2018) 115–123.
[263] Z. Hu, et al., Crystal structure of NLRC4 reveals its autoinhibition mechanism, Science 341 (6142) (2013) 172–175.
[264] L. Zhang, et al., Cryo-EM structure of the activated NAIP2-NLRC4 inflammasome reveals nucleated polymerization, Science 350 (6259) (2015) 404–409.
[265] C.A. Diebolder, et al., Cryoelectron tomography of the NAIP5/NLRC4 inflammasome: implications for NLR activation, Structure 23 (12) (2015) 2349–2357.
[266] Z. Hu, et al., Structural and biochemical basis for induced self-propagation of NLRC4, Science 350 (6259) (2015) 399–404.
[267] D.P. Zheng, T. Liwinski, E. Elinav, Inflammasome activation and regulation: toward a better understanding of complex mechanisms, Cell Discov. 6 (1) (2020) 36, https://doi.org/10.1038/s41421-020-0167-x.
[268] H. Xu, et al., Innate immune sensing of bacterial modifications of Rho GTPases by the Pyrin inflammasome, Nature 513 (7517) (2014) 237–241.
[269] N. Richards, et al., Interaction between pyrin and the apoptotic speck protein (ASC) modulates ASC-induced apoptosis, J. Biol. Chem. 276 (42) (2001) 39320–39329.
[270] O. Dufies, L. Boyer, RhoGTPases and inflammasomes: guardians of effector-triggered immunity, PLoS Pathog. 17 (4) (2021) e1009504.
[271] T. Jin, et al., Structures of the HIN domain:DNA complexes reveal ligand binding and activation mechanisms of the AIM2 inflammasome and IFI16 receptor, Immunity 36 (4) (2012) 561–571.
[272] T. Jin, et al., Structure of the absent in melanoma 2 (AIM2) pyrin domain provides insights into the mechanisms of AIM2 autoinhibition and inflammasome assembly, J. Biol. Chem. 288 (19) (2013) 13225–13235.

[273] A. Lu, et al., Plasticity in PYD assembly revealed by cryo-EM structure of the PYD filament of AIM2, Cell Discov. 1 (2015) 15013.
[274] S.R. Morrone, et al., Assembly-driven activation of the AIM2 foreign-dsDNA sensor provides a polymerization template for downstream ASC, Nat. Commun. 6 (2015) 7827.
[275] S.R. Morrone, et al., Cooperative assembly of IFI16 filaments on dsDNA provides insights into host defense strategy, Proc. Natl. Acad. Sci. USA 111 (1) (2014) E62–E71.
[276] N. Kayagaki, et al., Non-canonical inflammasome activation targets caspase-11, Nature 479 (7371) (2011) 117–121.
[277] N. Kayagaki, et al., Noncanonical inflammasome activation by intracellular LPS independent of TLR4, Science 341 (6151) (2013) 1246–1249.
[278] J.A. Hagar, et al., Cytoplasmic LPS activates caspase-11: implications in TLR4-independent endotoxic shock, Science 341 (6151) (2013) 1250–1253.
[279] J. Shi, et al., Inflammatory caspases are innate immune receptors for intracellular LPS, Nature 514 (7521) (2014) 187–192.
[280] H. Van Gorp, N. Van Opdenbosch, M. Lamkanfi, Inflammasome-dependent cytokines at the crossroads of health and autoinflammatory disease, Cold Spring Harb. Perspect. Biol. 11 (1) (2019), a028563.
[281] Y. Li, et al., Inflammasomes as therapeutic targets in human diseases, Signal Transduct. Target. Ther. 6 (1) (2021) 247.
[282] F.L. Zhong, et al., Germline NLRP1 mutations cause skin inflammatory and cancer susceptibility syndromes via inflammasome activation, Cell 167 (1) (2016) 187–202 e17.
[283] C.B. Levandowski, et al., NLRP1 haplotypes associated with vitiligo and autoimmunity increase interleukin-1beta processing via the NLRP1 inflammasome, Proc. Natl. Acad. Sci. USA 110 (8) (2013) 2952–2956.
[284] Y. Jin, et al., NALP1 in vitiligo-associated multiple autoimmune disease, N. Engl. J. Med. 356 (12) (2007) 1216–1225.
[285] F.L. Zhong, et al., Human DPP9 represses NLRP1 inflammasome and protects against autoinflammatory diseases via both peptidase activity and FIIND domain binding, J. Biol. Chem. 293 (49) (2018) 18864–18878.
[286] S. Grandemange, et al., A new autoinflammatory and autoimmune syndrome associated with NLRP1 mutations: NAIAD (NLRP1-associated autoinflammation with arthritis and dyskeratosis), Ann. Rheum. Dis. 76 (7) (2017) 1191–1198.
[287] S.B. Drutman, et al., Homozygous NLRP1 gain-of-function mutation in siblings with a syndromic form of recurrent respiratory papillomatosis, Proc. Natl. Acad. Sci. USA 116 (38) (2019) 19055–19063.
[288] F.S. Ong, et al., The M694V mutation in Armenian-Americans: a 10-year retrospective study of MEFV mutation testing for familial Mediterranean fever at UCLA, Clin. Genet. 84 (1) (2013) 55–59.
[289] A. Kanneganti, et al., GSDMD is critical for autoinflammatory pathology in a mouse model of Familial Mediterranean Fever, J. Exp. Med. 215 (6) (2018) 1519–1529.
[290] M.C. Heymann, et al., Human procaspase-1 variants with decreased enzymatic activity are associated with febrile episodes and may contribute to inflammation via RIP2 and NF-kappaB signaling, J. Immunol. 192 (9) (2014) 4379–4385.
[291] S. Winkler, A. Rosen-Wolff, Caspase-1: an integral regulator of innate immunity, Semin. Immunopathol. 37 (4) (2015) 419–427.
[292] F. Kapplusch, et al., CASP1 variants influence subcellular caspase-1 localization, pyroptosome formation, pro-inflammatory cell death and macrophage deformability, Clin. Immunol. 208 (2019) 108232.

[293] N. Van Opdenbosch, M. Lamkanfi, Caspases in cell death, inflammation, and disease, Immunity 50 (6) (2019) 1352–1364.
[294] A. Halle, et al., The NALP3 inflammasome is involved in the innate immune response to amyloid-beta, Nat. Immunol. 9 (8) (2008) 857–865.
[295] M.T. Heneka, et al., NLRP3 is activated in Alzheimer's disease and contributes to pathology in APP/PS1 mice, Nature 493 (7434) (2013) 674–678.
[296] M.T. Heneka, et al., Neuroinflammation in Alzheimer's disease, Lancet Neurol. 14 (4) (2015) 388–405.
[297] S. Li, et al., Gasdermin D in peripheral myeloid cells drives neuroinflammation in experimental autoimmune encephalomyelitis, J. Exp. Med. 216 (11) (2019) 2562–2581.
[298] S. Voet, et al., Inflammasomes in neuroinflammatory and neurodegenerative diseases, EMBO Mol. Med. 11 (6) (2019), e10248.
[299] S.L. Masters, et al., Activation of the NLRP3 inflammasome by islet amyloid polypeptide provides a mechanism for enhanced IL-1beta in type 2 diabetes, Nat. Immunol. 11 (10) (2010) 897–904.
[300] S.L. Masters, E. Latz, L.A. O'Neill, The inflammasome in atherosclerosis and type 2 diabetes, Sci. Transl. Med. 3 (81) (2011) 81ps17.
[301] H. Wen, et al., Fatty acid-induced NLRP3-ASC inflammasome activation interferes with insulin signaling, Nat. Immunol. 12 (5) (2011) 408–415.
[302] J.I. Shin, et al., Inflammasomes and autoimmune and rheumatic diseases: a comprehensive review, J. Autoimmun. 103 (2019) 102299.
[303] N. Ji, et al., Pyroptosis: a new regulating mechanism in cardiovascular disease, J. Inflamm. Res. 14 (2021) 2647–2666.
[304] D.P. Del Re, et al., Fundamental mechanisms of regulated cell death and implications for heart disease, Physiol. Rev. 99 (4) (2019) 1765–1817.
[305] C. Chang, Neonatal autoimmune diseases: a critical review, J. Autoimmun. 38 (2–3) (2012) J223–J238.
[306] J. Feldmann, et al., Chronic infantile neurological cutaneous and articular syndrome is caused by mutations in CIAS1, a gene highly expressed in polymorphonuclear cells and chondrocytes, Am. J. Hum. Genet. 71 (1) (2002) 198–203.
[307] H.M. Hoffman, et al., Mutation of a new gene encoding a putative pyrin-like protein causes familial cold autoinflammatory syndrome and Muckle-Wells syndrome, Nat. Genet. 29 (3) (2001) 301–305.
[308] J. Xiao, et al., Gasdermin D mediates the pathogenesis of neonatal-onset multisystem inflammatory disease in mice, PLoS Biol. 16 (11) (2018) e3000047.
[309] Y. Li, et al., GSDME-mediated pyroptosis promotes inflammation and fibrosis in obstructive nephropathy, Cell Death Differ. 28 (8) (2021) 2333–2350.
[310] W. Xia, et al., Gasdermin E deficiency attenuates acute kidney injury by inhibiting pyroptosis and inflammation, Cell Death Dis. 12 (2) (2021) 139.
[311] X. Shen, et al., Caspase 3/GSDME-dependent pyroptosis contributes to chemotherapy drug-induced nephrotoxicity, Cell Death Dis. 12 (2) (2021) 186.
[312] P. Liu, Z. Zhang, Y. Li, Relevance of the pyroptosis-related inflammasome pathway in the pathogenesis of diabetic kidney disease, Front. Immunol. 12 (2021) 603416.
[313] I. Jorgensen, E.A. Miao, Pyroptotic cell death defends against intracellular pathogens, Immunol. Rev. 265 (1) (2015) 130–142.
[314] S.M. Man, R. Karki, T.D. Kanneganti, Molecular mechanisms and functions of pyroptosis, inflammatory caspases and inflammasomes in infectious diseases, Immunol. Rev. 277 (1) (2017) 61–75.
[315] J.E. Galan, *Salmonella Typhimurium* and inflammation: a pathogen-centric affair, Nat. Rev. Microbiol. (2021), https://doi.org/10.1038/s41579-021-00561-4.

[316] I. Jorgensen, et al., Pyroptosis triggers pore-induced intracellular traps (PITs) that capture bacteria and lead to their clearance by efferocytosis, J. Exp. Med. 213 (10) (2016) 2113–2128.
[317] K.W. Chen, et al., Noncanonical inflammasome signaling elicits gasdermin D-dependent neutrophil extracellular traps, Sci. Immunol. 3 (26) (2018), eaar6676.
[318] G. Sollberger, et al., Gasdermin D plays a vital role in the generation of neutrophil extracellular traps, Sci. Immunol. 3 (26) (2018), eaar6689.
[319] C.L. Evavold, et al., The pore-forming protein gasdermin D regulates interleukin-1 secretion from living macrophages, Immunity 48 (1) (2018) 35–44 e6.
[320] M. Carty, et al., Cell survival and cytokine release after inflammasome activation is regulated by the Toll-IL-1R protein SARM, Immunity 50 (6) (2019) 1412–1424 e6.
[321] S. Xia, et al., Gasdermin D pore structure reveals preferential release of mature interleukin-1, Nature 593 (7860) (2021) 607–611.
[322] A.H. Chan, K. Schroder, Inflammasome signaling and regulation of interleukin-1 family cytokines, J. Exp. Med. 217 (1) (2020), e20190314.
[323] K.A. Deets, R.E. Vance, Inflammasomes and adaptive immune responses, Nat. Immunol. 22 (4) (2021) 412–422.
[324] K.W. Chen, et al., RIPK1 activates distinct gasdermins in macrophages and neutrophils upon pathogen blockade of innate immune signaling, Proc. Natl. Acad. Sci. USA 118 (28) (2021), e2101189118.
[325] S. Li, et al., Clinical and pathological investigation of patients with severe COVID-19, JCI Insight 5 (12) (2020), e138070.
[326] T.L. Freeman, T.H. Swartz, Targeting the NLRP3 inflammasome in severe COVID-19, Front. Immunol. 11 (2020) 1518.
[327] M.Z. Ratajczak, et al., SARS-CoV-2 entry receptor ACE2 is expressed on very small CD45(-) precursors of hematopoietic and endothelial cells and in response to virus spike protein activates the Nlrp3 inflammasome, Stem Cell Rev. Rep. 17 (1) (2021) 266–277.
[328] T.S. Rodrigues, et al., Inflammasomes are activated in response to SARS-CoV-2 infection and are associated with COVID-19 severity in patients, J. Exp. Med. 218 (3) (2021), e20201707.
[329] H. Xian, et al., Metformin inhibition of mitochondrial ATP and DNA synthesis abrogates NLRP3 inflammasome activation and pulmonary inflammation, Immunity 54 (7) (2021) 1463–1477 e11.
[330] M. Plassmeyer, et al., Caspases and therapeutic potential of caspase inhibitors in moderate-severe SARS CoV2 infection and long COVID, Allergy (2021), https://doi.org/10.1111/all.14907.
[331] A.C. Ferreira, et al., SARS-CoV-2 engages inflammasome and pyroptosis in human primary monocytes, Cell Death Discov. 7 (1) (2021) 43.
[332] R. Kolb, et al., Inflammasomes in cancer: a double-edged sword, Protein Cell 5 (1) (2014) 12–20.
[333] H. Van Gorp, M. Lamkanfi, The emerging roles of inflammasome-dependent cytokines in cancer development, EMBO Rep. 20 (6) (2019), e47575.
[334] S. Hamarsheh, R. Zeiser, NLRP3 inflammasome activation in cancer: a double-edged sword, Front. Immunol. 11 (2020) 1444.
[335] X. Lu, T. Guo, X. Zhang, Pyroptosis in cancer: friend or foe? Cancers 13 (14) (2021) 3620.
[336] H. Zhao, et al., Inflammation and tumor progression: signaling pathways and targeted intervention, Signal Transduct. Target. Ther. 6 (1) (2021) 263.
[337] Y. Tan, et al., Pyroptosis: a new paradigm of cell death for fighting against cancer, J. Exp. Clin. Cancer Res. 40 (1) (2021) 153.

[338] S.S. Faria, et al., NLRP3 inflammasome-mediated cytokine production and pyroptosis cell death in breast cancer, J. Biomed. Sci. 28 (1) (2021) 26.
[339] S. Missiroli, et al., Targeting the NLRP3 inflammasome as a new therapeutic option for overcoming cancer, Cancers 13 (10) (2021) 2297.
[340] Z. Zhang, Y. Zhang, J. Lieberman, Lighting a fire: can we harness pyroptosis to ignite antitumor immunity? Cancer Immunol. Res. 9 (1) (2021) 2–7.
[341] N. Ershaid, et al., NLRP3 inflammasome in fibroblasts links tissue damage with inflammation in breast cancer progression and metastasis, Nat. Commun. 10 (1) (2019) 4375.
[342] M. Bruchard, et al., Chemotherapy-triggered cathepsin B release in myeloid-derived suppressor cells activates the Nlrp3 inflammasome and promotes tumor growth, Nat. Med. 19 (1) (2013) 57–64.
[343] I. Kaplanov, et al., Blocking IL-1beta reverses the immunosuppression in mouse breast cancer and synergizes with anti-PD-1 for tumor abrogation, Proc. Natl. Acad. Sci. USA 116 (4) (2019) 1361–1369.
[344] B. Weichand, et al., S1PR1 on tumor-associated macrophages promotes lymphangiogenesis and metastasis via NLRP3/IL-1beta, J. Exp. Med. 214 (9) (2017) 2695–2713.
[345] D. Hofbauer, et al., Beta2-microglobulin triggers NLRP3 inflammasome activation in tumor-associated macrophages to promote multiple myeloma progression, Immunity 54 (8) (2021) 1772–1787.e9.
[346] J. Dupaul-Chicoine, et al., The Nlrp3 inflammasome suppresses colorectal cancer metastatic growth in the liver by promoting natural killer cell tumoricidal activity, Immunity 43 (4) (2015) 751–763.
[347] Y. Huang, et al., Myeloid PTEN promotes chemotherapy-induced NLRP3-inflammasome activation and antitumour immunity, Nat. Cell Biol. 22 (6) (2020) 716–727.
[348] R.J. North, et al., Interleukin 1-induced, T cell-mediated regression of immunogenic murine tumors. Requirement for an adequate level of already acquired host concomitant immunity, J. Exp. Med. 168 (6) (1988) 2031–2043.
[349] F. Ghiringhelli, et al., Activation of the NLRP3 inflammasome in dendritic cells induces IL-1beta-dependent adaptive immunity against tumors, Nat. Med. 15 (10) (2009) 1170–1178.
[350] O.A. Haabeth, et al., Interleukin-1 is required for cancer eradication mediated by tumor-specific Th1 cells, OncoImmunology 5 (1) (2016) e1039763.
[351] K.O. de Beeck, L. Van Laer, G. Van, Camp, DFNA5, a gene involved in hearing loss and cancer: a review, Ann. Otol. Rhinol. Laryngol. 121 (3) (2012) 197–207.
[352] X. Xia, et al., The role of pyroptosis in cancer: pro-cancer or pro-"host"? Cell Death Dis. 10 (9) (2019) 650.
[353] M. Wang, X. Chen, Y. Zhang, Biological functions of gasdermins in cancer: from molecular mechanisms to therapeutic potential, Front. Cell Dev. Biol. 9 (2021) 638710.
[354] Z. Zhang, et al., Gasdermin E suppresses tumour growth by activating anti-tumour immunity, Nature 579 (7799) (2020) 415–420.
[355] X. Liu, et al., Channelling inflammation: gasdermins in physiology and disease, Nat. Rev. Drug Discov. 20 (5) (2021) 384–405.
[356] Q. Wang, et al., A bioorthogonal system reveals antitumour immune function of pyroptosis, Nature 579 (7799) (2020) 421–426.
[357] C. Lee, et al., Inflammasome as a promising therapeutic target for cancer, Life Sci. 231 (2019) 116593.
[358] A. Zahid, et al., Pharmacological inhibitors of the NLRP3 inflammasome, Front. Immunol. 10 (2019) 2538.

[359] D. Chauhan, L. Vande Walle, M. Lamkanfi, Therapeutic modulation of inflammasome pathways, Immunol. Rev. 297 (1) (2020) 123–138.
[360] R. Cross, Could an NLRP3 inhibitor be the one drug to conquer common diseases? Chem. Eng. News 98 (7) (2020).
[361] Y. Yang, et al., Recent advances in the mechanisms of NLRP3 inflammasome activation and its inhibitors, Cell Death Dis. 10 (2) (2019) 128.
[362] D.G. Perregaux, et al., Identification and characterization of a novel class of interleukin-1 post-translational processing inhibitors, J. Pharmacol. Exp. Therapeut. 299 (1) (2001) 187–197.
[363] M. Lamkanfi, et al., Glyburide inhibits the Cryopyrin/Nalp3 inflammasome, J. Cell Biol. 187 (1) (2009) 61–70.
[364] R.C. Coll, et al., A small-molecule inhibitor of the NLRP3 inflammasome for the treatment of inflammatory diseases, Nat. Med. 21 (3) (2015) 248–255.
[365] R.C. Coll, et al., MCC950 directly targets the NLRP3 ATP-hydrolysis motif for inflammasome inhibition, Nat. Chem. Biol. 15 (6) (2019) 556–559.
[366] H. Jiang, et al., Identification of a selective and direct NLRP3 inhibitor to treat inflammatory disorders, J. Exp. Med. 214 (11) (2017) 3219–3238.
[367] Y. Huang, et al., Tranilast directly targets NLRP3 to treat inflammasome-driven diseases, EMBO Mol. Med. 10 (4) (2018).
[368] C. Juliana, et al., Anti-inflammatory compounds parthenolide and Bay 11-7082 are direct inhibitors of the inflammasome, J. Biol. Chem. 285 (13) (2010) 9792–9802.
[369] Y. He, et al., 3,4-methylenedioxy-beta-nitrostyrene inhibits NLRP3 inflammasome activation by blocking assembly of the inflammasome, J. Biol. Chem. 289 (2) (2014) 1142–1150.
[370] M. Cocco, et al., Design, synthesis, and evaluation of acrylamide derivatives as direct NLRP3 inflammasome inhibitors, ChemMedChem 11 (16) (2016) 1790–1803.
[371] H. He, et al., Oridonin is a covalent NLRP3 inhibitor with strong anti-inflammasome activity, Nat. Commun. 9 (1) (2018) 2550.
[372] C. Marchetti, et al., OLT1177, a beta-sulfonyl nitrile compound, safe in humans, inhibits the NLRP3 inflammasome and reverses the metabolic cost of inflammation, Proc. Natl. Acad. Sci. USA 115 (7) (2018) E1530–E1539.
[373] G. Bateman, et al., Great balls of fire: activation and signalling of inflammatory caspases, Biochem. Soc. Trans. 49 (3) (2021) 1311–1324.
[374] B.A. McKenzie, et al., Caspase-1 inhibition prevents glial inflammasome activation and pyroptosis in models of multiple sclerosis, Proc. Natl. Acad. Sci. USA 115 (26) (2018) E6065–E6074.
[375] J. Chen, et al., VX-765 reduces neuroinflammation after spinal cord injury in mice, Neural Regen. Res. 16 (9) (2021) 1836–1847.
[376] J. Flores, et al., Pre-symptomatic Caspase-1 inhibitor delays cognitive decline in a mouse model of Alzheimer disease and aging, Nat. Commun. 11 (1) (2020) 4571.
[377] X.E. Xu, et al., Caspase-1 inhibitor exerts brain-protective effects against sepsis-associated encephalopathy and cognitive impairments in a mouse model of sepsis, Brain Behav. Immun. 80 (2019) 859–870.
[378] J. Flores, et al., Caspase-1 inhibition alleviates cognitive impairment and neuropathology in an Alzheimer's disease mouse model, Nat. Commun. 9 (1) (2018) 3916.
[379] M. Xu, et al., Identification of small-molecule inhibitors of Zika virus infection and induced neural cell death via a drug repurposing screen, Nat. Med. 22 (10) (2016) 1101–1107.
[380] J.J. Hu, et al., FDA-approved disulfiram inhibits pyroptosis by blocking gasdermin D pore formation, Nat. Immunol. 21 (7) (2020) 736–745.

[381] J.K. Rathkey, et al., Chemical disruption of the pyroptotic pore-forming protein gasdermin D inhibits inflammatory cell death and sepsis, Sci. Immunol. 3 (26) (2018) eaat2738.
[382] T. Liu, et al., NF-kappaB signaling in inflammation, Signal Transduct. Target. Ther. 2 (2017) 17023.
[383] Q. Zhang, M.J. Lenardo, D. Baltimore, 30 Years of NF-kappaB: a blossoming of relevance to human pathobiology, Cell 168 (1–2) (2017) 37–57.
[384] K. Taniguchi, M. Karin, NF-kappaB, inflammation, immunity and cancer: coming of age, Nat. Rev. Immunol. 18 (5) (2018) 309–324.
[385] H. Yu, et al., Targeting NF-kappaB pathway for the therapy of diseases: mechanism and clinical study, Signal Transduct. Target. Ther. 5 (1) (2020) 209.
[386] N. Awasthee, et al., Targeting IkappaappaB kinases for cancer therapy, Semin. Cancer Biol. 56 (2019) 12–24.
[387] M. Labbozzetta, M. Notarbartolo, P. Poma, Can NF-kappaB Be considered a valid drug target in neoplastic diseases? Our point of view, Int. J. Mol. Sci. 21 (9) (2020) 3070.
[388] A. Mantovani, et al., Interleukin-1 and related cytokines in the regulation of inflammation and immunity, Immunity 50 (4) (2019) 778–795.
[389] C. Garlanda, A. Mantovani, Interleukin-1 in tumor progression, therapy, and prevention, Cancer Cell 39 (8) (2021) 1023–1027.
[390] H. Malcova, et al., Interleukin-1 blockade in polygenic autoinflammatory disorders: where are we now? Front. Pharmacol. 11 (2020) 619273.
[391] H.M. Hoffman, et al., Efficacy and safety of rilonacept (interleukin-1 Trap) in patients with cryopyrin-associated periodic syndromes: results from two sequential placebo-controlled studies, Arthritis Rheum. 58 (8) (2008) 2443–2452.
[392] D.S. Aschenbrenner, New indication for rilonacept, Am. J. Nurs. 121 (7) (2021) 24–25.
[393] P.M. Ridker, et al., Effect of interleukin-1beta inhibition with canakinumab on incident lung cancer in patients with atherosclerosis: exploratory results from a randomised, double-blind, placebo-controlled trial, Lancet 390 (10105) (2017) 1833–1842.
[394] E.J. Kooistra, et al., Anakinra treatment in critically ill COVID-19 patients: a prospective cohort study, Crit. Care 24 (1) (2020) 688.
[395] G. Cavalli, et al., Interleukin-1 blockade with high-dose anakinra in patients with COVID-19, acute respiratory distress syndrome, and hyperinflammation: a retrospective cohort study, Lancet Rheumatol. 2 (6) (2020) e325–e331.
[396] R. Caricchio, et al., Effect of canakinumab vs placebo on survival without invasive mechanical ventilation in patients hospitalized with severe COVID-19: a randomized clinical trial, J. Am. Med. Assoc. 326 (3) (2021) 230–239.
[397] E.A. McKie, et al., A study to investigate the efficacy and safety of an anti-interleukin-18 monoclonal antibody in the treatment of type 2 diabetes mellitus, PLoS One 11 (3) (2016) e0150018.
[398] E. Wlodek, et al., A pilot study evaluating GSK1070806 inhibition of interleukin-18 in renal transplant delayed graft function, PLoS One 16 (3) (2021) e0247972.
[399] C. Gabay, et al., Open-label, multicentre, dose-escalating phase II clinical trial on the safety and efficacy of tadekinig alfa (IL-18BP) in adult-onset Still's disease, Ann. Rheum. Dis. 77 (6) (2018) 840–847.
[400] S.W. Canna, et al., Life-threatening NLRC4-associated hyperinflammation successfully treated with IL-18 inhibition, J. Allergy Clin. Immunol. 139 (5) (2017) 1698–1701.

[401] P. Samir, R.K.S. Malireddi, T.D. Kanneganti, The PANoptosome: a deadly protein complex driving pyroptosis, apoptosis, and necroptosis (PANoptosis), Front. Cell. Infect. Microbiol. 10 (2020) 238.
[402] M.D. Stutz, et al., Macrophage and neutrophil death programs differentially confer resistance to tuberculosis, Immunity 54 (8) (2021), 1758.e7–1771.e7.

CHAPTER 2

Cell death: machinery and regulation

Daolin Tang and Rui Kang
Department of Surgery, UT Southwestern Medical Center, Dallas, TX, United States

1. Introduction

Cell death is not only a physiological process occurring during organism development and tissue homeostasis, but also a pathological event implicated in tissue injury and dysfunction. Over the last few decades, the Nomenclature Committee on Cell Death has published several guidelines on the definition, detection, and interpretation of various types of cell death, which accelerates the basic and translational research of cell death in human health and diseases [1–3]. Although existing classification criteria differ in many ways, cell death can be generally discriminated into accidental cell death and regulated cell death (Fig. 2.1) [3]. Accidental cell death is an uncontrolled process due to severe injury, such as chemical or mechanical damage. In contrast, regulated cell death is a controlled process that takes place during various types of stresses, such as hypoxia, oxidative stress, metabolic stress, and pathogen invasion, which can be modulated by gene- or drug-based approaches. Regulated cell death can be further divided into apoptotic and nonapoptotic forms, which may exhibit different morphology, genetics, biochemistry, and molecular biology [4]. In this chapter, we first introduce the molecular machinery of major types of nonapoptotic regulated cell death and then focus on a discussion of the pathways and core regulators of apoptosis. Understanding the process and function of regulated cell death, including apoptosis, is important for the development of drugs targeting human diseases.

2. Nonapoptotic cell death

The forms of nonapoptotic cell death are growing with the advancement of biomedicine research. Many experimental approaches and techniques, including chemical or RNAi screening, are used to identify new types of cell death. A general strategy for pharmacologically screening small

Mechanisms of Cell Death and Opportunities for Therapeutic Development
ISBN 978-0-12-814208-0
https://doi.org/10.1016/B978-0-12-814208-0.00003-8

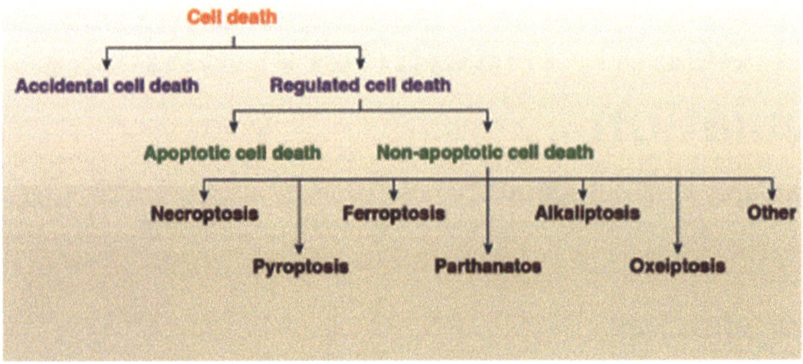

Figure 2.1 ***Classification of cell death.*** Cell death can be generally divided into accidental cell death and regulated cell death. Accidental cell death is an uncontrolled process caused by severe injury, such as chemical or mechanical damage. In contrast, regulated cell death is a controlled process that occurs during various types of stresses. Regulated cell death can be further divided into apoptotic and nonapoptotic forms, such as necroptosis, pyroptosis, ferroptosis, parthanatos, alkaliptosis, and oxeiptosis.

molecular compounds to see if they induce new types of cell death depends on confirming whether well-known cell death inhibitors block the cell death induced by a candidate drug. Below, we describe the molecular mechanisms of three main forms of nonapoptotic cell death, including necroptosis, pyroptosis, and ferroptosis, which generally are forms of regulated necrosis with plasma membrane injury and the release of cellular contents, such as damage-associated molecular patterns (DAMPs) [5], into the extracellular space.

2.1 Necroptosis

Necroptosis was first described in 1996 when pig kidney cells were infected by cowpox virus in the presence of a viral caspase (CASP1 and CASP8) inhibitor [6]. Later, the cytokine tumor necrosis factor (TNF, also known as TNFα) was found to also cause nonapoptotic cell death, namely, necroptosis, in mouse fibroblasts in the presence of a CASP8 inhibitor [7]. Indeed, many pro-inflammatory cytokines or agents targeting a cell death receptor or toll-like receptor (TLR) can induce necroptosis when caspases are inhibited by using caspase inhibitors (such as Z-VAD-FMK) or the knockdown of a caspase or other core pro-apoptotic regulator [8,9].

Although there are different upstream regulators of necroptosis, depending on the triggers, the downstream mediators of necroptosis generally require the activation of receptor-interacting serine/threonine kinase 3 (RIPK3) [10–12] and subsequent mixed-lineage kinase domain-like pseudokinase (MLKL) [13]. In particular, the ligands of cell death receptor or TLR can lead to RIPK3-dependent necroptosis through the activation of receptor-interacting serine/threonine kinase 1 (RIPK1) or TLR adaptor molecule 1 (TICAM1, also known as TRIF), respectively. The term "necroptosis" was originally based on the discovery of necrostatin-1, a RIPK1 inhibitor [14]. Phosphorylation is a key process in regulating the activation of RIPK1, RIPK3, and MLKL, which promotes the formation of a specific protein complex called the "necrosomes," which ultimately results in plasma membrane permeabilization [10–12,15,16]. Although early studies indicate that mitochondrial dysfunction, such as mitochondrial oxidative stress, the activation of mitochondrial phosphatase PGAM family member 5 (PGAM5), or an increase of mitochondrial membrane potential, is essential for the induction of necroptosis [17], recent findings suggest that mitochondria could play a context-dependent role in promoting necroptosis [18,19]. Thus, carefully dissecting these pathways will provide a better understanding of the central role of mitochondria in cell stress and death.

2.2 Pyroptosis

The term "pyroptosis" was first used by Susan Fink and Brad Cookson to describe bacterial infection-induced cell death in macrophages [20]. We now know that pyroptosis is an inflammasome-dependent regulated cell death characterized by the activation of pro-inflammatory caspases (CASP1 and CASP11 [also known as CASP4 and CASP5 in humans]) and the release of interleukin (IL) 1 family members (e.g., interleukin-1β [IL1B] and IL18) and DAMPs (e.g., high-mobility group box 1 [HMGB1] and lactate dehydrogenase) [21].

The inflammasome is a multiprotein intracellular complex that detects pathogen-associated molecular patterns (PAMPs, e.g., lipopolysaccharides [LPS] and pathogenic DNA) or DAMPs (e.g., adenosine triphosphate, uric acid, and host DNA) in innate immunity. In general, the activation of inflammasomes in macrophages or monocytes is a dynamic process requiring two steps [21]. Step one, also known as the priming stage, requires inducing the expression of inflammasome components through the activation of transcriptional factors, such as nuclear factor κB (NF-κB), via TLR ligands or other cytokines. Step two involves the formation and maturation of inflammasome structure and machinery, which leads to pro-inflammatory caspase activation and subsequent pyroptosis.

The inflammasome is divided into canonical or noncanonical inflammasomes, which rely on CASP1 or CASP11, respectively [21]. The main subtypes of canonical inflammasomes include NOD-like receptors (NLR, e.g., NLRP1, NLRP2, NLRP3, NLRP6, NLRP7, and NLRC4) and non-NLR (e.g., absent in melanoma 2 [AIM2])-mediated inflammasomes. LPS, the major component of the outer membrane of Gram-negative bacteria, can either bind TLR4 in the plasma membrane or CASP11 in the cytosol to trigger pyroptosis and inflammation through different pathways [22,23]. Gasdermin D (GSDMD) seems to play a central role in mediating pyroptosis through the formation of pores in the plasma membrane [24–29]. This process requires CASP1- or CASP11-mediated cleavage of GSDMD to produce an N-terminal fragment called GSDMD-N [26,27,29]. The formation of GSDMD-N can be further positively regulated by CASP8 or negatively regulated by CASP3 or CASP7 [30,31]. Moreover, the signals of reactive oxygen species (ROS) or calcium can accelerate or fix GSDMD-N-mediated injury during pyroptosis [32–34]. The cleavage pattern of GSDMD as well as GSDMD-N activity, however, may vary between different cell types [35]. In contrast to CASP1- or CASP11-mediated GSDMD-N production in macrophages, neutrophil elastase-mediated GSDMD-N production in neutrophils may play a distinct role in limiting bacterial infection [36,37], indicating that there is a competition between different host cells and pathogens to regulate pyroptotic cell death and infection.

2.3 Ferroptosis

The term "ferroptosis" was coined by Brent Stockwell and colleagues to describe an iron-dependent cell death in the study of targeted cancer therapies in 2012 [38]. Iron is an important trace element that not only promotes cell proliferation but also causes various types of cell death, including ferroptosis, which depends on the absorption, storage, and use of iron. Iron overload promotes ferroptosis through at least two mechanisms [39,40]. On the one hand, iron can increase ROS production through the Fenton reaction. On the other hand, iron can increase the activity of iron-containing enzymes (e.g., lipoxygenase family members [ALOX, e.g., ALOX5 and ALOX15]), which are responsible for lipid peroxidation, a core event in driving ferroptosis, mainly through the activation of the acyl-CoA synthetase long-chain family member 4 (ACSL4)-lysophosphatidylcholine acyltransferase 3 (LPCAT3)-ALOX pathway [41–45]. It is clear that iron metabolism is a complex process requiring a wide range of regulators of the balance in redox forms: ferrous cation (Fe^{2+}) and ferric cation (Fe^{3+}). In general, increased iron intracellular uptake (e.g., the overexpression of transferrin receptor [TFRC]) [46], decreased iron storage (knockdown of ferritin or overexpression of CDGSH iron sulfur domain 1 [CISD1]) [47,48], and impaired iron extracellular export (e.g., the knockdown of solute carrier family 40 member 1 [SLC40A1] [49] or inhibition of exosomes [50]) can increase free Fe^{2+} in the cytosol, which finally results in ferroptotic cell death through oxidative stress.

There are two major types of ferroptosis activators that target antioxidant system xc^- and glutathione peroxidase 4 (GPX4). System xc^- is composed of solute carrier family 7 member 11 (SLC7A11) and solute carrier family 3 member 2 (SLC3A2) that mediate the exchange between extracellular cysteine and intracellular glutamate. System xc^--mediated cystine uptake can be further used to produce glutathione, which is important for sustaining the activity and function of GPX4, an antioxidant enzyme that can reduce the production of phospholipid hydroperoxide during lipid peroxidation. Consequently, the pharmacological inhibition of system xc^- and GPX4 by erastin or RSL3, respectively, has been shown to induce ferroptosis in various cells, including cancer and noncancer cells [51]. GPX4 also protects against apoptosis [52], necroptosis [53], and pyroptosis [32], indicating that different downstream effectors of GPX4 inhibition are responsible for different types of regulated cell death. In addition to the classical system xc^--GPX4 antiferroptotic pathways, other alternative pathways, such as apoptosis-inducing factor mitochondria associated 2 (AIFM2)-mediated coenzyme Q10 production [54,55] and endosomal sorting complexes required for transport (ESCRT)-III-mediated membrane repair [56], can limit lipid peroxidation in ferroptosis. Moreover, several stress-associated regulators, such as tumor protein P53 (TP53) [57–59] and nuclear factor, erythroid 2-like 2 (NFE2L2, best known as NRF2) [60–62], play a context-dependent role in the regulation of ferroptosis depending on its transcription-dependent and -independent functions.

Notably, ferroptosis was originally described as a novel cell death different from apoptosis, necrosis, and autophagy [38]. This notion is challenged by recent studies. First, ferroptosis is generally a type of regulated necrosis with the characteristic morphological changes of necrosis [39]. Second, certain classical ferroptosis activators, such as erastin, can induce apoptosis or mixed cell death in cancer cells [63–65]. Third, ferroptosis requires autophagy machinery or components, especially selective types of autophagy (e.g., ferritinophagy [66,67], lipophagy [68], clockophagy [69,70], and chaperone-mediated autophagy [71]), to initiate or accelerate cell death [72–74]. These findings indicate that ferroptosis may occur in multiple forms sharing common regulators or stimuli of other nonapoptotic types of cell death [75,76].

2.4 Other

In addition to necroptosis, pyroptosis, and ferroptosis discussed above, other types of nonapoptotic cell death can occur under stress conditions within different signals and pathways. For example, parthanatos is a poly(ADP-Ribose) polymerase 1 (PARP1)-dependent form of cancer cell death via the activation of apoptosis-inducing factor mitochondria-associated 1 (AIFM1) and macrophage migration inhibitory factor (MIF) pathways [77,78]. Alkaliptosis is a pH-dependent cancer cell death driven by intracellular alkalinization, which is partly regulated by the NF-κB–dependent downregulation of carbonic anhydrase 9 (CA9), a member of the zinc metalloenzymes that catalyzes the reversible hydration of carbon dioxide [79,80]. Oxeiptosis is also an ROS-mediated cell death that requires the activation of the PGAM5-AIFM1 pathway in macrophages [81,82]. The selectivity of regulated cell death in different conditions must be further elucidated as it is important for understanding immune and metabolism responses to cell death in human diseases.

3. Apoptotic cell death

The term "apoptosis" is used in Greek to describe the falling process of leaves or flowers from trees or plants. In 1842, Karl Vogt was the first to describe an apoptotic event in midwife toad (*Alytes obstetricans*) embryos. In 1885, Walther Flemming further described this event in embryonic development as the process of programmed cell death. In 1972, the term "apoptosis" was coined by Drs. John Kerr, Alastair Currie, and Andrew Wyllie to indicate a form of programmed cell death with morphological changes that differ from necrosis [83]. Apoptosis was originally defined by the characteristic morphology of the dying cells. Unlike necrotic cells, apoptotic cells usually become smaller and generate apoptotic bodies with intact plasma membranes. In the following decades, with the use of biochemical and genetic approaches, and the clone of pro-apoptotic and antiapoptotic genes during *Caenorhabditis elegans* development [84–86] marked the beginning of a new era in molecular apoptosis research. In 2002, Sydney Brenner, H. Robert Horvitz, and John E. Sulston were awarded the Nobel Prize for their discoveries of the molecular mechanisms and function of apoptosis. It is clear that apoptosis is not only a physiological process in the regulation of organ development through removing interdigital cells or nonfunctional cells but also a pathological event implicated in disease development through the induction of an inflammation response or by causing tissue dysfunction [87]. In biochemistry, apoptosis is usually accompanied by the degradation of chromosomal DNA, rapid exposure of membrane phosphatidylserine, and loss of mitochondrial membrane potential [88]. Below, we discuss the pathways and regulators of apoptosis in mammalian cells.

3.1 Apoptotic pathways

In general, apoptosis can be triggered through two pathways, namely, extrinsic and intrinsic apoptosis (Fig. 2.2). Extrinsic apoptosis is mainly induced by ligands of death receptors, such as Fas ligand (FASLG, also known as FasL), TNF superfamily member 10 (TNFSF10, also known as TRAIL), and TNF [89]. These ligands share common structures and can be

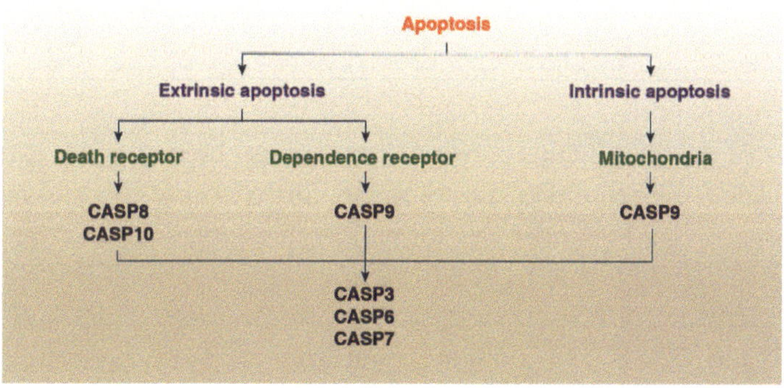

Figure 2.2 ***Classification of apoptosis.*** Apoptosis can be divided into extrinsic or intrinsic apoptosis, which use different caspases to cleave substrates. Extrinsic apoptosis is triggered by the activation of death receptors by ligands or the inhibition of dependence receptors by the withdrawing of trophic factors or the presence of antitrophic antibodies. The intrinsic pathway of apoptosis is also known as the mitochondria-mediated death pathway, which is triggered by mitochondrial outer membrane permeabilization.

cleaved from membrane-bound forms to generate soluble forms via specific protease enzymes, such as metalloprotease. Membrane-bound ligands of death receptors can bind to death receptors such as FAS (also known as CD95), and TNF receptor superfamily member 1A (TNFRSF1A, also known as TNFR1) in plasma membrane to generate death-inducing signaling complex (DISC), a protein complex containing the death receptor, Fas associated via death domain (FADD), CASP8 or CASP10. Subsequently, the formation of DISC leads to the cleavage and activation of CASP8 or CASP10 from pro-CASP8 or pro-CASP10, which are inactive single polypeptide chain zymogen procaspases [89]. Another pathway for extrinsic apoptosis induction is mediated by dependence receptors, such as unc-5 netrin receptor B (UNC5B, also known as UNC5H2) and DCC netrin 1 receptor (DCC) [90]. Trophic factors, such as fibroblast growth factor, nerve growth factor, and those of the epidermal growth factor family, are best known for their pro-survival functions in neurons. In contrast, the absence of these trophic factors or the presence of antitrophic antibodies can induce dependence receptor-mediated apoptosis through the cleavage and activation of CASP9 but not CASP8 or CASP10 [90].

The intrinsic pathway, namely, mitochondrial pathway, can be triggered by a number of factors or stresses (e.g., DNA damage, ischemia, and oxidative stress), which leads to mitochondrial outer membrane permeabilization (MOMP) and the subsequent release of pro-apoptotic mitochondrial proteins, such as cytochrome C, somatic (CYCS), diablo IAP-binding mitochondrial protein (DIABLO, also known as Smac), and HtrA serine peptidase 2 (HTRA2) [91]. MOMP is directly modulated by a member of the BCL2 family, which can increase or inhibit MOMP [87]. The released mitochondrial proteins can further cause the cleavage and activation of CASP9 under the formation of apoptosomes via apoptotic protease activating factor 1 (APAF1, also known as CED-4 in *C. elegans*). Finally, activated CASP9 can cleave effector caspases, such as CASP3, CASP6, and CASP7, to induce the intrinsic apoptosis pathway via the cleavage of cellular targets or the destruction of cellular structures [92].

Extrinsic apoptosis signals can activate the intrinsic apoptosis pathway via CASP8-mediated cleavage of BID, a member of the BCL-2 family [93]. Cleaved BID (also known as tBID) can translocate from cytosol to mitochondria, leading to MOMP and CYCS release [93]. Thus, BID is an important regulator of the interplay between extrinsic and intrinsic apoptosis.

3.2 Apoptotic regulators

3.2.1 Caspase family

Caspases are a family of cysteinyl-containing aspartate-specific proteinases that play key roles in cell death by apoptosis or pyroptosis [92,94]. Based on their function and structure, mammalian caspases can be divided into four subfamilies: initiator caspases (CASP2, CASP8, CASP9, and CASP10 [that exists in humans, not mice]), executioner caspases (CASP3, CASP6, and CASP7), inflammatory caspases (CASP1, CASP11 [exists in mice, corresponding to CASP4 or CASP5 in humans], and CASP12), and the keratinization-associated caspase (CASP14). Under normal conditions, caspases are usually found as monomeric forms of inactive proenzymes. Under apoptotic conditions, caspases can occur as dimerized forms of active enzymes through proteolytic processing of aspartic residues to produce two subunits, namely, the large and small subunits [92]. These subunits finally assemble into active enzymes responsible for substrate cleavage. Although apoptosis is a process of caspase cascade, caspases can be activated by three different routes. First, the previously activated initiator caspase (e.g., CASP9) can cleave its downstream caspase (e.g., CASP3). Second, a caspase (e.g., CASP8) can be activated via DISC. Third, the activation of a caspase (e.g., CASP9) requires a specific regulatory protein complex (e.g., apoptosome). The activation or inhibition of CASP8 appears to play a wide role in various kinds of regulated cell death, including apoptosis, necroptosis, and pyroptosis, but the underlying structure of this process remains largely unknown [8,95–99].

Caspases are fundamental components of the apoptotic machinery that can cause cell death through protein cleavages. Although the identification of all relevant protein substrates of caspases remains a challenge, studies have shown that over 1300 different proteins can be cleaved by caspases at 1706 sites in an apoptotic human cell [100]. Moreover, individual activated caspases have preferred protein substrates with overlap or distinct recognition sequences. For example, CASP3 and CASP7 share similar substrate specificity by recognizing the cleavage motif Asp–Glu–Val–Asp, whereas CASP8 and CASP9 can recognize cleavage motif Ile–Glu–Thr–Asp or Ser–Glu–Val–Asp, respectively. Caspase inhibitors are developed by binding the active site of caspases in either a reversible or irreversible manner [94].

3.2.2 BCL2 family

The BCL2 family is one of the most evolutionarily conserved regulators of apoptosis. The family is generally composed of antiapoptotic members, such as BCL2 and BCL2-like 1 (BCL2L1, also known as BCLXL) and pro-apoptotic members, such as BCL2-associated X, apoptosis regulator (BAX), and BCL2 antagonist/killer 1 (BAK1, also known as BAK) [87,101]. Although members of the BCL2 family display tissue-specific differences in expression, they have structural similarities. In some members, the C-terminal end has a hydrophobic transmembrane domain that determines the subcellular localization of BCL2 family proteins. They also share one or more of the four conserved BCL2 homology (BH) domains, namely, BH1, BH2, BH3, and BH4. Most antiapoptotic BCL2 members contain BH1 and BH2 domains and some of them (e.g., BCL2 and BCL2L1) have an extra BH4 domain that is important for their antiapoptotic activity, whereas pro-apoptotic proteins usually contain a BH3 domain responsible for dimerization with other members of the BCL2 family.

The BCL2 family regulates apoptosis mainly through modulation of the opening and closing of the mitochondrial permeability transition pore (MPTP), a transmembrane protein channel residing in the mitochondrial inner membrane for control of the transport of Ca^{2+}, pH, and voltage [87]. Although the structure of MPTP remains largely unknown, adenine nucleotide translocase, cyclophilin D, and voltage-dependent anion channels have been recognized as key components of the MPTP [102]. The interaction between BCL2 and MPTP components is multifaceted and determines MOMP in apoptosis. In a well-studied model, antiapoptotic BCL2 has been found to limit the arrangement of pro-apoptotic BAX and BAK1 in membrane-bound complexes, which finally block MOMP and CYCS release [103]. In contrast, activated BH3-only pro-apoptotic BCL2 proteins can enhance the formation of the BAX-BAK1 complex in mitochondria [103]. Moreover, the expression and location of several BCL2 family members is regulated by TP53, a sequence-specific transcription factor for cell cycle arrest or apoptosis [104]. These findings indicate that there exists a network that broadly modulates MOMP and MPTP during apoptosis.

3.2.3 IAP family

Inhibitor of apoptosis proteins (IAPs) are a group of negative regulators of apoptosis that were first discovered in baculoviruses [105,106]. In mammalian cells, a couple of IAP family proteins, such as X-linked inhibitor of IAP (XIAP), baculoviral IAP repeat containing-2 (BIRC2, also known as cIAP1), baculoviral IAP repeat-containing 3 (BIRC3, also known as cIAP2), NLR family apoptosis inhibitory protein (NAIP), baculoviral IAP repeat-containing 7 (BIRC7, also known as livin), and baculoviral IAP repeat-containing 5 (BIRC5, also known as survivin), have been identified. A common structural feature of IAPs is the presence of one to three baculovirus IAP repeat (BIR) domains responsible for the protein–protein interactions [105,106]. In addition, certain IAP family members contain a Ubiquitin-associated domain, a really interesting new gene domain, or a conserved caspase recruitment domain, which plays a role in the regulation of protein degradation as well as protein–protein interactions.

A common mechanism for the inactivation of apoptosis by the IAP family is to inhibit caspase activation [107]. For example, XIAP, the best-studied member of the IAP family, can block apoptosis through binding to CASP3, CASP7, and CASP9 [108]. In contrast, released mitochondrial proteins, such as DIABLO and HTRA2, can reverse the antiapoptotic activity of XIAP through protein–protein interactions. Elucidating the relationship between their binding partners may expand our understanding of the mechanism of caspase activation. Although the BIR domain is required for the antiapoptotic activity of all IAP family proteins, not all BIR-containing proteins have similar functions in the suppression of apoptosis [109,110]. Time will further tell the fundamental structure–function relationship between IAP family and apoptosis.

3.2.4 HSP family

Heat shock proteins (HSPs) are conserved proteins that are divided into several groups, including HSP100, HSP90, HSP70, HSP60, HSP40, and small HSPs (sHSPs), based on their molecular sizes [111]. They can be upregulated by cells in response to various environmental stresses, especially heat stress, and play a molecular-chaperone role in the regulation of protein folding, translocation, and degradation. Moreover, the HSP family is an important cell survival regulator during various types of cell death, including apoptosis [112,113], necroptosis [114,115], and ferroptosis [71,116,117]. Different members of the HSP family suppress apoptosis through different mechanisms. For example, HSPB1 (also known as HSP25 in mice or HSP27 in humans) plays a broad role in the inhibition of apoptosis through directly binding to BAX or by inhibiting CYCS activity. HSP60, a mitochondrial HSP, can bind to an MPTP component (e.g., cyclophilin D), limiting MOMP or enhancing BIRC5-mediated caspase inhibition [118,119]. Both HSP70 and HSP90 have been shown to bind APAF1 to inhibit CASP9 activation [120–122]. Moreover, HSP90 can enhance AKT activity to sustain antiapoptotic BCL2 function [123,124]. In addition to AKT, the activity of apoptosis-associated kinases (e.g., mitogen-activated protein kinase kinase kinase 5 [MAP3K5, also known as ASK1] and c-Jun N-terminal kinases [JNKs]) is fine-tuned by certain HSPs. Thus, the inhibition of HSPs could be a strategy for the enhancement of the anticancer activity of apoptosis inducers, although different HSP inhibitor may not block only one target.

4. Conclusions and perspectives

Deregulated apoptosis is a hallmark of many diseases and can result in cancer therapy chemoresistance. In general, apoptosis is a caspase-dependent cell death, which can be controlled by multiple factors, such as BCL2, IAP, and HSP protein families. Alternatively, caspase-independent apoptotic execution has been reported through the activation of crosstalk signals between mitochondria and the nucleus [125]. However, it remains a challenge to distinguish the difference between caspase-independent apoptosis and nonapoptotic cell death. Indeed, a mixed type of cell death appears to be more common in diseases, which may share the same regulators although the underlying molecular network, including upstream and downstream signals, may be variable [4]. Moreover, the clearance of apoptotic cells by phagocytic cells or the activation of autophagy is also important to sustain homeostasis [126]. Although apoptotic cell death is generally immune-silent, the release of nuclear DAMPs, such as DNA [127,128] and histone [129], may cause an inflammatory response in human diseases. To understand the context-dependent regulation of apoptotic cell death, we need ideal models, reliable measurement methods, and specific inhibitors to define the different forms of apoptosis, including classical and nonclassical apoptosis.

Acknowledgments

We thank Dave Primm (Department of Surgery, University of Texas Southwestern Medical Center) for his critical reading of the manuscript.

References

[1] L. Galluzzi, et al., Molecular definitions of cell death subroutines: recommendations of the Nomenclature Committee on Cell Death 2012, Cell Death Differ. 19 (1) (2012) 107–120.

[2] L. Galluzzi, et al., Molecular mechanisms of cell death: recommendations of the Nomenclature Committee on Cell Death 2018, Cell Death Differ. 25 (3) (2018) 486–541.
[3] L. Galluzzi, et al., Essential versus accessory aspects of cell death: recommendations of the NCCD 2015, Cell Death Differ. 22 (1) (2015) 58–73.
[4] D. Tang, et al., The molecular machinery of regulated cell death, Cell Res. 29 (5) (2019) 347–364.
[5] D. Tang, et al., PAMPs and DAMPs: signal 0s that spur autophagy and immunity, Immunol. Rev. 249 (1) (2012) 158–175.
[6] M. Pasparakis, P. Vandenabeele, Necroptosis and its role in inflammation, Nature 517 (7534) (2015) 311–320.
[7] S.M. Laster, J.G. Wood, L.R. Gooding, Tumor necrosis factor can induce both apoptic and necrotic forms of cell lysis, J. Immunol. 141 (8) (1988) 2629–2634.
[8] D. Vercammen, et al., Inhibition of caspases increases the sensitivity of L929 cells to necrosis mediated by tumor necrosis factor, J. Exp. Med. 187 (9) (1998) 1477–1485.
[9] L. Galluzzi, et al., Necroptosis: mechanisms and relevance to disease, Annu. Rev. Pathol. 12 (2017) 103–130.
[10] D.W. Zhang, et al., RIP3, an energy metabolism regulator that switches TNF-induced cell death from apoptosis to necrosis, Science 325 (5938) (2009) 332–336.
[11] S. He, et al., Receptor interacting protein kinase-3 determines cellular necrotic response to TNF-alpha, Cell 137 (6) (2009) 1100–1111.
[12] Y.S. Cho, et al., Phosphorylation-driven assembly of the RIP1-RIP3 complex regulates programmed necrosis and virus-induced inflammation, Cell 137 (6) (2009) 1112–1123.
[13] L. Sun, et al., Mixed lineage kinase domain-like protein mediates necrosis signaling downstream of RIP3 kinase, Cell 148 (1–2) (2012) 213–227.
[14] A. Degterev, et al., Chemical inhibitor of nonapoptotic cell death with therapeutic potential for ischemic brain injury, Nat. Chem. Biol. 1 (2) (2005) 112–119.
[15] Y. Xie, et al., Inhibition of aurora kinase A induces necroptosis in pancreatic carcinoma, Gastroenterology 153 (5) (2017) 1429e5–1443e5.
[16] W. Chen, et al., Ppm1b negatively regulates necroptosis through dephosphorylating Rip3, Nat. Cell Biol. 17 (4) (2015) 434–444.
[17] Z. Wang, et al., The mitochondrial phosphatase PGAM5 functions at the convergence point of multiple necrotic death pathways, Cell 148 (1–2) (2012) 228–243.
[18] J.M. Murphy, et al., The pseudokinase MLKL mediates necroptosis via a molecular switch mechanism, Immunity 39 (3) (2013) 443–453.
[19] Q. Remijsen, et al., Depletion of RIPK3 or MLKL blocks TNF-driven necroptosis and switches towards a delayed RIPK1 kinase-dependent apoptosis, Cell Death Dis. 5 (2014) e1004.
[20] S.L. Fink, B.T. Cookson, Apoptosis, pyroptosis, and necrosis: mechanistic description of dead and dying eukaryotic cells, Infect. Immun. 73 (4) (2005) 1907–1916.
[21] P. Broz, V.M. Dixit, Inflammasomes: mechanism of assembly, regulation and signalling, Nat. Rev. Immunol. 16 (7) (2016) 407–420.
[22] J.A. Hagar, et al., Cytoplasmic LPS activates caspase-11: implications in TLR4-independent endotoxic shock, Science 341 (6151) (2013) 1250–1253.
[23] N. Kayagaki, et al., Noncanonical inflammasome activation by intracellular LPS independent of TLR4, Science 341 (6151) (2013) 1246–1249.
[24] J. Ding, et al., Pore-forming activity and structural autoinhibition of the gasdermin family, Nature 535 (7610) (2016) 111–116.
[25] X. Liu, et al., Inflammasome-activated gasdermin D causes pyroptosis by forming membrane pores, Nature 535 (7610) (2016) 153–158.

[26] N. Kayagaki, et al., Caspase-11 cleaves gasdermin D for non-canonical inflammasome signalling, Nature 526 (7575) (2015) 666–671.
[27] J. Shi, et al., Cleavage of GSDMD by inflammatory caspases determines pyroptotic cell death, Nature 526 (7575) (2015) 660–665.
[28] X. Chen, et al., Pyroptosis is driven by non-selective gasdermin-D pore and its morphology is different from MLKL channel-mediated necroptosis, Cell Res. 26 (9) (2016) 1007–1020.
[29] W.T. He, et al., Gasdermin D is an executor of pyroptosis and required for interleukin-1beta secretion, Cell Res. 25 (12) (2015) 1285–1298.
[30] P. Orning, et al., Pathogen blockade of TAK1 triggers caspase-8-dependent cleavage of gasdermin D and cell death, Science 362 (6418) (2018) 1064–1069.
[31] J. Sarhan, et al., Caspase-8 induces cleavage of gasdermin D to elicit pyroptosis during *Yersinia* infection, Proc. Natl. Acad. Sci. USA 115 (46) (2018) E10888–E10897.
[32] R. Kang, et al., Lipid peroxidation drives gasdermin D-mediated pyroptosis in lethal polymicrobial sepsis, Cell Host Microbe 24 (1) (2018) 97–108. e4.
[33] S. Ruhl, et al., ESCRT-dependent membrane repair negatively regulates pyroptosis downstream of GSDMD activation, Science 362 (6417) (2018) 956–960.
[34] R. Chen, et al., AGER-mediated lipid peroxidation drives caspase-11 inflammasome activation in sepsis, Front. Immunol. 10 (2019) 1904.
[35] P. Broz, P. Pelegrin, F. Shao, The gasdermins, a protein family executing cell death and inflammation, Nat. Rev. Immunol. (2019).
[36] G. Sollberger, et al., Gasdermin D plays a vital role in the generation of neutrophil extracellular traps, Sci. Immunol. 3 (26) (2018).
[37] K.W. Chen, et al., Noncanonical inflammasome signaling elicits gasdermin D-dependent neutrophil extracellular traps, Sci. Immunol. 3 (26) (2018).
[38] S.J. Dixon, et al., Ferroptosis: an iron-dependent form of nonapoptotic cell death, Cell 149 (5) (2012) 1060–1072.
[39] B.R. Stockwell, et al., Ferroptosis: a regulated cell death nexus linking metabolism, redox biology, and disease, Cell 171 (2) (2017) 273–285.
[40] Y. Xie, et al., Ferroptosis: process and function, Cell Death Differ. 23 (3) (2016) 369–379.
[41] S. Doll, et al., ACSL4 dictates ferroptosis sensitivity by shaping cellular lipid composition, Nat. Chem. Biol. 13 (1) (2017) 91–98.
[42] V.E. Kagan, et al., Oxidized arachidonic and adrenic PEs navigate cells to ferroptosis, Nat. Chem. Biol. 13 (1) (2017) 81–90.
[43] H. Yuan, et al., Identification of ACSL4 as a biomarker and contributor of ferroptosis, Biochem. Biophys. Res. Commun. 478 (3) (2016) 1338–1343.
[44] S.E. Wenzel, et al., PEBP1 wardens ferroptosis by enabling lipoxygenase generation of lipid death signals, Cell 171 (3) (2017) 628e26–641e26.
[45] W.S. Yang, et al., Peroxidation of polyunsaturated fatty acids by lipoxygenases drives ferroptosis, Proc. Natl. Acad. Sci. USA 113 (34) (2016) E4966–E4975.
[46] W.S. Yang, B.R. Stockwell, Synthetic lethal screening identifies compounds activating iron-dependent, nonapoptotic cell death in oncogenic-RAS-harboring cancer cells, Chem. Biol. 15 (3) (2008) 234–245.
[47] H. Yuan, et al., CISD1 inhibits ferroptosis by protection against mitochondrial lipid peroxidation, Biochem. Biophys. Res. Commun. 478 (2) (2016) 838–844.
[48] Y.Q. Wang, et al., The protective role of mitochondrial ferritin on erastin-induced ferroptosis, Front. Aging Neurosci. 8 (2016) 308.
[49] N. Geng, et al., Knockdown of ferroportin accelerates erastin-induced ferroptosis in neuroblastoma cells, Eur. Rev. Med. Pharmacol. Sci. 22 (12) (2018) 3826–3836.
[50] C.W. Brown, et al., Prominin2 drives ferroptosis resistance by stimulating iron export, Dev. Cell (2019).

[51] W.S. Yang, et al., Regulation of ferroptotic cancer cell death by GPX4, Cell 156 (1–2) (2014) 317–331.
[52] Q. Ran, et al., Embryonic fibroblasts from Gpx4+/− mice: a novel model for studying the role of membrane peroxidation in biological processes, Free Radic. Biol. Med. 35 (9) (2003) 1101–1109.
[53] O. Canli, et al., Glutathione peroxidase 4 prevents necroptosis in mouse erythroid precursors, Blood 127 (1) (2016) 139–148.
[54] K. Bersuker, et al., The CoQ oxidoreductase FSP1 acts parallel to GPX4 to inhibit ferroptosis, Nature (2019).
[55] S. Doll, et al., FSP1 is a glutathione-independent ferroptosis suppressor, Nature (2019).
[56] E. Dai, et al., ESCRT-III-dependent membrane repair blocks ferroptosis, Biochem. Biophys. Res. Commun. (2019).
[57] L. Jiang, et al., Ferroptosis as a p53-mediated activity during tumour suppression, Nature 520 (7545) (2015) 57–62.
[58] Y. Xie, et al., The tumor suppressor p53 limits ferroptosis by blocking DPP4 activity, Cell Rep. 20 (7) (2017) 1692–1704.
[59] R. Kang, G. Kroemer, D. Tang, The tumor suppressor protein p53 and the ferroptosis network, Free Radic. Biol. Med. (2018).
[60] X. Sun, et al., Activation of the p62-Keap1-NRF2 pathway protects against ferroptosis in hepatocellular carcinoma cells, Hepatology 63 (1) (2016) 173–184.
[61] X. Sun, et al., Metallothionein-1G facilitates sorafenib resistance through inhibition of ferroptosis, Hepatology 64 (2) (2016) 488–500.
[62] D. Chen, et al., NRF2 is a major target of ARF in p53-independent tumor suppression, Mol. Cell 68 (1) (2017) 224e4–232e4.
[63] S.H. Hong, et al., Molecular crosstalk between ferroptosis and apoptosis: emerging role of ER stress-induced p53-independent PUMA expression, Oncotarget 8 (70) (2017) 115164–115178.
[64] C. Huang, et al., Upregulation and activation of p53 by erastininduced reactive oxygen species contribute to cytotoxic and cytostatic effects in A549 lung cancer cells, Oncol. Rep. 40 (4) (2018) 2363–2370.
[65] T. Muller, et al., Necroptosis and ferroptosis are alternative cell death pathways that operate in acute kidney failure, Cell. Mol. Life Sci. 74 (19) (2017) 3631–3645.
[66] W. Hou, et al., Autophagy promotes ferroptosis by degradation of ferritin, Autophagy 12 (8) (2016) 1425–1428.
[67] Z. Zhang, et al., Activation of ferritinophagy is required for the RNA-binding protein ELAVL1/HuR to regulate ferroptosis in hepatic stellate cells, Autophagy 14 (12) (2018) 2083–2103.
[68] Y. Bai, et al., Lipid storage and lipophagy regulates ferroptosis, Biochem. Biophys. Res. Commun. 508 (4) (2019) 997–1003.
[69] M. Yang, et al., Clockophagy is a novel selective autophagy process favoring ferroptosis, Sci. Adv. 5 (7) (2019) eaaw2238.
[70] J. Liu, et al., Autophagic degradation of the circadian clock regulator promotes ferroptosis, Autophagy 15 (11) (2019) 2033–2035.
[71] Z. Wu, et al., Chaperone-mediated autophagy is involved in the execution of ferroptosis, Proc. Natl. Acad. Sci. USA 116 (8) (2019) 2996–3005.
[72] R. Kang, et al., BECN1 is a new driver of ferroptosis, Autophagy 14 (12) (2018) 2173–2175.
[73] X. Song, et al., AMPK-mediated BECN1 phosphorylation promotes ferroptosis by directly blocking system xc(−) activity, Curr. Biol. 28 (15) (2018) 2388e5–2399e5.
[74] R. Kang, D. Tang, Autophagy and ferroptosis - what's the connection? Curr. Pathobiol. Rep. 5 (2) (2017) 153–159.

[75] H. Gao, et al., Ferroptosis is a lysosomal cell death process, Biochem. Biophys. Res. Commun. 503 (3) (2018) 1550–1556.
[76] B. Zhou, et al., Ferroptosis is a type of autophagy-dependent cell death, Semin. Cancer Biol. (2019).
[77] S.A. Susin, et al., Molecular characterization of mitochondrial apoptosis-inducing factor, Nature 397 (6718) (1999) 441–446.
[78] Y. Wang, et al., A nuclease that mediates cell death induced by DNA damage and poly(ADP-ribose) polymerase-1, Science 354 (6308) (2016).
[79] X. Song, et al., JTC801 induces pH-dependent death specifically in cancer cells and slows growth of tumors in mice, Gastroenterology 154 (5) (2018) 1480–1493.
[80] J. Liu, et al., Alkaliptosis: a new weapon for cancer therapy, Cancer Gene Ther. (2019).
[81] C. Holze, et al., Oxeiptosis, a ROS-induced caspase-independent apoptosis-like cell-death pathway, Nat. Immunol. 19 (2) (2018) 130–140.
[82] P. Scaturro, A. Pichlmair, Oxeiptosis: a discreet way to respond to radicals, Curr. Opin. Immunol. 56 (2019) 37–43.
[83] J.F. Kerr, A.H. Wyllie, A.R. Currie, Apoptosis: a basic biological phenomenon with wide-ranging implications in tissue kinetics, Br. J. Canc. 26 (4) (1972) 239–257.
[84] M.O. Hengartner, R.E. Ellis, H.R. Horvitz, *Caenorhabditis elegans* gene ced-9 protects cells from programmed cell death, Nature 356 (6369) (1992) 494–499.
[85] M.O. Hengartner, H.R. Horvitz, *C. elegans* cell survival gene ced-9 encodes a functional homolog of the mammalian proto-oncogene bcl-2, Cell 76 (4) (1994) 665–676.
[86] J. Yuan, H.R. Horvitz, The *Caenorhabditis elegans* cell death gene ced-4 encodes a novel protein and is expressed during the period of extensive programmed cell death, Development 116 (2) (1992) 309–320.
[87] R. Singh, A. Letai, K. Sarosiek, Regulation of apoptosis in health and disease: the balancing act of BCL-2 family proteins, Nat. Rev. Mol. Cell Biol. 20 (3) (2019) 175–193.
[88] M.O. Hengartner, The biochemistry of apoptosis, Nature 407 (6805) (2000) 770–776.
[89] K. Schulze-Osthoff, et al., Apoptosis signaling by death receptors, Eur. J. Biochem. 254 (3) (1998) 439–459.
[90] D.E. Bredesen, P. Mehlen, S. Rabizadeh, Apoptosis and dependence receptors: a molecular basis for cellular addiction, Physiol. Rev. 84 (2) (2004) 411–430.
[91] J.E. Chipuk, L. Bouchier-Hayes, D.R. Green, Mitochondrial outer membrane permeabilization during apoptosis: the innocent bystander scenario, Cell Death Differ. 13 (8) (2006) 1396–1402.
[92] D.R. McIlwain, T. Berger, T.W. Mak, Caspase functions in cell death and disease, Cold Spring Harb. Perspect. Biol. 7 (4) (2015).
[93] H. Li, et al., Cleavage of BID by caspase 8 mediates the mitochondrial damage in the Fas pathway of apoptosis, Cell 94 (4) (1998) 491–501.
[94] N. Van Opdenbosch, M. Lamkanfi, Caspases in cell death, inflammation, and disease, Immunity 50 (6) (2019) 1352–1364.
[95] R. Weinlich, et al., Necroptosis in development, inflammation and disease, Nat. Rev. Mol. Cell Biol. 18 (2) (2017) 127–136.
[96] A.M. Espinosa-Oliva, et al., Brainiac caspases: beyond the wall of apoptosis, Front. Cell. Neurosci. 13 (2019) 500.
[97] M. Fritsch, et al., Caspase-8 is the molecular switch for apoptosis, necroptosis and pyroptosis, Nature (2019).
[98] K. Newton, et al., Cleavage of RIPK1 by caspase-8 is crucial for limiting apoptosis and necroptosis, Nature 574 (7778) (2019) 428–431.

[99] K. Newton, et al., Activity of caspase-8 determines plasticity between cell death pathways, Nature (2019).
[100] E.D. Crawford, et al., The DegraBase: a database of proteolysis in healthy and apoptotic human cells, Mol. Cell. Proteomics 12 (3) (2013) 813–824.
[101] P.E. Czabotar, et al., Control of apoptosis by the BCL-2 protein family: implications for physiology and therapy, Nat. Rev. Mol. Cell Biol. 15 (1) (2014) 49–63.
[102] A. Rasola, P. Bernardi, The mitochondrial permeability transition pore and its involvement in cell death and in disease pathogenesis, Apoptosis 12 (5) (2007) 815–833.
[103] A. Pena-Blanco, A.J. Garcia-Saez, Bax, Bak and beyond - mitochondrial performance in apoptosis, FEBS J. 285 (3) (2018) 416–431.
[104] J.T. Zilfou, S.W. Lowe, Tumor suppressive functions of p53, Cold Spring Harb. Perspect. Biol. 1 (5) (2009) a001883.
[105] J. Silke, J. Vince, IAPs and cell death, Curr. Top. Microbiol. Immunol. 403 (2017) 95–117.
[106] R. Budhidarmo, C.L. Day, IAPs: modular regulators of cell signalling, Semin. Cell Dev. Biol. 39 (2015) 80–90.
[107] J. Berthelet, L. Dubrez, Regulation of apoptosis by inhibitors of apoptosis (IAPs), Cells 2 (1) (2013) 163–187.
[108] B.P. Eckelman, G.S. Salvesen, F.L. Scott, Human inhibitor of apoptosis proteins: why XIAP is the black sheep of the family, EMBO Rep. 7 (10) (2006) 988–994.
[109] J. Silke, P. Meier, Inhibitor of apoptosis (IAP) proteins-modulators of cell death and inflammation, Cold Spring Harb. Perspect. Biol. 5 (2) (2013).
[110] F. Cossu, et al., Targeting the BIR domains of inhibitor of apoptosis (IAP) proteins in cancer treatment, Comput. Struct. Biotechnol. J. 17 (2019) 142–150.
[111] H. Jee, Size dependent classification of heat shock proteins: a mini-review, J. Exerc. Rehabil. 12 (4) (2016) 255–259.
[112] H.M. Beere, D.R. Green, Stress management - heat shock protein-70 and the regulation of apoptosis, Trends Cell Biol. 11 (1) (2001) 6–10.
[113] K. Purandhar, et al., Understanding the role of heat shock protein isoforms in male fertility, aging and apoptosis, World J. Mens Health 32 (3) (2014) 123–132.
[114] A.V. Jacobsen, J. Silke, The importance of being chaperoned: HSP90 and necroptosis, Cell Chem. Biol. 23 (2) (2016) 205–207.
[115] X.M. Zhao, et al., Hsp90 modulates the stability of MLKL and is required for TNF-induced necroptosis, Cell Death Dis. 7 (2016) e2089.
[116] X. Sun, et al., HSPB1 as a novel regulator of ferroptotic cancer cell death, Oncogene 34 (45) (2015) 5617–5625.
[117] S. Zhu, et al., HSPA5 regulates ferroptotic cell death in cancer cells, Cancer Res. 77 (8) (2017) 2064–2077.
[118] Y.H. Huang, et al., Targeting HSP60 by subcutaneous injections of jetPEI/HSP60-shRNA destabilizes cytoplasmic survivin and inhibits hepatocellular carcinoma growth, Mol. Carcinog. 57 (9) (2018) 1087–1101.
[119] W. Kim, J. Ryu, J.E. Kim, CCAR2/DBC1 and Hsp60 positively regulate expression of survivin in neuroblastoma cells, Int. J. Mol. Sci. 20 (1) (2019).
[120] T. Gotoh, et al., hsp70-DnaJ chaperone pair prevents nitric oxide- and CHOP-induced apoptosis by inhibiting translocation of Bax to mitochondria, Cell Death Differ. 11 (4) (2004) 390–402.
[121] B. Jiang, et al., Increased stability of Bcl-2 in HSP70-mediated protection against apoptosis induced by oxidative stress, Cell Stress Chaperones 16 (2) (2011) 143–152.
[122] P. Pandey, et al., Negative regulation of cytochrome c-mediated oligomerization of Apaf-1 and activation of procaspase-9 by heat shock protein 90, EMBO J. 19 (16) (2000) 4310–4322.

[123] X. Li, et al., The role of the Hsp90/Akt pathway in myocardial calpain-induced caspase-3 activation and apoptosis during sepsis, BMC Cardiovasc. Disord. 13 (2013) 8.
[124] Y. Minami, et al., Selective apoptosis of tandemly duplicated FLT3-transformed leukemia cells by Hsp90 inhibitors, Leukemia 16 (8) (2002) 1535–1540.
[125] H.K. Lorenzo, et al., Apoptosis inducing factor (AIF): a phylogenetically old, caspase-independent effector of cell death, Cell Death Differ. 6 (6) (1999) 516–524.
[126] S. Nagata, Apoptosis and clearance of apoptotic cells, Annu. Rev. Immunol. 36 (2018) 489–517.
[127] P.C. Liaw, et al., DAMP and DIC: the role of extracellular DNA and DNA-binding proteins in the pathogenesis of DIC, Blood Rev. 30 (4) (2016) 257–261.
[128] K. McArthur, et al., BAK/BAX macropores facilitate mitochondrial herniation and mtDNA efflux during apoptosis, Science 359 (6378) (2018).
[129] R. Chen, et al., Release and activity of histone in diseases, Cell Death Dis. 5 (2014) e1370.

CHAPTER 3

Molecular mechanisms of cell death: a brief overview

Muthu K. Shanmugam and Gautam Sethi
Department of Pharmacology, Yong Loo Lin School of Medicine, National University of Singapore, Singapore, Singapore

1. Introduction

The word apoptosis is derived from the Greek language meaning "falling off" or "dropping off" of leaves from trees or petals from flowers [1]. Historically, naturally occurring cell death without an inflammatory response during the development process was considered a passive phenomenon that accounted for end point in all biological systems [2]. Incidentally during the development of silkworm and tadpole, a timed or programmed cell death was observed [3–6]. Subsequent ultrastructural studies by Kerr, Wylie, and Currie defined a series of distinct morphological changes that characterize cell death under normal physiological conditions [1]. Together with associated cell debris and fragments such as morphotypes disposal mechanisms cell death has been subdivided into three different forms: (1) apoptosis or type I cell death exhibiting cytoplasmic shrinkage, pyknosis (chromatin condensation), karyorrhexis (nuclear fragmentation), blebbing of plasma membrane resulting in the formation of apoptotic bodies containing intact small vesicles that undergo phagocytosis and are degraded by lysosomes; (2) autophagy or type II cell death, which is marked by widespread cytoplasmic vacuolization and similar to type I phagocytosis and degradation by lysosomes; (3) necrosis or type III cell death, which has no distinct morphological alterations as seen in type I or II cell death and is associated with the absence of any obvious phagocytosis or lysosome involvement [7–9].

Recently, Nomenclature Committee on Cell Death (NCDD) has provided additional guidelines by including genetic, biochemical, pharmacological, and function basis to identify consensus criteria in addition to the currently existing morphological classification [9–14]. Several novel advancements in this field have provided novel signaling mechanisms in the initiation, execution, and progression of cell death and provide molecularly

Mechanisms of Cell Death and Opportunities for Therapeutic Development
ISBN 978-0-12-814208-0
https://doi.org/10.1016/B978-0-12-814208-0.00002-6

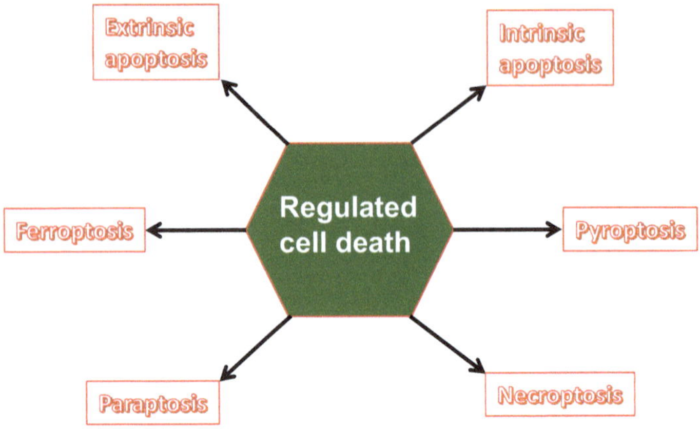

Figure 3.1 Various modes of regulated cell death in inflammation and cancer.

oriented definitions of terms including intrinsic apoptosis, extrinsic apoptosis, mitochondrial permeability transition (MPT)-driven necrosis, necroptosis, ferroptosis, pyroptosis, parthanatos, entotic cell death, NETotic cell death, lysosome-dependent cell death, autophagy-dependent cell death, immunogenic cell death, cellular senescence, and mitotic catastrophe, some of which are discussed comprehensively in this book chapter. Fig. 3.1 illustrates various modes of regulated cell death encountered under physiological and pathological conditions.

2. Intrinsic apoptotic pathway

Intrinsic apoptosis is mainly regulated by the mitochondrial proteins that are released upon stimulation by various factors such as ultra violet radiation, osmotic stress and growth factor withdrawal stress, treatment with chemotherapeutic agents, natural compounds, etc. [15–21]. In addition, internal stimuli such as very high concentrations of cytosolic Ca^{2+}, oxidative stress, hypoxia, and DNA damage also trigger the initiation of the mitochondrial pathway-mediated apoptosis [22]. Disruption in mitochondrial outer membrane permeability often leads to the release of apoptogenic factors such as cytochrome *c* that can regulate the mitochondria-dependent cell death process [23].

Upon release cytochrome *C* can attach to the cytosolic Apaf-1 (apoptosis protease activating factor-1) and this complex can recruit procaspase-9 to the caspase recruitment domain (CARD) which in turn activates downstream executor caspases 3, 6, and 7 for breakdown of cellular substrates such as PARP leading to apoptotic cell death [24,25]. The main drivers of the intrinsic pathway are the B-cell lymphoma 2 (Bcl-2) family of proteins. The Bcl-2 gene family encodes for approximately 20 proteins that can regulate the intrinsic apoptotic pathway by distinct mechanisms [26–29]. The main proteins involved are the pro-apoptotic proteins Bax and Bak belonging to the Bcl-2 family, which form pores in the outer membrane of mitochondria. The pro-apoptotic activity of Bax and Bak can be neutralized by antiapoptotic proteins Bcl-2, Bcl-xL, and Mcl-1 [30]. Patients with diffuse large B-cell lymphoma and follicular lymphoma are often linked with t(14;18) chromosome translocation that leads to transcriptional upregulation of Bcl-2 proteins [31,32]. Initially Bcl-2 family proteins were believed to be oncogene that can predominantly drive growth. However, subsequent studies have demonstrated that Bcl-2 drives tumor cells survival by abrogating apoptosis [33,34].

The antiapoptotic Bcl-2 family contain four conserved Bcl-2 homology (BH) domains (BH1-4). BH1 and BH2 domains of Bcl-2 are required for dimerization with pro-apoptotic proteins, while BH3 domain is required for interaction between pro-apoptotic and antiapoptotic proteins and present in all the members. The amino-terminal BH4 domain is associated with antiapoptotic activity but is also present in pro-apoptotic molecules. The antiapoptotic proteins containing four BH domains include Bcl-2, Bcl-xL (Bcl-2L1), Bcl-W (Bcl-2L2), myeloid cell leukemia 1 (MCL1), A1 (Bcl-2A1), and Bcl-B (Bcl-2L10) [35]. A subfamily of Bcl-2 proteins contains only BH3 domain that includes BH3 interacting-domain death agonist (Bid), Bcl-2-like protein 11 (Bim), Bcl-2 associated death promoter (Bad), p53 upregulated modulator of apoptosis (Puma), harakiri (Hrk), Noxa, BMF, and Bik [36]. These BH3-only proteins promote apoptosis by antagonizing the antiapoptotic Bcl-2 family proteins or activating the pro-apoptotic multidomain proteins Bax and Bak. Consequently, the antiapoptotic proteins prevent the release of cytochrome *c* from mitochondria while the pro-apoptotic proteins can actively promote its release. Thus, any deregulation in Bcl-2 functions that alters the homeostasis between pro- and antiapoptotic proteins will have direct implications in cellular survival or induction of apoptosis [37].

The cytosolic pro-apoptotic protein Bax can translocate into mitochondria during apoptosis and oligomerizes with Bak (an integral mitochondrial membrane protein), thereby promoting initiation of mitochondrial outer membrane permeabilization (MOMP) culminating in apoptosis [38,39]. Some BH3 domain only containing pro-apoptotic proteins act upstream of this event by binding to Bcl-2 proteins, thereby allowing Bax/Bak to provoke MOMP and apoptosis [27,40,41]. However, in cancer cells with overexpression of Bcl-2 proteins sequester pro-apoptotic proteins by binding to their BH3 domain thereby inhibiting apoptosis. Bcl-2-mediated resistance to apoptosis is a hallmark of malignant cells exhibiting chemoresistance. Thus, targeting Bcl-2 proteins remains an attractive and effective therapeutic strategy in cancer treatment [42]. Several other pro-apoptotic factors that are released from the mitochondrial intermembrane spaces include apoptosis-inducing factor, second mitochondria-derived activator of caspase (Smac, also known as direct IAP binding protein with low pI or DIABLO), and Omi/high temperature requirement protein A (HtrA2) [43].

3. Extrinsic apoptotic pathway

The extrinsic apoptotic pathway is activated by the interaction between death ligands and death receptors (DRs). Death ligands include tumor necrosis factor α (TNFα), the Fas ligand (FASLG), and TNF-related apoptosis-inducing ligand (TRAIL, TNFSF10). These death ligands bind to their cognate DRs TNF receptors (TNFRs), Fas, and DR4/DR5, respectively, that are present on cell surface [44]. The TNF receptor superfamily comprises of 20 proteins containing a type-I transmembrane topology with 2 to 4 cysteine-rich extracellular domains. DRs are characterized by intracellular protein–protein interaction domain known as the death domain (DD) containing 80 amino acid in the cytoplasmic part, which is necessary to activate apoptosis-inducing signals [45–47]. The major function of DRs and their ligands has been implicated in the regulation and activity of innate and adaptive immune system [48]. The various cells in the body may be primed for either apoptotic or necrotic cell death as a result of their natural differentiation program or due to external factors such as infections, cellular damage, or oncogenic transformation. There are six DRs that have been implicated in the activation of apoptosis. They are TNFR1 (TNFRSF1A; cognate ligands: TNFα/TNFSF2 and lymphotoxin α/TNFSF1), Fas (APO1/CD95/TNFRSF6; cognate ligand: FasL/CD95L/TNFSF6), DR3 (APO3/TNFRSF25; cognate ligand: TL1A/TNFSF15), DR4 (TRAILR1/TNFRSF10A; cognate ligand: Apo2L/TRAIL/TNFSF10), DR5 (TRAILR2/TNFRSF10B; cognate ligand: Apo2L/TRAIL/TNFSF10), and DR6 (TNFRSF21) [48–55]. Typically, ligand binding to DRs can trigger apoptosis in responsive cells. Crystallographic studies have shown that the death ligands FasL and Apo2L/TRAIL are homotrimeric proteins forming a basic ligand receptor complex consisting of the trimeric ligands and three receptor molecules [56,57]. Upon ligand binding, DRs such as Fas, DR4, and DR5 can multimerize thereby leading to conformational alterations in the cytosolic domain of the receptor [57,58]. This in turn induces apoptosis by forming the death-inducing signaling complex (DISC) containing the adaptor protein Fas-associated DD (FADD) that contains both DD and death effector domain (DED) and recruit apoptosis-inducing protease caspases 8 or 10 and/or cFLIP, a caspase 8-like protein lacking protease activity [59–63].

Novel posttranslational modifications have been reported on DRs such as palmitoylation of Fas [64] and O-linked glycosylation of DR4 and DR5 in their ectodomains [65] that enable ligand-induced clustering and apoptosis activation. Interestingly, clathrin-mediated DR5 internalization upon ligand binding may attenuate caspase activation; however, caspase 8-mediated cleavage of clathrin adaptor AP2α can negatively regulate DR5 internalization thereby driving apoptosis activation [66,67]. Intracellular endoplasmic stress has also been shown to activate DR5 receptor even without ligand binding [68]. In epithelial cells, E-cadherin (a homotypic adhesion protein) may bind to ligated DR4 and DR5, tethering actin cytoskeleton through E-cadherin's interacting partner α-catenin thereby promoting DISC assembly [69]. In addition, cullin 3 (an E3 ubiquitin ligase) recruitment to the death ligand/DR complex induces caspase 8 ubiquitination to enhance the crosstalk between E-cadherin and DR4/5 [69].

Furthermore, an atypical TNFR superfamily member, p75 neurotrophin receptor, and the ectodysplasin A receptor EDAR (TNFRSF19) also have DD motifs. In addition, decoy receptors such as DcR1 (TNFRSF10C) and DcR2 (TNFRSF10D) antagonize ligand engagement with DRs by binding to Apo2L/TRAIL, while DcR3 (TNFRSF6B) interacts with FasL [70]. The fundamental role of DRs is to induce apoptosis upon ligand binding. However, depending upon key cytosolic signaling molecules, these DRs also induce necroptosis or even promote cell survival. Some of the key pro-survival pathways may include NF-κB, JNK/c-Jun, p38 MAPK/AP1 pathways when apoptosis signal is intercepted [71–79]. DR3 has been shown to regulate the noncanonical NF-κB signaling pathway; however, it can also induce apoptosis indirectly via TNFα [80,81]. Likewise, TNFR1 has also been shown to play a dual role of either inducing apoptosis or activating pro-survival signaling pathways such as IKK/NF-κB, JNK/c-Jun, and p38/MAPK [82]. Apoptosis is activated when the downstream surviving signaling is blocked, for example, upon cellular depletion of inhibitor of apoptosis proteins (cIAPs) [83] or severe DNA damage [84]. Furthermore, inhibition of caspases in conjunction with a blockade of the cell survival signaling of TNFR1/TNF binding was found to induce necroptotic cell death [85]. Fig. 3.2 summarizes the intrinsic and extrinsic branches of the apoptosis pathway.

4. Caspases

Caspases are cysteinyl, aspartic-specific proteases that function as proteolytic enzymes to cleave specific substrates in cell death [86]. The role of caspases in apoptosis was identified more than 2 decades ago with the discovery of ced-3 as a final executioner caspase during the development of the nematode *Caenorhabditis elegans* [87]. The mammalian homologs of ced-3 are interleukin-1-β converting enzyme and Nedd2, which were later renamed as caspase 1 and caspase 2, respectively [88–90]. Caspases can exist in the cytoplasm as pro-caspases or zymogens in inactive forms. They become active caspases through an auto-proteolysis process that is facilitated by specific adaptor molecules [91,92]. Once the initiator caspases are activated, they in-turn activate executioner caspases that cleave off specific cellular substrates resulting in apoptotic cell death characterized by chromatin condensation and fragmentation as well as plasma membrane blebbing [91–93].

The apoptotic process is primarily initiated by the initiator caspases 2, 8, 9, and 10, and the final execution is done by the execution caspases 3, 6, and 7 by cleaving specific substrates in the apoptotic pathway [30,94,95]. The initiator caspases contain specific protein interaction domains such as DED (caspases 8 and 10) or CARD (caspases 2 and 9) in their N-terminal part that is responsible for their recruitment into large complexes thereby facilitating their activation [95,96]. Importantly the initiator caspases are activated by both the Bcl-2 regulated intrinsic and the DR-mediated extrinsic pathways [95,96]. As mentioned above, cellular stress such as cytotoxic drugs or DNA damage initiates the intrinsic apoptotic pathway by the activation of Bcl-2 family proteins Bax and Bak, induction of MOMP, and cytochrome *C* release [38,39,95].

Released cytochrome *C* associates with apoptotic protease-activating factor 1 (Apaf-1) into large multimeric complexes known as apoptosome,

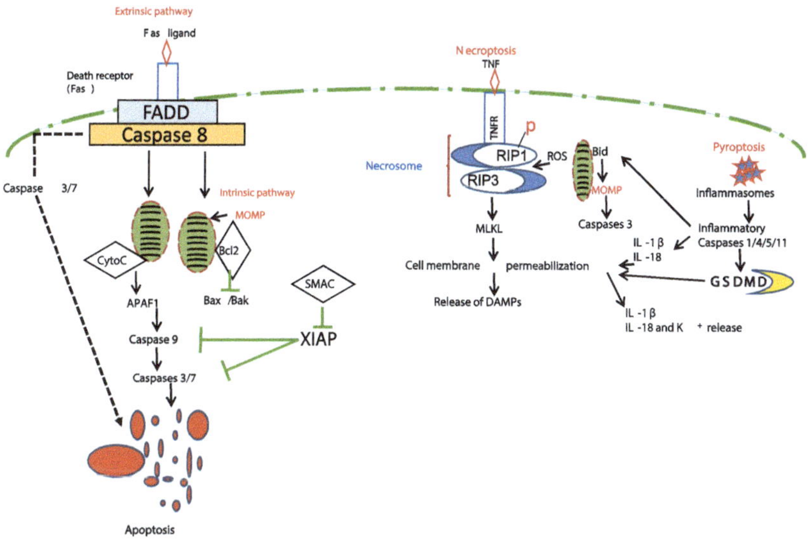

Figure 3.2 ***Cell death signaling pathways activated in inflammation and cancer.*** Two major pathways of apoptosis have been extensively studied. First, extrinsic pathway involves the receptor–ligand interaction. Fas-ligand binding to Fas death receptor initiates apoptosis by recruiting Fas-associated death domain (FADD) and initiator caspase 8 with subsequent activation of executioner caspases 3 and 7 thus inducing apoptotic mode of cell death. The second mechanism is the intrinsic pathway also known as the mitochondrial apoptotic pathway that is activated by a wide variety of stimuli such as DNA damage, growth factor withdrawal, and mitotic arrest. In this pathway, BH3-only members inhibit the antiapoptotic members of the B-cell lymphoma 2 (BCL-2) family to activate the pro-apoptotic Bax and Bak with the subsequent MOMP. The permeabilized mitochondria releases cytochrome *C*, which in turn binds to apoptotic peptidase-activating factor 1 (APAF1). This apoptosome complex initiates caspase 9, which cleaves and activates caspases 3 and 7. MOMP also releases SMAC that binds to caspase inhibitor X inhibitor of apoptosis (XIAP) and induces apoptosis. Necroptosis can be triggered by tumor necrosis factor (TNF) and is a pro-inflammatory cell death process with the release of DAMPs. Upon caspase inhibition, TNF promotes the phosphorylation of receptor interacting protein kinase 1 (RIP1) and RIP3 with the formation of necrosome. Sequentially the necrosome phosphorylates mixed-lineage kinase domain-like pseudokinase (MLKL), which permeabilizes the cell membrane and releases DAMPs. Mitochondrial ROS is also involved in the formation of necrosome in this cell death pathway. Pyroptosis is another mode of regulated cell death, which is characterized by the formation of inflammasome complexes, activation of inflammatory caspases that cleave and activate gasdermin D (GSDMD), which form a pore in the plasma membrane to release IL-1β, IL-18, and K^+ ion.

which recruits and activates the initiator caspase-9 [97]. As discussed above, the extrinsic apoptosis is activated by the engagement of death ligand with DRs [48–50,98]. Ligand-receptor complex recruits and activates caspases 8 or 10 through the so-called DISC. The activated initiator caspases in turn activate the downstream executioner caspases, namely, caspases 3, 6, and 7. Additionally, activated caspase 8 also cleaves Bid to a shortened Bid (tBid) that engages mitochondria and amplifies the apoptotic response [99]. Consequently, the activated executioner caspases 3, 6, and 7 cleave many cellular components, ultimately leading to the classical morphological changes associated with apoptosis and cell death [93]. Recently, it has become evident that several members of the caspase family of proteases are involved in several other modes of cell death mechanism in addition to apoptosis such as necroptosis, pyroptosis, autophagy, ferroptosis, NETosis, and paraptosis [95,100,101].

5. The inhibitors of apoptosis

The process of proteolysis is an irreversible event; therefore, a stringent control of the caspase-mediated proteolytic cleavage is of key importance to avoid inappropriate cell destruction. Negative regulation of caspases is mediated by a family of highly conserved proteins known as inhibitors of apoptosis (IAPs). IAP family of proteins is comprised of 8 members [102]. They are neuronal apoptosis inhibitor protein (NAIP) [BIRC1] [103], cellular IAPs (cIAP1 and cIAP2) [BIRC2 and BIRC3] [104–106], X-chromosome linked IAP (XIAP) [BIRC4] [105–107], survivin [BIRC5] [108], Apollon [BRUCE, BIRC6] [109,110], melanoma IAP (ML-IAP, also known as livin) [BIRC7] [111–113], and IAP-like protein 2 (ILP2) [BIRC8] [114]. All these proteins carry the conserved baculovirus IAP repeat (BIR) domain that contains 80 amino-acid residues [115]. The BIR domain interacts with various proteins including caspases thus enabling IAPs to bind and inactivate caspases. XIAP is the most versatile member of IAPs that block caspases mediated apoptotic signals [116]. For example, the second BIR of XIAP favorably binds to caspases 3 and 7 while the third BIR specifically binds to caspase 9 [117]. Notably, cIAPs, XIAP, ML-IAP, and ILP2 possess a RING domain that can function as an E3 ubiquitin ligase in C-terminal region and is involved in ubiquitinating caspases for degradation [118]. In cancer cells, IAPs are found in abundance thereby inhibiting caspase activation and enabling evasion of apoptosis [119].

Overexpression of IAPs has also been implicated in the development of resistance to apoptosis induced by chemotherapeutic agents. Therefore, IAPs represent an attractive target for cancer therapy [119]. Remarkably, mitochondrial proteins Omi/HtrA2 and Smac/DIABLO were shown to inhibit the activities of IAPs during apoptosis by antagonizing the BIR domain of IAPs thus inhibiting their capability to bind to caspases, consequently reestablishing their pro-apoptotic activity [120]. Several novel Smac mimetics have been developed by designing compounds that mimic the N-terminal 4-amino acid sequence of Ala-Val-Pro-Ile that interacts with BIR2 and BIR3 domains. These Smac mimetics potently inhibit the functions of IAPs [121,122] (Fig. 3.3).

Interestingly, Smac mimetics activate the E3 ubiquitin ligase activity of IAPs for their auto-ubiquitination and thus primes them for proteasomal degradation. This in turn activates noncanonical NF-κB pathway by

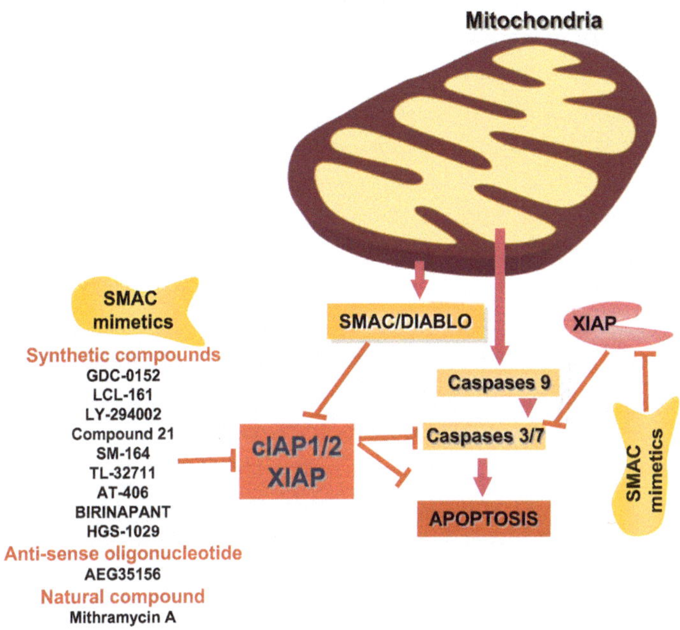

Figure 3.3 Under normal conditions, caspases are active and can induce apoptosis. However, cancer cells escape apoptosis and thus forms tumors. IAPs are endogenous proteins that inactivate caspases via direct binding thereby contributing to tumorigenesis and resistance to therapy. Endogenous SMAC/DIABLO proteins inhibit IAPs. In addition, there are several approaches to target IAPs for anticancer therapy. They are antisense oligonucleotides, small molecule inhibitors, and SMAC mimetics that target either XIAP or cIAP1/2 to promote apoptosis.

activating NF-κB inducing kinase (NIK) that promotes proteolytic processing of NFKB2/p100, leading to NF-κB transcriptional activation of downstream target genes such as TNF-α. TNF-α binds to its receptor TNFR1 in an autocrine or paracrine fashion, thus inducing cell death [118,123–127]. Binding of IAP by Smac mimetics has also been shown to potently prevent XIAP from binding to caspases 3, 7 and, 9, thus leading to subsequent caspase activation and apoptosis [119]. Smac mimetics have been tested as monotherapy agents for the treatment of hematological malignancies such as acute lymphoblastic leukemia (ALL), chronic lymphocytic leukemia, and acute myeloid leukemia (AML) [128–130].

In addition, Smac mimetics have also been used in combination therapy with classical chemotherapeutic agents and cellular signaling pathway modulators. Smac mimetics in combination with TRAIL synergistically induced apoptosis in childhood acute leukemia cells and overcame Bcl-2 mediated resistance [131]. This combination was also tested in a patient-derived ALL xenograft model in mice with similar synergistic activity [131]. Furthermore, a Smac mimetic in combination with TRAIL-induced apoptosis in cells derived from subgroups of CLL patients with poor prognosis. These findings suggest profound implications toward the development of apoptosis-based therapies for hematological malignancies [132]. In colorectal cancer, the Smac mimetic SM83 in combination with a diabody consisting of a TRAIL variant conjugated to Cetuximab that binds epidermal growth factor receptor can induce substantial apoptotic cell death [133].

In pancreatic cancer cells, Smac mimetics in combination with TRAIL receptor agonists such as antibodies that bind to the receptor, soluble TRAIL or TNFα can also cause apoptosis [134,135]. Interestingly, Hannes, et al. reported that Smac mimetics in combination with TRAIL-induced necroptosis when caspase activation was blocked in pancreatic cancer cells [136]. Temozolomide (TMZ), the first-line choice for glioblastoma treatment, in combination with Smac mimetics augmented TMZ efficacy by inducing RIP1 and NF-κB dependent apoptosis [137]. Smac mimetics have also been found to upregulate the production of IFNγ and sensitized glioblastoma cells to TMZ-induced cell death [138]. The role of DRs in cell death can be context dependent; however, Smac mimetics increased cancer cell response to chemotherapeutic agents and induced apoptosis in a TNFα dependent manner [139]. In another study, Smac mimetics have been reported to sensitize both primary glioblastoma and glioblastoma stem cells to γ-irradiation-induced apoptosis [140,141]. In another study, Smac mimetics in combination either with TMZ or radiotherapy abrogated tumor growth in a mouse model of glioblastoma [142]. Several studies have demonstrated that targeting IAPs by Smac mimetics augmented radiosensitivity to breast, colorectal, head and neck squamous cell carcinoma, and pancreatic adenocarcinoma cells [143–146]. Epigenetic demethylating agents such as 5-Azacitadine or 5-Aza-2-deoxycytidine (decitabine) in combination with IAP antagonist birinapant-induced potent apoptosis of AML stem/progenitor cells in vitro and in vivo in an AML mouse xenograft model [147]. In combination with proteasome inhibitor bortezomib or with TRAIL, Smac mimetics induced potent apoptosis in multiple myeloma or melanoma cells [148,149]. In the presence of tyrosine kinase inhibitors such as the BCR-ABL inhibitor nilotinib, the FLT3 inhibitor PKC412 or platelet-derived growth factor inhibitors, Smac mimetics induced potent antitumor activity [150–152]. In another study, siRNA induced depletion of XIAP or Smac mimetics that target multiple IAPs-induced potent apoptosis in combination with ErbB antagonists such as trastuzumab, lapatinib, or gefitinib in Her2-overexpressing BT474 breast cancer cells or in combination with gefitinib in epidermal growth factor receptor (EGFR)-overexpressing MDA-MB-468 breast cancer cells [153].

Apart from IFNs, there are other immune stimuli that have been reported to augment the antitumor activity of Smac mimetics. A study by Beug et al. showed that oncolytic viruses or noninfectious immune stimulatory molecules including poly(I:C) or CpG oligonucleotides induced bystander cancer cell death treated with Smac mimetics, which appears to be mediated by IFNβ, TNFα, and TRAIL. This combinatorial approach resulted in rapid tumor regression and prolonged the life span in two mouse tumor models [154]. In a preclinical model of bladder cancer, Smac mimetics improved the efficacy of Bacillus Calmette-Guérin immunotherapy, which appears to be mediated through TNFα without the involvement of TRAIL and FasL [155]. The two major apoptosis pathways, namely, the intrinsic and extrinsic apoptosis mechanisms that can be activated by various stimuli are depicted in Fig. 3.3.

6. Other types of programmed cell death

6.1 Necroptosis (programmed necrosis)

Cell death due to sudden changes in extracellular and/or intracellular microenvironment and an unregulated inflammatory response is often referred to as necrosis [156]. A form of necrosis with defined molecular mechanisms is known as necroptosis, which is often seen in chronic inflammation-driven diseases, and in ischemic injury. Several external factors that induce necroptosis have been identified including viral infections and lipopolysaccharides that trigger Toll receptor signaling [157–160]. However, necroptosis is best characterized by the involvement of TNFα signaling [9,161–164]. The basic scenario is that under caspase 8 deficiency, TNFα/TNFR interaction leads to the activation of receptor interaction protein kinase 1 (RIPK1) and RIPK3 and formation of necrosomes. Several ligands belonging to the TNF superfamily are capable of causing necroptosis, especially when caspase activation has been limited. The works by Degterev and Yuan showed the importance of RIPK1 in necroptosis. They have discovered necrostatin-1 (Nec-1), a small molecule compound, inhibited necroptosis, and RIPK1 was found to be its intracellular molecular target [165–167]. These and other published studies have provided insights into the specific molecular pathway in inducing and regulating necroptosis.

At the molecular level, necroptosis depends on the sequential activation of RIPK3 through phosphorylation followed by mixed-lineage kinase domain-like pseudokinase (MLKL) activation [156,168,169]. Active RIPK3 catalyzes MLKL phosphorylation and oligomerization (trimer or tetramers) that can translocate to the plasma membrane thereby inducing membrane permeabilization and cell death [9,169–173]. In addition, it has been reported that chemical induced activation of RIPK3 and dimerization of MLKL along with concomitant depletion of mitochondria by mitophagy induced necroptosis. This indicates that irrespective of mitochondria activation, MLKL is shown to be the final executioner mechanism of necroptosis [174].

However, in certain cell types, mitochondrial ROS is involved in the initiation of necroptosis by activating RIPK1 autophosphorylation, activation, and necrosome formation [175,176]. In addition, in another mechanism RIPK3 activates pyruvate dehydrogenase complex to induce improved aerobic respiration with concomitant ROS generation [177]. ROS is considered an important determinant as to whether a cell initiates necroptosis. For example, progressive mitochondrial dysfunction especially during aging has been shown to increase the propensity of cells to undergo necroptosis [178]. In addition, lysosomal leakage and lipid peroxidation may also control necroptotic cell demise [167]. Moreover, damage-associated molecular patterns (DAMPs) such as bacterial lipopolysaccharide or viral nucleic acid products and their associated receptors, that is, certain Toll-like receptors, NOD-like receptors, and viral DNA receptors such as ZBP1 (Z-DNA-binding protein-1, also known as DNA-dependent activator of IFN-regulatory factor or DAI) [85,157,158], as well as interferons [179,180] can also initiate necroptotic signaling upon ligand engagement.

The importance of necroptotic cell death in cancer chemotherapy has been recently reviewed [181—183]. Several lines of evidence suggest that both activation and/or suppression of tumor growth were observed upon necroptosis in cancer development and progression. Similar to apoptosis resistance, necroptosis resistance by cancer cells is also considered as one of the hallmarks of cancer [181]. The tumor suppressive effect of necroptosis is observed in tumors expressing RIPK3 [184,185]. The expression levels of RIPK3 and MLKL as well as RIPK3 polymorphism are implicated in cancer patient prognosis [184—187]. Several studies have reported that the expression of RIPK3 and MLKL varies in tumor tissues depending upon the stage of cancer and its subtypes and is associated with downregulation of necroptosis mediators. Clinicopathological analysis showed low MLKL expression in gastric cancer and significantly correlated with older age, histologic type, differentiation grade, depth of invasion, and lymph node metastasis. In gastric cancer, low expression of MLKL was significantly associated with decreased overall survival [188—190].

On the other hand, necroptosis is also implicated in tumor-promoting activities. Fox example, RIPK3 is shown to be upregulated in late-stage MMTV-PyMT breast cancer thus possibly contributing to tumor growth and metastasis [191]. Necroptosis-induced inflammation promotes pancreatic tumorigenesis and is also associated with an immunosuppressive tumor microenvironment. Interestingly, it was reported that necroptosis was associated with increased expression of CXCL1 and that CXCL1 blockade suppressed pancreatic tumor progression [188,192]. Moreover, tumor cells were shown to induce necroptosis of endothelial cells, which may promote tumor cell extravasation and metastasis [193]. Anticancer agents have been shown to induce innate and adaptive immune system to attack transformed cells leading to the release of DAMPs and necroptotic cells determine cross-priming of CD^{8+} T cells [194,195]. Therefore, clearly detailed mechanistic studies are warranted to understand the precise or context-dependent roles of necroptosis in tumorigenesis. However, the ability of necroptosis to modulate the adaptive and innate immune response makes it an attractive target for cancer therapy. Fig. 3.2 illustrates the mechanism underlying necroptosis. Necroptosis is also discussed in Chapter 1 of this volume.

6.2 Ferroptosis

Ferroptosis is another form of regulated cell death that is activated by lipid peroxides which attack the lipid membranes leading to loss of cell membrane integrity and cell death [9,196,197]. Ferroptosis was first discovered as an iron-dependent oxidative cell death with reactive iron (Fe^2) as it is required for the Fenton reaction responsible for lipid peroxidation. Under normal cellular homeostasis, glutathione peroxidase 4 (GPX4) catalyzes the conversion of peroxidized lipids into nontoxic lipid alcohol. For this reaction, glutathione (GSH), a key cellular antioxidant that scavenges ROS through reduction, is a cofactor for GPX4 to convert the peroxidized lipids into lipid alcohol. Cysteine, via cystine, an oxidized cysteine dimer, is imported into the cells by exchanging for glutamate through the antiporter System Xc^-, which can be inhibited by erastin and other small molecule compounds. Blocking of System Xc^- by erastin leads to decreased levels of glutathione, thus impairing GPX function to neutralize lipid peroxides [198].

Remarkably, triple-negative breast cancer cells have been reported to have increased addiction to glutamine and their ability to import cystine via the System Xc^- provides a pivotal therapeutic target [199,200]. The FDA-approved tyrosine kinase inhibitor sorafenib was shown to induce ferroptosis in specific cellular models by depleting GSH when System Xc^- is inhibited [201–204]. Another FDA-approved alkylating agent, altretamine, has been identified through a regulatory network genome-wide system screen as a potential inhibitor of GPX4 [205]. Thus, sorafenib and altretamine may in part induce cancer cell death by inducing ferroptosis. The role of mitochondria in ferroptosis is contentious [206]. For example, mitochondrial generation of ROS can contribute to ferroptosis; however, loss of mtDNA or loss of mitochondria does not affect sensitivity to ferroptosis [197,207]. Cytosolic as well as mitochondrial ferritin was shown to chelate iron thereby preventing accumulating iron-dependent lipid peroxidation by the Fenton reaction [208]. Ferroptosis is discussed in some detail in Chapter 8 of this volume.

6.3 Pyroptosis

Pyroptosis is another form of regulated cell death [209]. Similar to apoptosis, the word pyroptosis is also derived from the Greek language “pyro” and “ptosis,” which mean fever and falling, respectively, to emphasize on the newly discovered inflammatory programmed cell death [210]. Mitochondria have been implicated in several modalities of cell death; however, the mechanisms are poorly defined. The gasdermin family of proteins especially gasdermin D and E have been implicated in the development of MOMP, also known as pore-formation upon cleavage [211–213]. In particular, gasdermin D (GSDMD) has been found to induce inflammatory type of cell death known as pyroptosis independently of BAX and BAK [178]. Pyroptosis is primarily driven by inflammatory caspases 1, 4, 5, and 11 (caspase 11 is only found in rodents, which lack caspases 4 and 5), which cleave GSDMD [214–216]. During pyroptosis, the GSDMD amino-terminal cleavage fragment was found to oligomerize to form a pore that permeabilizes plasma membrane and releases of proinflammatory cytokines IL-1β and IL-18 [178,213]. In the absence of GSDMD, caspase 1 can cleave and thus activate executioner caspases and also Bid to induce MOMP. Pyroptosis has been observed in various human diseases including cancers. In cancer cells, pyroptosis is associated with activation of multiple pro-survival signaling pathways and is also implicated in cancer cell chemoresistance to therapeutic drugs [217–219]. Several studies have reported that pyroptosis can be pro-tumorigenic as well as antitumorigenic in different tumor cells [209]. NLRP3 (NOD-, LRR-, and pyrin domain-containing protein 3) inflammasomes can be activated by several stimuli such as bacteria, viruses, fungi, uric acid, and ATP [209]. Wei et al. reported that downregulated expression of NLRP3 in hepatic tissue contributed to the development of hepatocellular carcinoma (HCC) [220]. In a recent study, they reported that 17β-estradiol inhibited the growth of HCC by inducing pyroptosis by activating NLRP3 inflammasome [221]. Mice lacking active inflammasomes such as *Nlrp3−/−* or *Casp1−/−* mice showed increased sensitivity to azoxymethane/dextran sulfate sodium (AOM/DSS)-induced colitis-associated colon cancer [222–225]. Eukaryotic elongation factor-2 kinase (eEF-2K) has been shown to be a negative regulator of protein synthesis and plays a key role in pyroptosis and autophagy in melanoma cells. Silencing of eEF-2K was found to promote doxorubicin-induced pyroptosis in association with suppression of autophagy, suggesting that pyroptosis plays a critical role in sensitization of melanoma cells to doxorubicin [226]. In gastric cancer cells, it was reported that chemotherapeutic drugs can convert caspase-3 dependent apoptosis to pyroptosis via DFNA5/GSDME [227]. Promoter methylation has been found to downregulate DFNA5/GSDME expression. However, the demethylating agent decitabine activated DFNA5/GSDME expression, thereby inducing pyroptosis and sensitizing them to chemotherapeutic agents [211,228]. Similarly, in lung cancer cells, DFNA5/GSDME knockdown is shown to reduce pyroptosis and is associated with drug resistance. In contrast, overexpression of DFNA5/GSDME leads to increased drug sensitivity [229]. Reactive oxygen species (ROS) induced activation of NLRP3 inflammasome in cervical carcinoma cells induced pyroptosis [209]. In Barrett's esophageal cancer cells, lipopolysaccharide induced upregulation of NLRP3, mitochondrial ROS, IL-18, IL-1β, and induced pyroptosis [230]. A PLK1 kinase inhibitor was shown to enhance the chemosensitivity of cisplatin by inducing pyroptosis in esophageal squamous cell carcinoma cells [231]. By contrast, microRNA miR-214 inhibited the growth and migration of glioma cells by targeting caspase 1 associated pyroptosis [232]. In pancreatic adenocarcinoma cells, the mammalian STE20-like kinase 1 (MST1) is found to be downregulated. Upregulation of MST1 suppresses pancreatic cancer cell proliferation, migration, and invasion by inducing ROS-mediated pyroptosis [233]. Fig. 3.2 briefly summarizes the mechanism underlying pyroptosis. The mechanisms underlying pyroptosis, its involvement in diseases, and efforts on therapeutic development targeting this form of regulated cell death are also covered in Chapter 1 of this volume.

7. Conclusion

Cell death either due to external exposure to toxic stimuli or due to internal physiochemical environment is an essential component during the embryonic development process of all multicellular organisms and immune homeostasis. Apoptosis or necroptosis were considered the two major forms of cell death mechanisms. Several decades of research on diverse mechanisms of apoptosis have led to the identification of several key players involved in both the intrinsic and extrinsic cascades. The intrinsic apoptotic pathways were found to be regulated by MOMP and mitochondria-associated proteins such as the Bcl-2 family of proteins that are either pro-apoptotic or antiapoptotic, while the extrinsic apoptotic pathways are controlled by receptor—ligand interaction that leads to the activation of several intracellular signaling pathways, converging at the level of executioner caspases. Activation of caspases is frequently impaired in human cancers, contributing to cancer formation, progression, and therapy resistance. Thus, triggering caspase activation is fundamental in increasing the therapeutic efficacy in cancer therapy. Upregulated caspase activation leads to induction of apoptosis, which augments anticancer activity of chemotherapeutic agents. Several lines of evidence suggest that both activation and/or suppression of tumor growth are observed upon necroptosis in cancer development and progression. Similar to apoptosis, resistance necroptosis resistance by cancer cells is also considered as one of the hallmarks of cancer. While there are no approved drugs in the market that directly target necroptosis, various agents have the potential to both up- and downregulate necroptosis, and to induce different levels of necroptosis signaling.

Mitochondria have also been implicated in other forms of cell death including necroptosis, pyroptosis, ferroptosis, and paraptosis. It is of note that these various forms of cell death communicate with one another with the involvement of mitochondria. Some of these forms of cell death are pro-inflammatory in nature and their modalities of initiation, propagation, and final execution of cell death can have key consequences in cellular homeostasis and in various disease settings. Another type of regulated cell death mediated by Gasdermin family of proteins known as pyroptosis is associated with inflammasomes in cancer cells. Cancer therapy activating pyroptosis holds promise. Some chemotherapeutic drugs and miRNA have been shown to induce pyroptosis to inhibit tumor growth. Upregulating the levels of DFNA5/GSDME and GSDMD expression in cancer cells enhances chemotherapeutic drug sensitivity and thus overcoming drug resistance. More preclinical experiments and clinical trials are needed to explore the potential application of anticancer therapy based on pyroptosis. It is expected that many small molecule drugs with enhanced anticancer activities will be available for the treatment of cancer in the future.

References

[1] J.F. Kerr, A.H. Wyllie, A.R. Currie, Apoptosis: a basic biological phenomenon with wide-ranging implications in tissue kinetics, Br. J. Cancer 26 (4) (1972) 239–257.

[2] A. Glucksmann, Cell deaths in normal vertebrate ontogeny, Biol. Rev. Camb. Phil. Soc. 26 (1) (1951) 59–86.

[3] R.A. Lockshin, C.M. Williams, Programmed cell death–I. Cytology of degeneration in the intersegmental muscles of the *Pernyi* silkmoth, J. Insect Physiol. 11 (1965) 123–133.

[4] J.R. Tata, Requirement for RNA and protein synthesis for induced regression of the tadpole tail in organ culture, Dev. Biol. 13 (1) (1966) 77–94.

[5] M.C. Raff, et al., Programmed cell death and the control of cell survival: lessons from the nervous system, Science 262 (5134) (1993) 695–700.

[6] M.D. Jacobson, M. Weil, M.C. Raff, Programmed cell death in animal development, Cell 88 (3) (1997) 347–354.

[7] J.U. Schweichel, H.J. Merker, The morphology of various types of cell death in prenatal tissues, Teratology 7 (3) (1973) 253–266.

[8] L. Galluzzi, et al., Cell death modalities: classification and pathophysiological implications, Cell Death Differ. 14 (7) (2007) 1237–1243.
[9] L. Galluzzi, et al., Molecular mechanisms of cell death: recommendations of the Nomenclature Committee on Cell Death 2018, Cell Death Differ. 25 (3) (2018) 486–541.
[10] L. Galluzzi, et al., Guidelines for the use and interpretation of assays for monitoring cell death in higher eukaryotes, Cell Death Differ. 16 (8) (2009) 1093–1107.
[11] L. Galluzzi, et al., Molecular definitions of cell death subroutines: recommendations of the Nomenclature Committee on Cell Death 2012, Cell Death Differ. 19 (1) (2012) 107–120.
[12] L. Galluzzi, et al., Essential versus accessory aspects of cell death: recommendations of the NCCD 2015, Cell Death Differ. 22 (1) (2015) 58–73.
[13] G. Kroemer, et al., Classification of cell death: recommendations of the Nomenclature Committee on Cell Death, Cell Death Differ. 12 (Suppl. 2) (2005) 1463–1467.
[14] G. Kroemer, et al., Classification of cell death: recommendations of the Nomenclature Committee on Cell Death 2009, Cell Death Differ. 16 (1) (2009) 3–11.
[15] D.R. Green, G. Kroemer, The pathophysiology of mitochondrial cell death, Science 305 (5684) (2004) 626–629.
[16] M.K. Shanmugam, R. Kannaiyan, G. Sethi, Targeting cell signaling and apoptotic pathways by dietary agents: role in the prevention and treatment of cancer, Nutr. Cancer 63 (2) (2011) 161–173.
[17] S.M. Tan, et al., Identification of beta-escin as a novel inhibitor of signal transducer and activator of transcription 3/Janus-activated kinase 2 signaling pathway that suppresses proliferation and induces apoptosis in human hepatocellular carcinoma cells, J. Pharmacol. Exp. Ther. 334 (1) (2010) 285–293.
[18] P. Rajendran, et al., Honokiol inhibits signal transducer and activator of transcription-3 signaling, proliferation, and survival of hepatocellular carcinoma cells via the protein tyrosine phosphatase SHP-1, J. Cell. Physiol. 227 (5) (2012) 2184–2195.
[19] X. Dai, et al., Targeting TNF-related apoptosis-inducing ligand (TRAIL) receptor by natural products as a potential therapeutic approach for cancer therapy, Exp. Biol. Med. 240 (6) (2015) 760–773.
[20] S.K. Manna, et al., Morin (3,5,7,2′,4′-Pentahydroxyflavone) abolishes nuclear factor-kappaB activation induced by various carcinogens and inflammatory stimuli, leading to suppression of nuclear factor-kappaB-regulated gene expression and up-regulation of apoptosis, Clin. Cancer Res. 13 (7) (2007) 2290–2297.
[21] A.S. Nair, et al., Deguelin, an Akt inhibitor, suppresses IkappaBalpha kinase activation leading to suppression of NF-kappaB-regulated gene expression, potentiation of apoptosis, and inhibition of cellular invasion, J. Immunol. 177 (8) (2006) 5612–5622.
[22] G. Kroemer, L. Galluzzi, C. Brenner, Mitochondrial membrane permeabilization in cell death, Physiol. Rev. 87 (1) (2007) 99–163.
[23] N.N. Danial, S.J. Korsmeyer, Cell death: critical control points, Cell 116 (2) (2004) 205–219.
[24] E.A. Slee, C. Adrain, S.J. Martin, Serial killers: ordering caspase activation events in apoptosis, Cell Death Differ. 6 (11) (1999) 1067–1074.
[25] K. Kuribayashi, P.A. Mayes, W.S. El-Deiry, What are caspases 3 and 7 doing upstream of the mitochondria? Cancer Biol. Ther. 5 (7) (2006) 763–765.
[26] Z. Jin, W.S. El-Deiry, Overview of cell death signaling pathways, Cancer Biol. Ther. 4 (2) (2005) 139–163.
[27] A. Shamas-Din, et al., BH3-only proteins: orchestrators of apoptosis, Biochim. Biophys. Acta 1813 (4) (2011) 508–520.
[28] J.E. Chipuk, et al., The BCL-2 family reunion, Mol. Cell 37 (3) (2010) 299–310.

[29] P.E. Czabotar, et al., Control of apoptosis by the BCL-2 protein family: implications for physiology and therapy, Nat. Rev. Mol. Cell Biol. 15 (1) (2014) 49–63.
[30] G. Pistritto, et al., Apoptosis as anticancer mechanism: function and dysfunction of its modulators and targeted therapeutic strategies, Aging (Albany NY) 8 (4) (2016) 603–619.
[31] Y. Tsujimoto, et al., Cloning of the chromosome breakpoint of neoplastic B cells with the t(14;18) chromosome translocation, Science 226 (4678) (1984) 1097–1099.
[32] Y. Tsujimoto, et al., Molecular cloning of the chromosomal breakpoint of B-cell lymphomas and leukemias with the t(11;14) chromosome translocation, Science 224 (4656) (1984) 1403–1406.
[33] D.L. Vaux, S. Cory, J.M. Adams, Bcl-2 gene promotes haemopoietic cell survival and cooperates with c-myc to immortalize pre-B cells, Nature 335 (6189) (1988) 440–442.
[34] T.J. McDonnell, et al., Bcl-2-immunoglobulin transgenic mice demonstrate extended B cell survival and follicular lymphoproliferation, Cell 57 (1) (1989) 79–88.
[35] K.L. O'Neill, et al., Inactivation of prosurvival Bcl-2 proteins activates Bax/Bak through the outer mitochondrial membrane, Genes Dev. 30 (8) (2016) 973–988.
[36] R.J. Youle, A. Strasser, The BCL-2 protein family: opposing activities that mediate cell death, Nat. Rev. Mol. Cell Biol. 9 (1) (2008) 47–59.
[37] M.H. Kang, C.P. Reynolds, Bcl-2 inhibitors: targeting mitochondrial apoptotic pathways in cancer therapy, Clin. Cancer Res. 15 (4) (2009) 1126–1132.
[38] G. Dewson, et al., To trigger apoptosis, Bak exposes its BH3 domain and homodimerizes via BH3:groove interactions, Mol. Cell 30 (3) (2008) 369–380.
[39] S. Bleicken, et al., Molecular details of Bax activation, oligomerization, and membrane insertion, J. Biol. Chem. 285 (9) (2010) 6636–6647.
[40] E. Gavathiotis, et al., BAX activation is initiated at a novel interaction site, Nature 455 (7216) (2008) 1076–1081.
[41] J.K. Brunelle, A. Letai, Control of mitochondrial apoptosis by the Bcl-2 family, J. Cell Sci. 122 (Pt 4) (2009) 437–441.
[42] S.A. Amundson, et al., An informatics approach identifying markers of chemosensitivity in human cancer cell lines, Cancer Res. 60 (21) (2000) 6101–6110.
[43] R. Hegde, et al., Identification of Omi/HtrA2 as a mitochondrial apoptotic serine protease that disrupts inhibitor of apoptosis protein-caspase interaction, J. Biol. Chem. 277 (1) (2002) 432–438.
[44] M.E. Guicciardi, G.J. Gores, Life and death by death receptors, FASEB J. 23 (6) (2009) 1625–1637.
[45] S. Fulda, K.M. Debatin, Death receptor signaling in cancer therapy, Curr. Med. Chem. Anticancer Agents 3 (4) (2003) 253–262.
[46] A. Ashkenazi, V.M. Dixit, Death receptors: signaling and modulation, Science 281 (5381) (1998) 1305–1308.
[47] S. Nagata, Apoptosis by death factor, Cell 88 (3) (1997) 355–365.
[48] N.S. Wilson, V. Dixit, A. Ashkenazi, Death receptor signal transducers: nodes of coordination in immune signaling networks, Nat. Immunol. 10 (4) (2009) 348–355.
[49] A. Ashkenazi, G. Salvesen, Regulated cell death: signaling and mechanisms, Annu. Rev. Cell Dev. Biol. 30 (2014) 337–356.
[50] A. Ashkenazi, Targeting the extrinsic apoptotic pathway in cancer: lessons learned and future directions, J. Clin. Invest. 125 (2) (2015) 487–489.
[51] G. Sethi, et al., Targeting TNF for treatment of cancer and autoimmunity, Adv. Exp. Med. Biol. 647 (2009) 37–51.
[52] G. Sethi, et al., Pinitol targets nuclear factor-kappaB activation pathway leading to inhibition of gene products associated with proliferation, apoptosis, invasion, and angiogenesis, Mol. Cancer Ther. 7 (6) (2008) 1604–1614.

[53] G. Sethi, B. Sung, B.B. Aggarwal, TNF: a master switch for inflammation to cancer, Front. Biosci. 13 (2008) 5094–5107.
[54] B.B. Aggarwal, et al., TNF blockade: an inflammatory issue, Ernst Schering Res. Found. Workshop (56) (2006) 161–186.
[55] K.S. Ahn, G. Sethi, B.B. Aggarwal, Reversal of chemoresistance and enhancement of apoptosis by statins through down-regulation of the NF-kappaB pathway, Biochem. Pharmacol. 75 (4) (2008) 907–913.
[56] S.G. Hymowitz, et al., Triggering cell death: the crystal structure of Apo2L/TRAIL in a complex with death receptor 5, Mol. Cell 4 (4) (1999) 563–571.
[57] F.L. Scott, et al., The Fas-FADD death domain complex structure unravels signalling by receptor clustering, Nature 457 (7232) (2009) 1019–1022.
[58] L. Wang, et al., The Fas-FADD death domain complex structure reveals the basis of DISC assembly and disease mutations, Nat. Struct. Mol. Biol. 17 (11) (2010) 1324–1329.
[59] M.P. Boldin, et al., Involvement of MACH, a novel MORT1/FADD-interacting protease, in Fas/APO-1- and TNF receptor-induced cell death, Cell 85 (6) (1996) 803–815.
[60] F.C. Kischkel, et al., Cytotoxicity-dependent APO-1 (Fas/CD95)-associated proteins form a death-inducing signaling complex (DISC) with the receptor, EMBO J. 14 (22) (1995) 5579–5588.
[61] F.C. Kischkel, et al., Apo2L/TRAIL-dependent recruitment of endogenous FADD and caspase-8 to death receptors 4 and 5, Immunity 12 (6) (2000) 611–620.
[62] M.R. Sprick, et al., FADD/MORT1 and caspase-8 are recruited to TRAIL receptors 1 and 2 and are essential for apoptosis mediated by TRAIL receptor 2, Immunity 12 (6) (2000) 599–609.
[63] R.M. Siegel, et al., Fas preassociation required for apoptosis signaling and dominant inhibition by pathogenic mutations, Science 288 (5475) (2000) 2354–2357.
[64] C. Feig, et al., Palmitoylation of CD95 facilitates formation of SDS-stable receptor aggregates that initiate apoptosis signaling, EMBO J. 26 (1) (2007) 221–231.
[65] K.W. Wagner, et al., Death-receptor O-glycosylation controls tumor-cell sensitivity to the proapoptotic ligand Apo2L/TRAIL, Nat. Med. 13 (9) (2007) 1070–1077.
[66] C.D. Austin, et al., Death-receptor activation halts clathrin-dependent endocytosis, Proc. Natl. Acad. Sci. USA 103 (27) (2006) 10283–10288.
[67] S.L. Kohlhaas, et al., Receptor-mediated endocytosis is not required for tumor necrosis factor-related apoptosis-inducing ligand (TRAIL)-induced apoptosis, J. Biol. Chem. 282 (17) (2007) 12831–12841.
[68] M. Lu, et al., Opposing unfolded-protein-response signals converge on death receptor 5 to control apoptosis, Science 345 (6192) (2014) 98–101.
[69] M. Lu, et al., E-cadherin couples death receptors to the cytoskeleton to regulate apoptosis, Mol. Cell 54 (6) (2014) 987–998.
[70] A. Ashkenazi, Targeting death and decoy receptors of the tumour-necrosis factor superfamily, Nat. Rev. Cancer 2 (6) (2002) 420–430.
[71] R. Imamura, et al., Fas ligand induces cell-autonomous NF-kappaB activation and interleukin-8 production by a mechanism distinct from that of tumor necrosis factor-alpha, J. Biol. Chem. 279 (45) (2004) 46415–46423.
[72] E. Varfolomeev, et al., Molecular determinants of kinase pathway activation by Apo2 ligand/tumor necrosis factor-related apoptosis-inducing ligand, J. Biol. Chem. 280 (49) (2005) 40599–40608.
[73] F. Li, et al., NF-kappaB in cancer therapy, Arch. Toxicol. 89 (5) (2015) 711–731.
[74] Y.R. Puar, et al., Evidence for the involvement of the master transcription factor NF-kappaB in cancer initiation and progression, Biomedicines 6 (3) (2018) 82.

[75] E.M. Shin, et al., DEAD-box helicase DP103 defines metastatic potential of human breast cancers, J. Clin. Invest. 124 (9) (2014) 3807–3824.
[76] L. Liu, et al., Oleuropein induces apoptosis via abrogating NF-kappaB activation cascade in estrogen receptor-negative breast cancer cells, J. Cell. Biochem. 120 (3) (2019) 4504–4513.
[77] M.K. Shanmugam, et al., Celastrol attenuates the invasion and migration and augments the anticancer effects of bortezomib in a xenograft mouse model of multiple myeloma, Front. Pharmacol. 9 (2018) 365.
[78] C.D. Mohan, et al., Novel 1,3,4-oxadiazole induces anticancer activity by targeting NF-kappaB in hepatocellular carcinoma cells, Front. Oncol. 8 (2018) 42.
[79] R. Ningegowda, et al., A novel 4,6-disubstituted-1,2,4-triazolo-1,3,4-thiadiazole derivative inhibits tumor cell invasion and potentiates the apoptotic effect of TNFalpha by abrogating NF-kappaB activation cascade, Apoptosis 22 (1) (2017) 145–157.
[80] A. Ikner, A. Ashkenazi, TWEAK induces apoptosis through a death-signaling complex comprising receptor-interacting protein 1 (RIP1), Fas-associated death domain (FADD), and caspase-8, J. Biol. Chem. 286 (24) (2011) 21546–21554.
[81] P. Schneider, et al., TWEAK can induce cell death via endogenous TNF and TNF receptor 1, Eur. J. Immunol. 29 (6) (1999) 1785–1792.
[82] D. Wallach, et al., Tumor necrosis factor receptor and Fas signaling mechanisms, Annu. Rev. Immunol. 17 (1999) 331–367.
[83] D. Vucic, V.M. Dixit, I.E. Wertz, Ubiquitylation in apoptosis: a post-translational modification at the edge of life and death, Nat. Rev. Mol. Cell Biol. 12 (7) (2011) 439–452.
[84] S. Biton, A. Ashkenazi, NEMO and RIP1 control cell fate in response to extensive DNA damage via TNF-alpha feedforward signaling, Cell 145 (1) (2011) 92–103.
[85] N. Vanlangenakker, T. Vanden Berghe, P. Vandenabeele, Many stimuli pull the necrotic trigger, an overview, Cell Death Differ. 19 (1) (2012) 75–86.
[86] J. Li, J. Yuan, Caspases in apoptosis and beyond, Oncogene 27 (48) (2008) 6194–6206.
[87] H.M. Ellis, H.R. Horvitz, Genetic control of programmed cell death in the nematode *C. elegans*, Cell 44 (6) (1986) 817–829.
[88] J. Yuan, et al., The *C. elegans* cell death gene ced-3 encodes a protein similar to mammalian interleukin-1 beta-converting enzyme, Cell 75 (4) (1993) 641–652.
[89] S. Kumar, Y. Tomooka, M. Noda, Identification of a set of genes with developmentally down-regulated expression in the mouse brain, Biochem. Biophys. Res. Commun. 185 (3) (1992) 1155–1161.
[90] S. Kumar, et al., Induction of apoptosis by the mouse Nedd2 gene, which encodes a protein similar to the product of the *Caenorhabditis elegans* cell death gene ced-3 and the mammalian IL-1 beta-converting enzyme, Genes Dev. 8 (14) (1994) 1613–1626.
[91] D.W. Nicholson, Caspase structure, proteolytic substrates, and function during apoptotic cell death, Cell Death Differ. 6 (11) (1999) 1028–1042.
[92] H.R. Stennicke, G.S. Salvesen, Caspases - controlling intracellular signals by protease zymogen activation, Biochim. Biophys. Acta 1477 (1–2) (2000) 299–306.
[93] A. Degterev, M. Boyce, J. Yuan, A decade of caspases, Oncogene 22 (53) (2003) 8543–8567.
[94] N.A. Thornberry, Y. Lazebnik, Caspases: enemies within, Science 281 (5381) (1998) 1312–1316.
[95] S. Shalini, et al., Old, new and emerging functions of caspases, Cell Death Differ. 22 (4) (2015) 526–539.
[96] S. Kumar, Caspase function in programmed cell death, Cell Death Differ. 14 (1) (2007) 32–43.

[97] D.R. McIlwain, T. Berger, T.W. Mak, Caspase functions in cell death and disease, Cold Spring Harb. Perspect. Biol. 5 (4) (2013) a008656.
[98] U. Gaur, B.B. Aggarwal, Regulation of proliferation, survival and apoptosis by members of the TNF superfamily, Biochem. Pharmacol. 66 (8) (2003) 1403–1408.
[99] H. Li, et al., Cleavage of BID by caspase 8 mediates the mitochondrial damage in the Fas pathway of apoptosis, Cell 94 (4) (1998) 491–501.
[100] G. Pistritto, et al., Expression and transcriptional regulation of caspase-14 in simple and complex epithelia, Cell Death Differ. 9 (9) (2002) 995–1006.
[101] G. Pistritto, et al., Divergent modulation of neuronal differentiation by caspase-2 and -9, PLoS One 7 (5) (2012) e36002.
[102] G.S. Salvesen, C.S. Duckett, IAP proteins: blocking the road to death's door, Nat. Rev. Mol. Cell Biol. 3 (6) (2002) 401–410.
[103] N. Roy, et al., The gene for neuronal apoptosis inhibitory protein is partially deleted in individuals with spinal muscular atrophy, Cell 80 (1) (1995) 167–178.
[104] M. Rothe, et al., The TNFR2-TRAF signaling complex contains two novel proteins related to baculoviral inhibitor of apoptosis proteins, Cell 83 (7) (1995) 1243–1252.
[105] A.G. Uren, et al., Cloning and expression of apoptosis inhibitory protein homologs that function to inhibit apoptosis and/or bind tumor necrosis factor receptor-associated factors, Proc. Natl. Acad. Sci. USA 93 (10) (1996) 4974–4978.
[106] P. Liston, et al., Suppression of apoptosis in mammalian cells by NAIP and a related family of IAP genes, Nature 379 (6563) (1996) 349–353.
[107] C.S. Duckett, et al., A conserved family of cellular genes related to the baculovirus iap gene and encoding apoptosis inhibitors, EMBO J. 15 (11) (1996) 2685–2694.
[108] G. Ambrosini, C. Adida, D.C. Altieri, A novel anti-apoptosis gene, survivin, expressed in cancer and lymphoma, Nat. Med. 3 (8) (1997) 917–921.
[109] H.P. Hauser, et al., A giant ubiquitin-conjugating enzyme related to IAP apoptosis inhibitors, J. Cell Biol. 141 (6) (1998) 1415–1422.
[110] Z. Chen, et al., A human IAP-family gene, apollon, expressed in human brain cancer cells, Biochem. Biophys. Res. Commun. 264 (3) (1999) 847–854.
[111] D. Vucic, et al., ML-IAP, a novel inhibitor of apoptosis that is preferentially expressed in human melanomas, Curr. Biol. 10 (21) (2000) 1359–1366.
[112] J.H. Lin, et al., KIAP, a novel member of the inhibitor of apoptosis protein family, Biochem. Biophys. Res. Commun. 279 (3) (2000) 820–831.
[113] G.M. Kasof, B.C. Gomes, Livin, a novel inhibitor of apoptosis protein family member, J. Biol. Chem. 276 (5) (2001) 3238–3246.
[114] B.W. Richter, et al., Molecular cloning of ILP-2, a novel member of the inhibitor of apoptosis protein family, Mol. Cell Biol. 21 (13) (2001) 4292–4301.
[115] R. Budhidarmo, C.L. Day, IAPs: modular regulators of cell signalling, Semin. Cell Dev. Biol. 39 (2015) 80–90.
[116] B.P. Eckelman, G.S. Salvesen, F.L. Scott, Human inhibitor of apoptosis proteins: why XIAP is the black sheep of the family, EMBO Rep. 7 (10) (2006) 988–994.
[117] J. Berthelet, L. Dubrez, Regulation of apoptosis by inhibitors of apoptosis (IAPs), Cells 2 (1) (2013) 163–187.
[118] R. Feltham, et al., Smac mimetics activate the E3 ligase activity of cIAP1 protein by promoting RING domain dimerization, J. Biol. Chem. 286 (19) (2011) 17015–17028.
[119] S. Fulda, D. Vucic, Targeting IAP proteins for therapeutic intervention in cancer, Nat. Rev. Drug Discov. 11 (2) (2012) 109–124.
[120] E.C. LaCasse, et al., IAP-targeted therapies for cancer, Oncogene 27 (48) (2008) 6252–6275.
[121] Z. Liu, et al., Structural basis for binding of Smac/DIABLO to the XIAP BIR3 domain, Nature 408 (6815) (2000) 1004–1008.

[122] G. Wu, et al., Structural basis of IAP recognition by Smac/DIABLO, Nature 408 (6815) (2000) 1008–1012.
[123] S. Fulda, Therapeutic opportunities based on caspase modulation, Semin. Cell Dev. Biol. 82 (2018) 150–157.
[124] J.E. Vince, et al., IAP antagonists target cIAP1 to induce TNFalpha-dependent apoptosis, Cell 131 (4) (2007) 682–693.
[125] E. Varfolomeev, et al., IAP antagonists induce autoubiquitination of c-IAPs, NF-kappaB activation, and TNFalpha-dependent apoptosis, Cell 131 (4) (2007) 669–681.
[126] E.C. Dueber, et al., Antagonists induce a conformational change in cIAP1 that promotes autoubiquitination, Science 334 (6054) (2011) 376–380.
[127] M. Darding, et al., Molecular determinants of Smac mimetic induced degradation of cIAP1 and cIAP2, Cell Death Differ. 18 (8) (2011) 1376–1386.
[128] D. Opel, et al., Targeting inhibitor of apoptosis proteins by Smac mimetic elicits cell death in poor prognostic subgroups of chronic lymphocytic leukemia, Int. J. Cancer 137 (12) (2015) 2959–2970.
[129] L. Steinhart, K. Belz, S. Fulda, Smac mimetic and demethylating agents synergistically trigger cell death in acute myeloid leukemia cells and overcome apoptosis resistance by inducing necroptosis, Cell Death Dis. 4 (2013) e802.
[130] S. Steinwascher, et al., Identification of a novel synergistic induction of cell death by Smac mimetic and HDAC inhibitors in acute myeloid leukemia cells, Cancer Lett. 366 (1) (2015) 32–43.
[131] M. Fakler, et al., Small molecule XIAP inhibitors cooperate with TRAIL to induce apoptosis in childhood acute leukemia cells and overcome Bcl-2-mediated resistance, Blood 113 (8) (2009) 1710–1722.
[132] S. Loeder, et al., A novel paradigm to trigger apoptosis in chronic lymphocytic leukemia, Cancer Res. 69 (23) (2009) 8977–8986.
[133] Y. Moller, et al., EGFR-targeted TRAIL and a Smac mimetic synergize to overcome apoptosis resistance in KRAS mutant colorectal cancer cells, PLoS One 9 (9) (2014) e107165.
[134] D. Stadel, et al., TRAIL-induced apoptosis is preferentially mediated via TRAIL receptor 1 in pancreatic carcinoma cells and profoundly enhanced by XIAP inhibitors, Clin. Cancer Res. 16 (23) (2010) 5734–5749.
[135] M. Vogler, et al., Small molecule XIAP inhibitors enhance TRAIL-induced apoptosis and antitumor activity in preclinical models of pancreatic carcinoma, Cancer Res. 69 (6) (2009) 2425–2434.
[136] S. Hannes, B.A. Abhari, S. Fulda, Smac mimetic triggers necroptosis in pancreatic carcinoma cells when caspase activation is blocked, Cancer Lett. 380 (1) (2016) 31–38.
[137] L. Wagner, et al., Smac mimetic sensitizes glioblastoma cells to Temozolomide-induced apoptosis in a RIP1- and NF-kappaB-dependent manner, Oncogene 32 (8) (2013) 988–997.
[138] V. Marschall, S. Fulda, Smac mimetic-induced upregulation of interferon-beta sensitizes glioblastoma to temozolomide-induced cell death, Cell Death Dis. 6 (2015) e1888.
[139] B.L. Probst, et al., Smac mimetics increase cancer cell response to chemotherapeutics in a TNF-alpha-dependent manner, Cell Death Differ. 17 (10) (2010) 1645–1654.
[140] S.H. Vellanki, et al., Small-molecule XIAP inhibitors enhance gamma-irradiation-induced apoptosis in glioblastoma, Neoplasia 11 (8) (2009) 743–752.
[141] R. Berger, et al., NF-kappaB is required for Smac mimetic-mediated sensitization of glioblastoma cells for gamma-irradiation-induced apoptosis, Mol. Cancer Ther. 10 (10) (2011) 1867–1875.

[142] D.S. Ziegler, et al., A small-molecule IAP inhibitor overcomes resistance to cytotoxic therapies in malignant gliomas in vitro and in vivo, Neuro Oncol. 13 (8) (2011) 820–829.

[143] S. Hehlgans, et al., The SMAC mimetic BV6 sensitizes colorectal cancer cells to ionizing radiation by interfering with DNA repair processes and enhancing apoptosis, Radiat. Oncol. 10 (2015) 198.

[144] J. Yang, et al., Radiosensitization of head and neck squamous cell carcinoma by a SMAC-mimetic compound, SM-164, requires activation of caspases, Mol. Cancer Ther. 10 (4) (2011) 658–669.

[145] D. Yang, et al., Smac-mimetic compound SM-164 induces radiosensitization in breast cancer cells through activation of caspases and induction of apoptosis, Breast Cancer Res. Treat. 133 (1) (2012) 189–199.

[146] S. Giagkousiklidis, et al., Sensitization of pancreatic carcinoma cells for gamma-irradiation-induced apoptosis by XIAP inhibition, Oncogene 26 (49) (2007) 7006–7016.

[147] B.Z. Carter, et al., Synergistic targeting of AML stem/progenitor cells with IAP antagonist birinapant and demethylating agents, J. Natl. Cancer Inst. 106 (2) (2014) djt440.

[148] D. Chauhan, et al., Targeting mitochondrial factor Smac/DIABLO as therapy for multiple myeloma (MM), Blood 109 (3) (2007) 1220–1227.

[149] D. Lecis, et al., Novel SMAC-mimetics synergistically stimulate melanoma cell death in combination with TRAIL and Bortezomib, Br. J. Cancer 102 (12) (2010) 1707–1716.

[150] E. Weisberg, et al., Potentiation of antileukemic therapies by Smac mimetic, LBW242: effects on mutant FLT3-expressing cells, Mol. Cancer Ther. 6 (7) (2007) 1951–1961.

[151] E. Weisberg, et al., Smac mimetics: implications for enhancement of targeted therapies in leukemia, Leukemia 24 (12) (2010) 2100–2109.

[152] D.S. Ziegler, et al., Resistance of human glioblastoma multiforme cells to growth factor inhibitors is overcome by blockade of inhibitor of apoptosis proteins, J. Clin. Invest. 118 (9) (2008) 3109–3122.

[153] F.M. Foster, et al., Targeting inhibitor of apoptosis proteins in combination with ErbB antagonists in breast cancer, Breast Cancer Res. 11 (3) (2009) R41.

[154] S.T. Beug, et al., Smac mimetics and innate immune stimuli synergize to promote tumor death, Nat. Biotechnol. 32 (2) (2014) 182–190.

[155] G.G. Jinesh, S. Chunduru, A.M. Kamat, Smac mimetic enables the anticancer action of BCG-stimulated neutrophils through TNF-alpha but not through TRAIL and FasL, J. Leukoc. Biol. 92 (1) (2012) 233–244.

[156] R. Weinlich, et al., Necroptosis in development, inflammation and disease, Nat. Rev. Mol. Cell Biol. 18 (2) (2017) 127–136.

[157] W.J. Kaiser, J.W. Upton, E.S. Mocarski, Viral modulation of programmed necrosis, Curr. Opin. Virol. 3 (3) (2013) 296–306.

[158] W.J. Kaiser, et al., Toll-like receptor 3-mediated necrosis via TRIF, RIP3, and MLKL, J. Biol. Chem. 288 (43) (2013) 31268–31279.

[159] J.W. Upton, W.J. Kaiser, E.S. Mocarski, DAI/ZBP1/DLM-1 complexes with RIP3 to mediate virus-induced programmed necrosis that is targeted by murine cytomegalovirus vIRA, Cell Host Microbe 11 (3) (2012) 290–297.

[160] J.W. Upton, W.J. Kaiser, E.S. Mocarski, Virus inhibition of RIP3-dependent necrosis, Cell Host Microbe 7 (4) (2010) 302–313.

[161] D. Vercammen, et al., Tumour necrosis factor-induced necrosis versus anti-Fas-induced apoptosis in L929 cells, Cytokine 9 (11) (1997) 801–808.

[162] D. Vercammen, et al., Dual signaling of the Fas receptor: initiation of both apoptotic and necrotic cell death pathways, J. Exp. Med. 188 (5) (1998) 919–930.
[163] D. Vercammen, et al., Inhibition of caspases increases the sensitivity of L929 cells to necrosis mediated by tumor necrosis factor, J. Exp. Med. 187 (9) (1998) 1477–1485.
[164] L. Galluzzi, et al., Molecular mechanisms of regulated necrosis, Semin. Cell Dev. Biol. 35 (2014) 24–32.
[165] A. Degterev, et al., Chemical inhibitor of nonapoptotic cell death with therapeutic potential for ischemic brain injury, Nat. Chem. Biol. 1 (2) (2005) 112–119.
[166] A. Degterev, et al., Identification of RIP1 kinase as a specific cellular target of necrostatins, Nat. Chem. Biol. 4 (5) (2008) 313–321.
[167] P. Vandenabeele, et al., The role of the kinases RIP1 and RIP3 in TNF-induced necrosis, Sci. Signal. 3 (115) (2010) re4.
[168] A. Linkermann, D.R. Green, Necroptosis, N. Engl. J. Med. 370 (5) (2014) 455–465.
[169] J.M. Murphy, et al., The pseudokinase MLKL mediates necroptosis via a molecular switch mechanism, Immunity 39 (3) (2013) 443–453.
[170] D.A. Rodriguez, et al., Characterization of RIPK3-mediated phosphorylation of the activation loop of MLKL during necroptosis, Cell Death Differ. 23 (1) (2016) 76–88.
[171] Z. Cai, et al., Plasma membrane translocation of trimerized MLKL protein is required for TNF-induced necroptosis, Nat. Cell Biol. 16 (1) (2014) 55–65.
[172] G. Quarato, et al., Sequential engagement of distinct MLKL phosphatidylinositol-binding sites executes necroptosis, Mol. Cell 61 (4) (2016) 589–601.
[173] Y. Dondelinger, et al., MLKL compromises plasma membrane integrity by binding to phosphatidylinositol phosphates, Cell Rep. 7 (4) (2014) 971–981.
[174] S.W. Tait, et al., Widespread mitochondrial depletion via mitophagy does not compromise necroptosis, Cell Rep. 5 (4) (2013) 878–885.
[175] B. Schenk, S. Fulda, Reactive oxygen species regulate Smac mimetic/TNFalpha-induced necroptotic signaling and cell death, Oncogene 34 (47) (2015) 5796–5806.
[176] Y. Zhang, et al., RIP1 autophosphorylation is promoted by mitochondrial ROS and is essential for RIP3 recruitment into necrosome, Nat. Commun. 8 (2017) 14329.
[177] Z. Yang, et al., RIP3 targets pyruvate dehydrogenase complex to increase aerobic respiration in TNF-induced necroptosis, Nat. Cell Biol. 20 (2) (2018) 186–197.
[178] F.J. Bock, S.W.G. Tait, Mitochondria as multifaceted regulators of cell death, Nat. Rev. Mol. Cell Biol. 21 (2) (2020) 85–100.
[179] C.P. Dillon, et al., RIPK1 blocks early postnatal lethality mediated by caspase-8 and RIPK3, Cell 157 (5) (2014) 1189–1202.
[180] R.J. Thapa, et al., Interferon-induced RIP1/RIP3-mediated necrosis requires PKR and is licensed by FADD and caspases, Proc. Natl. Acad. Sci. USA 110 (33) (2013) E3109–E3118.
[181] X. Qin, et al., The role of necroptosis in cancer: a double-edged sword? Biochim. Biophys. Acta Rev. Cancer 1871 (2) (2019) 259–266.
[182] M.N. Messmer, A.G. Snyder, A. Oberst, Comparing the effects of different cell death programs in tumor progression and immunotherapy, Cell Death Differ. 26 (1) (2019) 115–129.
[183] T. Molnar, et al., Current translational potential and underlying molecular mechanisms of necroptosis, Cell Death Dis. 10 (11) (2019) 860.
[184] N.V. Conev, et al., RIPK3 expression as a potential predictive and prognostic marker in metastatic colon cancer, Clin. Invest. Med. 42 (1) (2019) E31–E38.
[185] S.A. Dziedzic, et al., ABIN-1 regulates RIPK1 activation by linking Met1 ubiquitylation with Lys63 deubiquitylation in TNF-RSC, Nat. Cell Biol. 20 (1) (2018) 58–68.

[186] B. Hu, et al., Prognostic and clinicopathological significance of MLKL expression in cancer patients: a meta-analysis, BMC Cancer 18 (1) (2018) 736.
[187] J.R. Cerhan, et al., Genetic variation in 1253 immune and inflammation genes and risk of non-Hodgkin lymphoma, Blood 110 (13) (2007) 4455–4463.
[188] L. Seifert, et al., The necrosome promotes pancreatic oncogenesis via CXCL1 and Mincle-induced immune suppression, Nature 532 (7598) (2016) 245–249.
[189] Z. Ertao, et al., Prognostic value of mixed lineage kinase domain-like protein expression in the survival of patients with gastric caner, Tumour Biol. 37 (10) (2016) 13679–13685.
[190] L. He, et al., Low expression of mixed lineage kinase domain-like protein is associated with poor prognosis in ovarian cancer patients, OncoTargets Ther. 6 (2013) 1539–1543.
[191] D. Jiao, et al., Necroptosis of tumor cells leads to tumor necrosis and promotes tumor metastasis, Cell Res. 28 (8) (2018) 868–870.
[192] W. Wang, et al., RIP1 kinase drives macrophage-mediated adaptive immune tolerance in pancreatic cancer, Cancer Cell 34 (5) (2018) 757e7–774e7.
[193] B. Strilic, et al., Tumour-cell-induced endothelial cell necroptosis via death receptor 6 promotes metastasis, Nature 536 (7615) (2016) 215–218.
[194] N. Yatim, et al., RIPK1 and NF-kappaB signaling in dying cells determines cross-priming of CD8(+) T cells, Science 350 (6258) (2015) 328–334.
[195] A.G. Snyder, et al., Intratumoral activation of the necroptotic pathway components RIPK1 and RIPK3 potentiates antitumor immunity, Sci. Immunol. 4 (36) (2019).
[196] B.R. Stockwell, et al., Ferroptosis: a regulated cell death nexus linking metabolism, redox biology, and disease, Cell 171 (2) (2017) 273–285.
[197] S.J. Dixon, et al., Ferroptosis: an iron-dependent form of nonapoptotic cell death, Cell 149 (5) (2012) 1060–1072.
[198] W.S. Yang, et al., Regulation of ferroptotic cancer cell death by GPX4, Cell 156 (1–2) (2014) 317–331.
[199] L.A. Timmerman, et al., Glutamine sensitivity analysis identifies the xCT antiporter as a common triple-negative breast tumor therapeutic target, Cancer Cell 24 (4) (2013) 450–465.
[200] A. Muir, et al., Environmental cystine drives glutamine anaplerosis and sensitizes cancer cells to glutaminase inhibition, Elife 6 (2017) e27713.
[201] S.J. Dixon, et al., Pharmacological inhibition of cystine-glutamate exchange induces endoplasmic reticulum stress and ferroptosis, Elife 3 (2014) e02523.
[202] C. Louandre, et al., The retinoblastoma (Rb) protein regulates ferroptosis induced by sorafenib in human hepatocellular carcinoma cells, Cancer Lett. 356 (2 Pt B) (2015) 971–977.
[203] E. Lachaier, et al., Sorafenib induces ferroptosis in human cancer cell lines originating from different solid tumors, Anticancer Res. 34 (11) (2014) 6417–6422.
[204] C. Louandre, et al., Iron-dependent cell death of hepatocellular carcinoma cells exposed to sorafenib, Int. J. Cancer 133 (7) (2013) 1732–1742.
[205] J.H. Woo, et al., Elucidating compound mechanism of action by network perturbation analysis, Cell 162 (2) (2015) 441–451.
[206] M. Gao, et al., Role of mitochondria in ferroptosis, Mol. Cell 73 (2) (2019) 354e3–363e3.
[207] M.M. Gaschler, et al., Determination of the subcellular localization and mechanism of action of ferrostatins in suppressing ferroptosis, ACS Chem. Biol. 13 (4) (2018) 1013–1020.
[208] Y.Q. Wang, et al., The protective role of mitochondrial ferritin on erastin-induced ferroptosis, Front. Aging Neurosci. 8 (2016) 308.

[209] X. Xia, et al., The role of pyroptosis in cancer: pro-cancer or pro-"host"? Cell Death Dis. 10 (9) (2019) 650.
[210] B.T. Cookson, M.A. Brennan, Pro-inflammatory programmed cell death, Trends Microbiol. 9 (3) (2001) 113–114.
[211] Y. Wang, et al., Chemotherapy drugs induce pyroptosis through caspase-3 cleavage of a gasdermin, Nature 547 (7661) (2017) 99–103.
[212] C. Rogers, et al., Cleavage of DFNA5 by caspase-3 during apoptosis mediates progression to secondary necrotic/pyroptotic cell death, Nat. Commun. 8 (2017) 14128.
[213] C. Rogers, et al., Gasdermin pores permeabilize mitochondria to augment caspase-3 activation during apoptosis and inflammasome activation, Nat. Commun. 10 (1) (2019) 1689.
[214] P. Broz, V.M. Dixit, Inflammasomes: mechanism of assembly, regulation and signalling, Nat. Rev. Immunol. 16 (7) (2016) 407–420.
[215] N. Kayagaki, et al., Caspase-11 cleaves gasdermin D for non-canonical inflammasome signalling, Nature 526 (7575) (2015) 666–671.
[216] J. Shi, et al., Cleavage of GSDMD by inflammatory caspases determines pyroptotic cell death, Nature 526 (7575) (2015) 660–665.
[217] H.T.H. Thi, S. Hong, Inflammasome as a therapeutic target for cancer prevention and treatment, J. Cancer Prev. 22 (2) (2017) 62–73.
[218] C.B. Zhou, J.Y. Fang, The role of pyroptosis in gastrointestinal cancer and immune responses to intestinal microbial infection, Biochim. Biophys. Acta Rev. Cancer 1872 (1) (2019) 1–10.
[219] Z. Wei, et al., Advances of research in cancer-associated inflammation and tumor microenvironments, Chin. J. Clin. Oncol. 45 (2018) 1117–1121.
[220] Q. Wei, et al., Deregulation of the NLRP3 inflammasome in hepatic parenchymal cells during liver cancer progression, Lab. Invest. 94 (1) (2014) 52–62.
[221] Q. Wei, et al., E2-Induced activation of the NLRP3 inflammasome triggers pyroptosis and inhibits autophagy in HCC cells, Oncol. Res. 27 (7) (2019) 827–834.
[222] J. Dupaul-Chicoine, et al., Control of intestinal homeostasis, colitis, and colitis-associated colorectal cancer by the inflammatory caspases, Immunity 32 (3) (2010) 367–378.
[223] B. Hu, et al., Inflammation-induced tumorigenesis in the colon is regulated by caspase-1 and NLRC4, Proc. Natl. Acad. Sci. USA 107 (50) (2010) 21635–21640.
[224] I.C. Allen, et al., The NLRP3 inflammasome functions as a negative regulator of tumorigenesis during colitis-associated cancer, J. Exp. Med. 207 (5) (2010) 1045–1056.
[225] M.H. Zaki, M. Lamkanfi, T.D. Kanneganti, The Nlrp3 inflammasome: contributions to intestinal homeostasis, Trends Immunol. 32 (4) (2011) 171–179.
[226] P. Yu, et al., Eukaryotic elongation factor-2 kinase regulates the cross-talk between autophagy and pyroptosis in doxorubicin-treated human melanoma cells in vitro, Acta Pharmacol. Sin. 40 (9) (2019) 1237–1244.
[227] Y. Wang, et al., GSDME mediates caspase-3-dependent pyroptosis in gastric cancer, Biochem. Biophys. Res. Commun. 495 (1) (2018) 1418–1425.
[228] S. Feng, D. Fox, S.M. Man, Mechanisms of gasdermin family members in inflammasome signaling and cell death, J. Mol. Biol. 430 (18 Pt B) (2018) 3068–3080.
[229] H. Lu, et al., Molecular targeted therapies elicit concurrent apoptotic and GSDME-dependent pyroptotic tumor cell death, Clin. Cancer Res. 24 (23) (2018) 6066–6077.
[230] Y. Nadatani, et al., NOD-like receptor protein 3 inflammasome priming and activation in Barrett's epithelial cells, Cell. Mol. Gastroenterol. Hepatol. 2 (4) (2016) 439–453.

[231] M. Wu, et al., A PLK1 kinase inhibitor enhances the chemosensitivity of cisplatin by inducing pyroptosis in oesophageal squamous cell carcinoma, EBioMedicine 41 (2019) 244–255.
[232] Z. Jiang, et al., miRNA-214 inhibits cellular proliferation and migration in glioma cells targeting caspase 1 involved in pyroptosis, Oncol. Res. 25 (6) (2017) 1009–1019.
[233] J. Cui, et al., MST1 suppresses pancreatic cancer progression via ROS-induced pyroptosis, Mol. Cancer Res. 17 (6) (2019) 1316–1325.

CHAPTER 4

Major methods and technologies for assessing cell death

Tsz-Leung To[1] and Xiaokun Shu[2]
[1]Broad Institute, Cambridge, MA, United States; [2]Department of Pharmaceutical Chemistry and Cardiovascular Research Institute, University of California San Francisco, San Francisco, CA, United States

1. Introduction

Cells actively make decisions between proliferation and cell death in order to maintain homeostasis. For instance, defective cell death is a hallmark in cancers, whereas excessive cell death can lead to atrophy, cardiovascular, and neurodegenerative diseases. Deep understanding of the cell death process can lead to important fundamental insights in biology, as well as new approaches to control cell physiology and diseases. There are many ways for a cell to die [1]. Cell death can be programmed and regulated, or accidental. Apoptosis, also referred to as type I programmed cell death, is the best understood regulated cell death process and it serves many morphological or biochemical functions [2]. Necrosis is usually considered accidental cell death, although it has been suggested that necrosis can also be regulated. The programmed form of necrosis is termed necroptosis. Necroptosis is thought to be a line of defense against viral infection, and it plays a role in inflammatory diseases [3]. In contrast to necrosis, leaking of the membrane during necroptosis is regulated. Autophagy is an intriguing cellular process that primarily disposes of defective organelles and recycles nutrients. As such, it is thought to operate as a survival pathway. Yet, autophagy is also referred to as type II programmed cell death. For instance, some anticancer drugs induce cytotoxicity through autophagy [4]. Nevertheless, it should be noted that although autophagic cell death has been observed in immortalized cell lines, it is not clear whether autophagy is a common path to cell destruction in vivo.

By definition, death requires cells to trespass a point-of-no-return. Several physiological processes have been proposed as the key irreversible step for commitment during apoptosis: massive caspase-3 activation, complete loss of mitochondrial membrane potential, exposure of phosphatidylserine (PS), nuclear fragmentation, and cytoskeletal collapse.

Mechanisms of Cell Death and Opportunities for Therapeutic Development
ISBN 978-0-12-814208-0
https://doi.org/10.1016/B978-0-12-814208-0.00001-4

While some of the above steps, such as PS exposure, can potentially be reversed, a cell is definitely dead when it loses its plasma membrane integrity or undergoes fragmentation into discrete bodies. Upon this point-of-no-return, other cells can clean up the "corpse" of the dead cell [1]. Table 4.1 summarizes various methods for detecting apoptosis, necrosis, and autophagy.

2. Morphological changes during cell death

The morphological aspects of cell death are often used to distinguish between apoptosis, necrosis, and autophagy [5]. The following morphological features define apoptosis: rounding-up of the cell, projection of cytoplasm as pseudopods, chromatin condensation, nuclear fragmentation, and plasma membrane blebbing. The plasma membrane integrity is maintained until the final stage, when the cell implodes to enable disintegration of apoptotic bodies. The apoptotic bodies are then recognized and engulfed rapidly by phagocytes. Necrosis is morphologically defined by an increase in cell volume, swelling of organelles, osmatic shock, lysis of plasma membrane, and uncontrolled leakage of intracellular substances. The released content can often induce inflammatory responses [6]. Autophagy, and specifically macroautophagy, is defined by engulfment of cytoplasm by the autophagosomes, fusion of autophagosomes and lysosomes to create autolysosomes, and degradation by acidic hydrolases in lysosomes [7]. By the same token, autophagic cell death is characterized by the presence of large vacuoles, massive autophagosome engulfment of the cytoplasm, and a lack of chromatin condensation and nuclear fragmentation. Autophagic cell death may be reverted until the very end [8].

2.1 Detection of morphological changes by microscopy

Cell morphology can be directly observed using bright field or phase contrast microscopy. Time-lapse microscopy can be used to observe morphological changes that are specific for apoptosis or necrosis. The duration and order of the distinct morphological changes can be faithfully recorded. To identify apoptotic cell death, one can pay attention to features such as cell rounding-up, membrane blebbing, formation of apoptotic bodies, etc. To identify necrotic cell death, one can look for cellular swelling and lysis of plasma membrane, as well as absence of blebbing and apoptotic bodies.

Table 4.1 Conventional methods for monitoring cell death.

Mode	Features	Methods
Apoptosis	Membrane blebbing and formation of apoptotic bodies	DIC/fluorescence microscopy, FACS
	Clearance of dead cells by phagocytosis	DIC/fluorescence microscopy, TEM, transwell migration assay
	Loss of mitochondrial membrane potential	Lipophilic cationic fluorescent dyes
	Mitochondrial outer member permeabilization	Calcein-cobalt assay
	Cytochrome *c* release	Fluorescence microscopy localization, immunoblotting
	Activation of caspases	Fluorogenic substrate assay, immunoblotting, and immunofluorescence of active caspases
	PARP-1 cleavage	Immunoblotting and immunofluorescence
	DNA fragmentation and chromatin condensation	Gel electrophoresis, TUNEL assay, fluorescence microscopy
	Phosphatidylserine exposure	Annexin V staining
	Cytosolic acidification	pH sensitive fluorescent indicators
Necrosis	Swelling of cell and organelles	DIC/fluorescence microscopy, TEM
	Lysis of plasma membrane, absence of blebbing	DIC/fluorescence microscopy, propidium iodide staining
	LDH release	Cytosolic LDH activity by colorimetrical measurement
	Lack of caspase activation	Fluorogenic substrate assay, immunoblotting, and immunofluorescence of active caspases
	Lack of PARP-1 cleavage	Immunoblotting and immunofluorescence
	Activation of calpains and cathepsins	Fluorogenic substrate assay
Autophagy	Autophagosome formation	Fluorescence microscopy (LC3 localization), TEM
	Fusion of autophagosome and lysosome	TEM, fluorogenic substrate assay
	LC3 processing	Immunoblotting (LC3-I to LC3-II conversion)
	Degradation by lysosomes	Lysosomal markers, lysosomal pH indicators, immunobloting

Differential interference contrast (DIC) microscopy is especially useful to indicate morphological changes in the ultrastructure of the cells, such as organelle swelling and cytoplasmic granules. For example, necroptosis, but not accidental necrosis, is often characterized by cytoplasmic granulation which can be observed using DIC [9]. A combination of bright field and fluorescence imaging allows linking specific morphological changes during cell death to the underlying biochemical events [5]. Important fluorescence probes for this purpose include propidium iodide (PI), DAPI, and SYTOX (Life Technologies) for plasma membrane integrity; Annexin-V fluorophore conjugates for PS exposure; LysoTracker (Life Technologies) for lysosomal function; MitoTracker (Life Technologies) and JC-1 for mitochondrial quality. These probes will be discussed in more detail in the following sections. Nevertheless, attentions should be paid to reduce phototoxicity and photobleaching of fluorescence markers whenever possible. This can be achieved by optimization of parameters such as light intensity, number of z sections and time frame, exposure time, and opening of iris and diaphragm.

Transmission electron microscopy (TEM) is a powerful way to assess cell morphology during cell death. Its high resolving power of down to 0.1 nm is required for detecting changes in organelle ultrastructure. It is particularly useful for the detection of autophagy, for example, in which submicroscopic ultrastructures like autophagosomes and autolysosomes are prevalent [7]. It can also be used together with immunogold staining to detect the precise localization of the protein of interest. TEM can also be used to distinguish between apoptotic and necrotic cells, as it can identify ultrastructures like protrusion of pseudopods, condensed chromatin, presence of nucleolus, swollen mitochondria, and convolution of cell surface. However, there are multiple challenges in utilizing TEM for cell death detection. First, it is time-consuming, expensive, and requires technical expertise on sample preparation and handling. Second, proper identification of the ultrastructures of interest on the TEM morphometry requires significant experience. Finally, large-scale quantitative analysis of cell death, in which many cells need to be observed over a period of time, is not possible with TEM. Light microscopy is better suited for measuring of a larger number of cell death events to enable statistical analysis. For the largest-scale quantification, flow cytometry can be used to probe the statistics of cell death events of a much larger sample size (thousands to millions of cells). Significant changes in the forward and side scatter parameters, which are correlated with cell size and cell granularity respectively, can serve as rough proxies for cell death events [10]. Dying cells typically have lower forward scatter and higher side scatter than viable cells. However, these changes in light scatter do not distinguish between apoptosis and necrosis. To determine the mode of cell death, the light scatter parameters are best combined with fluorescence detection of viability dyes such as PI and Annexin V.

2.2 Observation of the clearance of dead cells

Macrophages are involved in the clearance of dead cells in order to maintain normal tissue homeostasis and orchestrate immune responses. In particular, phagocytes are responsible for selective and rapid clearance of apoptotic cells, and such clearance is considered an antiinflammatory response. On the other hand, clearance of necrotic cells by macrophages is considered inflammatory [11]. Phagocytosis of dead cells can be observed using microscopy. Common methods to image include confocal fluorescence and light microscopy, transmission and scanning electron microscopy, and flow fluorocytometry.

The phagocytosis of apoptotic cells consists of several key steps [6]. In chronological order these steps are (1) recognition; (2) binding or tethering; (3) signaling events such as PS exposure and integrin activation to initiate engulfment; and (4) phagocytosis to clean up the apoptotic cell or body, as well as other phagocyte responses such as release of antiinflammatory cytokines and promotion of intercellular adhesion. While most of these steps can be directly observed using light and fluorescence microscopy, the first step of sensing requires careful characterization of cell migration [12]. The Boyden chamber of transwell system [13] has been designed to study the sensing process specifically. In this migration assay, the macrophages are placed above a membrane, and chemoattractant (apoptotic cells in this case) is placed below in order to generate a concentration gradient. The membrane pores are large enough to the macrophages to "squeeze" through but small enough to block them from falling freely through the membrane. The number of macrophage cells below the membrane is counted after some time is given to the cells to migrate. Such transwell migration assay has been used to demonstrate the attraction of monocyte-derived macrophages to supernatants of apoptotic cells [14].

3. Assays for apoptosis

Apoptosis is essential in multicellular organisms for the maintenance of cellular homeostasis as well as embryogenesis and metamorphosis. The defining signaling events for apoptosis include cytochrome *c* release, formation of apoptosome, activation of initiator caspases such as caspase-8 (extrinsic pathway) or caspase-9 (intrinsic pathway), and activation of executioner caspases such as caspase-3 that leads to organized destruction of the cell [15]. No single biochemical parameter fully defines apoptotic death. Methods have been developed to distinguish between live cells, early apoptosis, late apoptosis, and necrosis. Loss of mitochondrial membrane potential and caspase-3 activation apparently precedes nuclear fragmentation and PS exposure during apoptosis, whereas permeability to PI and cytoskeletal collapse occurs later. Mitochondria play an important role in the apoptotic process as they act as central regulators and effectors of apoptotic pathways [16]. As such, mitochondrial responses such as collapse of membrane potential and opening of permeability transition pore can be used as a proxy for apoptotic cell signaling. During the cell destruction phase, the biochemical characteristics of apoptosis are fragmentation of chromatin and degradation of nuclear envelope. This breakdown of the nucleus leads to nuclear blebbing and the formation of micronuclei, and eventually apoptotic bodies.

3.1 Mitochondrial physiology

Mitochondria play crucial roles in energy metabolism and are indispensable for the survival and proliferation of eukaryotic cells. Mitochondria are at the heart of apoptosis: they are central regulators and effectors of apoptotic pathways [16]. Under physiological conditions, mitochondria maintain a membrane potential across the inner membrane. The membrane potential and the selective permeability of the inner membrane allow specific proteins to be maintained within the intermembrane space (IMS). Apoptotic signaling events lead to destabilization of the mitochondrial outer membrane and opening of the mitochondrial permeability transition pore (mPTP) via binding of Bax, Bak, and BH-3 only proteins. This event, known as mitochondrial outer membrane permeabilization (MOMP), leads to dissipation of mitochondrial membrane potential and release of cytochrome *c* and other IMS protein. In the cytosplasm, cytochrome *c* activates the apoptosome and the caspase cascade, which ultimately results in the irreversible destruction of a cell.

3.2 Collapse of mitochondrial membrane potential

The detection of a reduced mitochondrial membrane potential can be used to predict the irreversible cascade of apoptotic events. During respiration, a proton motive force is maintained by the electron transport chain, which pumps proton from the matrix to the IMS and thus creates an electrical and pH gradient across the inner membrane. The mitochondrial membrane potential, which is typically around 150 mV, represents the electrical component of the proton motive force [17].

Lipophilic cationic fluorescent dyes with high membrane permeability and low membrane binding can be used to measure the membrane potential in live cells [18]. Cations are attracted to the negative potential across the inner membrane and they accumulate in the matrix of energized mitochondria. JC-1 is commonly used for the study of mitochondrial membrane potential using flow cytometry and fluorescence spectrophotometry [19]. The dye exhibits a membrane potential-dependent accumulation in mitochondria, which results in a fluorescence emission shift from green to red fluorescence. Mitochondrial depolarization due to apoptotic events can be detected by a reduction in the red to green fluorescence intensity ratio.

Cationic rhodamine and carbocyanine derivatives such as TMRE, TMRM [20], and DiOC6(3) [21] are also commonly used for more quantitative measurement of membrane potential. Like JC-1, these dyes accumulate primarily in energized mitochondria, although they do not undergo fluorescence emission shift. Instead, the fluorescence intensity of these dyes is directly correlated with the mitochondrial polarization status. Once the mitochondrial membrane potential is disrupted by apoptotic events, the dyes are re-equilibrated between cytosol and mitochondria and their partition can be predicted by the Nernst equation [17]. Depolarization of mitochondria results in loss of fluorescence signal in the mitochondria, and reduction in overall fluorescence within a cell.

It should be noted that a loss of mitochondrial membrane potential is not an absolute indicator of apoptosis. The transmembrane potential can be dissipated by protonophores, inhibitors of oxidative phosphorylation, or as a physiological response during quiescence or under cellular stress. As such, the presence of apoptotic cell signaling needs to be confirmed by a more specific technique such as the detection of activated caspases.

3.3 Mitochondrial outer membrane permeabilization

The onset of MOMP can be considered a point of no return in the cellular commitment to apoptosis [16]. Once MOMP occurs, caspase activation, fragmentation of genomic DNA, and PS exposure follow. MOMP is orchestrated by the opening of mitochondrial permeability transition pore (mPTP). Collapse of mitochondrial membrane potential can be an indirect method for measuring mPTP opening. However, nonapoptotic events, such as insults to the respiratory chain, can also trigger mitochondrial depolarization. To directly detect mPTP opening, one can load the cells with acetoxymethyl ester of calcein (calcein AM) and its quencher cobalt(II) chloride ($CoCl_2$) [22]. The AM moiety helps calcein to be loaded into the cytosol and mitochondria. Once in the cells, the AM moiety is cleaved by intracellular esterases, and this cleavage causes the calcein to be trapped in the compartments. The quencher $CoCl_2$ is then loaded to quench the calcein fluorescence in the cytosol but not in the mitochondria. Loss of mitochondrial calcein fluorescence can be observed upon mPTP opening, as calcein is released into the cytoplasmic space. As a positive control for mPTP opening, cells loaded with calcein AM and $CoCl_2$ can be treated with ionomycin that allows excessive Ca^{2+} entry into mitochondria to induce mPTP opening. Ionomycin-induced mPTP opening can be blocked by cyclosporine A, which prevents the formation of the pore by binding to cyclophilin D [23].

3.4 Cytochrome *c* release

Cytoplasmic cytochrome *c* acts as a caspase activator and may serve as a marker for apoptotic cells [16]. Cytochrome *c* releasing from mitochondria into cytosol can be determined by Western blotting using the cytochrome *c* antibody. Detection of immunolabeled cells by fluorescence microscopy can also be used for the identification of apoptotic cells. The release of this mitochondrial component can be highlighted by the staining pattern: a punctiform staining in normal cells corresponding to the mitochondrial localization of cytochrome *c*, and a more diffuse staining in apoptotic cells due to the release of cytochrome *c* from the mitochondria to the cytosol [24]. To visualize the dynamics of cytochrome *c* release in living cells, a cytochrome *c*-GFP fusion protein can be expressed. Time-lapse microscopy enables kinetic study of the apoptotic process and provides spatiotemporal information that is inaccessible by immunostaining. Similar strategies can be applied to detect the release of other IMS proteins involved in apoptosis, such as Smac/Diablo. However, one should note that microscopy is required for such localization-based detection and as such, measurement of a large number of cells is difficult.

3.5 Caspase activation

An important feature of early apoptosis is the activation of caspases, a family of cysteine-aspartic specific proteases. Caspase activation is the key event that discriminates between apoptosis and necrosis. Caspases recognize specific sequences of three to four amino acids followed by an aspartic acid residue, and cleavage occurs after the aspartate. Caspases typically exist as inactive precursors. The activation, either by autocatalysis or by an upstream protease, leads to cleavage of internal aspartates within the caspase [15]. They can also be activated by the release of inhibitors or by binding of cofactors. In particular, caspase-3, caspase-6, and caspase-7 are key executioner (or effector) caspases within the apoptosis pathway and they relay the signal from initiator caspases such as caspase-2, caspase-8, caspase-9, and caspase-10 to implement a full cellular disassembly. There are many downstream substrates of caspases 3 and 7, including poly(ADP-ribose) polymerase (PARP), actin, protein kinases, and other proteases.

To detect the activation of executioner caspases, one can develop fluorogenic reagents that incorporate the four-amino acid peptide DEVD—which contains the nominal recognition site for caspase-3 and caspase-7. As an example, the CellEvent Caspase-3/7 detection reagent (Life Technologies) contains a nucleic acid-binding dye whose binding is inhibited by the DEVD peptide, and therefore the reagent is intrinsically nonfluorescent. Activated caspase-3/7 cleaves the DEVD peptide away from the dye and allows the dye to bind DNA and produce a bright whole-cell fluorescence signal. The CellEvent reagent provides advantages such as no requirement of washing steps and compatibility with formaldehyde fixation. Similar reagents have been developed for initiator caspases, such as caspase-8. Caspase-8 activity can be detected by fluorogenic substrates that contain the recognition sequence IETD (which is also recognized by caspase-6) or LETD. Other known caspases substrates include LEHD for caspase-5, YVAD for caspase-1/4, and WEHD for caspase-1/4/5 [25].

A similar type of reagent known as Image-iT LIVE caspase-3 and -7 detection kit (Life Technologies) have been developed based on the FLICA method. FLICA utilizes a fluorescent inhibitor of caspases. The carboxyfluorescein (FAM) labeled FLICA inhibitor contains a fluoromethyl ketone moiety that can covalently attach its cysteine to the enzymatic center of an activated caspase [26]. The FLICA inhibitor is designed to be an affinity label that recognizes specific sequences of various caspases—DEVD for caspase-3 and caspase-7, LETD for caspase-8. The FLICA reagent is cell permeant and able to measure the amount of active caspase directly via whole-cell fluorescence signal. It is also common to perform time-course experiments with this reagent in order to probe the dynamics of apoptosis.

Standard Western blots using antibodies against caspase-3 or caspase-7 can also be used to detect caspase activation. Apoptotic events result in the cleavage of precursor procaspase to yield the processed *p*20 subunit, which can be readily identified using immunoblotting [5]. In addition, antibodies that specifically recognize the activated form of caspase-3 or caspase-7 are also commercially available for immunofluorescence detection of apoptotic cells.

3.6 Cleavage of PARP-1

PARP-1 is an enzyme involved in DNA damage repair. Mild DNA damage leads to a (NAD+)-dependent PARP-1 activation that results in relaxation of chromatin structure and recruitment of repair enzymes. More severe DNA damage induces apoptotic cell death, in which PARP-1 is cleaved by the DEVDase activity of caspase-3 [27]. In necrotic cell death, the absence of DEVDase activity results in no PARP-1 proteolysis. Apoptotic cleavage of PARP-1 can be detected by Western blots. A shift from the native 116 kDa to 85 kDa is a hallmark of apoptosis [11]. Anticleaved PARP-1 antibodies are also commercially available for immunofluorescence detection of apoptotic cells.

3.7 DNA fragmentation

A landmark of cellular self-destruction by apoptosis is the activation of nucleases that degrade the higher order chromatin structure of the DNA into fragments of 50—300 kilobases and subsequently into smaller DNA pieces of about 200 base pairs in length [28]. The break-up of the genomic DNA by cellular nucleases is also one of the most straightforward events to detect. DNA fragmentation is a feature of apoptosis that is not present in necrosis. Unlike in cells undergoing apoptosis, DNA is degraded randomly in necrotic cells.

Caspase-activating DNase (CAD) is capable of degradation of chromosomal DNA [23]. Nucleic acid stains have been developed to detect apoptotic cells by fluorescent means. The simplest way of detecting DNA fragmentation is by using electrophoresis [29]. In this in vitro assay, the DNA of apoptotic cells are extracted and separated by gel electrophoresis. DNA fragments exhibit a ladder pattern of low molecular weight. Standard nucleic acid stains, such as ethidium bromide and SYBR (Life Technologies), allow detection of DNA fragments by gel electrophoresis.

Chromatin condensation provides another simple mean to detect apoptosis. The blue-fluorescent Hoechst dye stains the condensed chromatin of apoptotic cells more brightly due to an increased uptake of the dye. The red-fluorescent PI is permeant only to late apoptotic or necrotic cells with highly compromised membranes [5]. Special dyes, such as the carbocyanine-based YO-PRO-1 (Life Technologies) [30], can selectively pass through the plasma membrane of early apoptotic cells, when the membrane is still impermeant to the late apoptotic/necrotic marker PI stain. The relatively large size of YO-PRO-1 prevents it from penetrating the plasma membrane of live cells. During apoptosis, the opening of cation channels allows the dye to enter the dying cells and bind to the DNA.

To definitively detect damaged DNA, the Comet assay, or single-cell gel electrophoresis assay, can be used to detect and quantify DNA damage in single cells [31]. The assay is based on the alkaline lysis of labile DNA at the damage sites. Cells are immobilized (e.g., on thin agarose) and lysed. The damaged DNA in apoptotic cells migrates out of the cells under electrophoresis. The cells are then stained with a nucleic acid dye such as ethidium bromide or SYBR. Cells with DNA fragmentation appear as fluorescent comets, whereas cells with undamaged DNA appear as round dots as intact DNA does not migrate out of the cells. The Comet assay can be used in combination with fluorescent in situ hybridization to identify specific sequences with damaged DNA.

A common method to detect DNA in situ strand breaks during apoptosis is the terminal deoxynucleotidyl transferase-dUTP nick end labeling (TUNEL) assay. The assay is based on terminal deoxynucleotidyl transferase (TdT) that attaches labeled deoxynucleotides (e.g., dUTPs) to the 3′-hydroxyl termini fragmented DNA caused by double-strand breaks [32]. Initially, this method was subjected to debate about its accuracy, as the assay also labeled necrotic cells. Subsequent improvement of the assay has greatly increased its specificity for the late apoptotic cells. The TUNEL assay can be done together with a DNA counter stain such as DAPI for the determination of the cell cycle phase where apoptosis is occurring. It can also be used together with necrosis-specific, impermeant dyes such as PI and SYTOX [5].

The key to an effective TUNEL assay is the use of an appropriate deoxynucleotide substrate that allows high sensitivity and reliability. The simplest form of the assay is the one-step incorporation of dye-modified nucleotides (fluorescein-12-2′-dUTP). However, the one-step incorporation leads to nonspecific labeling and significant false-positive results [32]. Moreover, detachment and subsequent loss of the dye can also yield false-negative results. Two-step incorporation provides additional flexibility and specificity. For example, cells can be rapidly fixed after first-step addition of minimally modified dNTPs, in order to preserve the late-stage apoptotic cells. Its applicability on fixed samples gives TUNEL a distinct advantage over many other methods of detecting apoptosis (e.g., Annexin V labeling).

Several options for two-step labeling are available. The thymidine analog BrdU (5-Bromo-2′-Deoxyuridine) can be used to derivatize dUTP to create BrdUTP, which can then be attached to the exposed 3′-hydroxyl termini [32]. Once incorporated into the DNA, BrdU can be immunocytochemically detected using an anti-BrdU monoclonal antibody conjugated to a fluorophore. For detecting DNA fragments, BrdUTP is preferred over biotin or fluorochrome tagged dUTPs because of lower reagent cost and higher sensitivity of BrdU. An alkyne-modified dUTP can be used to detect an azide-derivatized fluorophore through copper-catalyzed azide-alkyne cycloaddition [33]. This "click" chemistry enables a high degree of specificity since the azide and alkyne moieties are not found in endogenous biological molecules (unlike the biotin-avidin pair) [34]. Alkyne-modified dUTP molecules (e.g., EdUTP) are commercially available and they can be coupled with a choice of azide-derivatized fluorophores (e.g., Alexa Fluor dyes).

Finally, formation of double-strand breaks in genomic DNA is a strong signal of genotoxicity. Phosphorylation of H2A histones is an early response to double strand breaks formation. A specific site of DSB induced phosphorylation is at Ser139 of histone variant H2AX [35]. As such, DNA damage can be indirectly identified using antibody-based detection of phosphorylated H2AX (Ser139).

3.8 Phosphatidylserine exposure

Loss of plasma membrane integrity is one of the best markers for apoptotic cell death. Under normal condition, the phospholipid PS is on the cytoplasmic surface of the cell membrane. During apoptosis, the aminophospholipid translocases (aka flippases, e.g., ATP11A and ATP11C) are inactivated by activated caspase 3, which leads to phospholipid scrambling. The typically inner leaflet facing PS is translocated to the outer leaflet of the plasma membrane due to the activation of XKR8 by activated caspase 3 and becomes extracellularly exposed [36,37]. The human vascular anticoagulant Annexin V is a Ca^{2+}-dependent protein with a high affinity for the anionic PS. As such, Annexin V conjugates provide an easy and reliable way to detect the externalization of PS, which represents the intermediate stage of apoptosis [38]. The exposure of PS sends an eat-me signal on the surface of the apoptotic cell to attract rapid engulfment by phagocytes.

Using highly fluorescent Annexin V conjugates (e.g., Alexa Fluor dyes), the large difference in fluorescence intensity between apoptotic and non-apoptotic cells allows rapid detection by flow cytometry and other imaging-based assays. It should be noted that Annexin V staining cannot reliably distinguish among apoptotic, necrotic, and autophagic cells, since PS exposure can happen in all three types of cell death [5]. As such, Annexin V conjugates should be used simultaneously with other dyes such as Hoechst and PI to allow detection of multiple stages during the cell death process within the same experiment. The impermeant red-fluorescent SYTOX stain can also be used instead of PI to label specifically the necrotic cells. One potential challenge with Annexin V labeling is that it does not work well with trypsinized cells. Moreover, unlike the TUNEL assay, Annexin V labeling cannot be done on fixed cells. Finally, it should be noted that PS exposure can be reversed, as in the case of blood granulocytes [39].

3.9 Cytosolic acidification

Acidification of the cytosol (typically 0.3–0.4 intracellular pH units) is often associated with the early stage of apoptosis [40,41] and has been observed in both caspase-dependent and caspase-independent apoptosis. The origin of such acidification is unclear [41], although it has been reported that mitochondrial dysfunction may underlie this pH change [42]. Several physiological consequences of cytosolic acidification have been proposed. For example, PS externalization has been found to be concurrent with a decrease in pH, possibly due to the pH sensitivity of the flipping process of PS [43]. In addition, key apoptotic enzymes such as caspase-3 may have been regulated by pH. A pH-dependent safety catch mechanism has been proposed for caspase-3 [44], in which the three consecutive aspartate residues located next to the processing site of caspase-3 block access until the pH is lowered. Cytosolic acidification can be readily detected using fluorescent pH indicators. Small molecule dyes such as BCECF [45] and SNARF-1 [46] have been optimized for measuring physiological pH values. In addition, pH-sensitive fluorescent protein variants, such as pHluorin [47], have been developed for measurement of intracellular pH. A key advantage of fluorescent protein-based pH indicators is that they can be targeted to specific subcellular locations to provide spatial resolution [48].

4. Assays for necrosis

For many years, cell death processes have been divided into those that are regulated and those that are accidental. The first-discovered form of regulated cell death, apoptosis, is used to describe regulated cell death in the context of development and homeostasis. On the other hand, the term necrosis is often reserved for accidental, disordered mode of cell death. Necrosis describes cell death due to acute severe injuries to the cell, such as withdrawal of nutrients and extreme physiochemical insults. Typically, necrosis is morphologically characterized by swelling of cells and organelles, as well as rupture of plasma membrane that leads to inflammatory responses. One should note that a particular difficulty in defining necrosis is that in the absence of phagocytosis, apoptotic cells become secondary necrotic cells with many morphological features of primary necrosis [6]. As previously mentioned, the commonly used red-fluorescent PI can detect both primary and secondary necrotic cells with highly compromised membranes [5].

4.1 Detection through lack of apoptotic signatures

Unlike apoptosis and autophagic cell death that can be distinguished by certain biochemical events, necrosis is often characterized mostly in negative terms by the absence of caspase activation, PARP-1 cleavage, cytochrome *c* release, DNA fragmentation, and the requirement of ATP. Although not firmly established, severely lysosomal alkalinization is required for necrotic cell death [49]. Tumor necrosis factor (TNF) signaling in the absence of caspase activation provides a biochemical route for necrosis [23]. The signaling pathway of necrotic cell death begins at the TNF receptor 1 (TNF-R1), which contains an intracellular death domain (DD). TNF-R1 DD trimerization has been shown to be sufficient to activate the necrotic pathway [50]. Owing to its regulated nature, this necrotic form of cell death was termed necroptosis. Necroptosis is linked to pathological conditions that have a strong inflammatory signature, as in Crohn's disease [51]. It should be noted that the TNF receptor family also induce the extrinsic pathway of apoptosis depending on the ligands.

To discriminate, several physiological parameters have been considered distinctive for necroptosis. For instance, TNF-induced necroptosis requires functional oxidative phosphorylation [52]. Moreover, proteases other than caspases, such as cytosolic calpains and lysosomal cathepsins, may be involved in TNF-induced cell death [23]. Detection of calpain activation is possible by using the t-BOC-Leu-Met-CMAC fluorogenic substrate [53].

4.2 Release of lactate dehydrogenase

Permeabilization of plasma membrane is a distinct signature of necrotic death. One way to detect and quantify membrane permeabilization is to measure the lactate dehydrogenase (LDH) level in the culture supernatant [54,55]. LDH is a soluble cytoplasmic enzyme expressed in most cell types, and it is released into the extracellular growth medium once the plasma membrane is broken. Extracellular LDH converts NAD+ into NADH with concomitant conversion of lactate into pyruvate. The NADH is then used to convert tetrazolium salt into red formazan product, which can be quantified using colorimetrical methods (e.g., absorbance measurement using spectrophotometer). The level of formazan level in the culture supernatant can be used to estimate the fraction of necrotic cells. This assay allows high-throughput experiments using microtiter plates and does not require labeling of cells prior to the measurement. The main limitation of the LDH assay is that it does not discriminate between primary and secondary necrosis. As such, it is recommended that the samples be simultaneously tested for apoptosis. For instance, after harvesting the culture supernatant, the cells can be tested for caspase-3 activity.

Another extracellular marker for cell death is cytokeratin-18 (CK18). CK18 is cleaved by caspases specifically during apoptosis. Primary or secondary necrosis causes the plasma membrane to rupture and the release of CK18 into the serum. The caspase-cleaved, fragmented form of CK18 indicates apoptosis followed by subsequent secondary necrosis, whereas soluble CK18 indicates primary necrosis [56]. Both forms of CK18 can be detected by immunoassays such as ELISA. However, it should be noted that hematopoietic and neuronal cells do not express CK18.

Any biochemical assay on its own, be it PI staining or LDH release, is not sufficiently discriminative between apoptosis and necrosis. As such, it is best to combine multiple techniques to firmly establish the mode of cell death [5].

5. Assays for autophagy

Autophagy, the process in which cells recycle cytoplasm and dispose of defective organelles, plays critical roles in cellular homeostasis and regulation of macromolecules. Autophagy is defined by the segregation and delivery of cytoplasmic cargo, including proteins and organelles, for degradation by acidic hydrolases in the lysosomes. The process of autophagy starts with the formation and elongation of phagophores. The cytoplasmic cargo is then engulfed by the phagophores to create an autophagosome, and the double-membrane autophagosome fuses with a lysosome to generate the autolysosome. Finally, degradation is achieved through the action of hydrolytic enzymes within the autolysosome [57].

Although autophagy is referred to as type II programmed cell death, it is often considered a cell survival strategy. Under conditions of nutrient deprivation, autophagy serves as a survival pathway by recycling nutrients and macromolecules. Moreover, even though autophagy has been shown to participate in the destruction of cells and to promote cell death, it is unclear whether autophagy can by itself lead to a well-defined point-of-no-return. In fact, cells can recover from having displayed features of autophagic cell death. As such, it has been speculated that autophagic cell death occurs with but not through autophagy [1].

5.1 Observation of LC3

One of the key protein components involved in the initial steps of the autophagy is LC3.

LC3 exists as a precursor protein (proLC3), which is subsequently cleaved to form the nonlipidated form LC3-I. Upon the initiation of autophagy, LC3-I is lipidated with the addition of a phosphatidylethanolamine (PE) group to the C-terminus. The lipidated form LC3-II migrates faster on the SDS-PAGE gels due to the hydrophobic nature of the PE group. As such, Western blots can be performed using an LC3 antibody to determine the relative abundances of LC3-I and LC3-II. As autophagy occurs, the band intensity of LC3-II increases and that of LC3-I decreases [7].

The LC3 protein is localized in the cytosol under normal circumstances. During autophagy, the lipophilic character of PE enables LC3-II to be inserted into the outer and inner autophagosomal membranes, thus relocating the LC3 protein to the autophagosomes. Hence, fluorescence imaging can be used to detect the initiation of autophagy. For example, LC3-GFP fluorescent protein chimera can enable visualization of the LC3 distribution in live cells [58]. An accumulation of autophagosomal green fluorescent puncta indicates an increase in autophagic flux. Likewise, immunofluorescence technique using LC3 antibody can also be used.

There are caveats on the use of LC3 as a marker for autophagy [7]. Technical challenges, such as difference in LC3 expression in various cell types, and difference in antibody affinities for LC3-I and LC3-II, can limit LC3 measurement as a stand-alone autophagic marker. Ultimately, the "flux" through the entire autophagic pathway is the parameter of interest. LC3-II is static readout of autophagosome abundance. An increase in LC3-II band intensity on the Western blot can be due to an increase in autophagosome synthesis, or a defect in autophagosome recycling. The latter may actually reduce the autophagic flux. To measure the autophagic flux, dynamic measurements need to be made on LC3. Samples should be harvested over a period of time to allow detection of LC3-I to LC3-II shift following the stress applied to induce autophagy. Various inhibitors of autophagy can also be used to help determine the magnitude of the flux [59].

5.2 Monitoring the lysosomes

Importantly, autophagic flux represents the entire process of autophagy, which encompasses autophagosome formation, the delivery of cargo to lysosomes, and breakdown and release of the autophagic substrates back into the cytosol [59]. Autophagic activity cannot be fully captured by the LC3 profile. Direct monitoring of the breakdown of autophagic substrates, such as organelles and proteins, provides complementary information to LC3 detection. Mitochondria and lysosomes play important roles in autophagy. Mitophagy refers to the removal of damaged or surplus mitochondria through the autophagic pathway [60]. Recycling of key nutrients such as amino acids can be achieved through mitophagy. Detection of mitophagy can be achieved by using MitoTracker, mitochondria-targeted fluorescent proteins, or antibodies of mitochondrial proteins to monitor mitochondrial abundance and dynamics. Formation of autolysosome through the fusion of autophagophore and lysosome is a key step of the autophagic pathway. Various dyes such as LysoTracker and LysoSensor [61] have been developed for monitoring the abundance and acidity of lysosomes, both of which are critical for basal and induced autophagy [62]. To more directly detect the formation of autolysosome, LC3-GFP can be imaged together with a fluorogenic substrate known as DQ BSA. DQ BSA is bovine serum albumin conjugated to a self-quenched fluorophore. The docking of autophagosome onto the lysosome leads to dequenching of DQ BSA and strong fluorescence signal [63]. The autolysosomes can be identified by colocalization of LC3-GFP and DQ-BSA fluorescence.

Keima, a coral-derived acid-stable fluorescent protein, emits different-colored signals at acidic and neutral pHs [64]. When accumulated inside lysosomal compartments during autophagy, Keima's fluorescent readout can be used to quantify the dynamics of autophagy. Due to the large Stokes shift of Keima, this autophagy probe can be visualized in combination with other fluorescent markers, providing multiplexing capabilities.

Finally, TEM is a very powerful method for detecting and monitoring autophagy and is the only technique that provides insight into the structural and morphological aspects of autophagosomes and autolysosomes. Compared to other techniques such as Western blots and fluorescence imaging, however, TEM is time-consuming and requires strong expertise on sample preparation.

6. Novel genetically encoded reporters of apoptosis

The two main types of cell death, namely, apoptosis and necrosis, are defined primarily by morphological changes in the nucleus and cytoplasm. These morphological distinctions are often used to characterize results from studies performed in vitro using immortalized cell lines. However, observations of cell death in vitro may not reflect the in vivo physiology. For instance, cells in their native environment are usually engulfed and cleared by phagocytosis before showing detectable signs of advanced apoptosis or necrosis [1]. Moreover, morphological features of cell death can be different depending on the tissue or cell type. For in vivo studies, it may be best to combine morphological and biochemical interrogations to classify cell death (Fig. 4.1A). A good example is to perform fluorescence imaging on the dynamics of a key biochemical phenomenon, such as caspase activation (Fig. 4.1B) or chromatin condensation, in living animals. To visualize spatiotemporal dynamics of caspase activation in vivo, genetically encoded fluorogenic reporters would be ideal as they utilize the animal's own protein synthesis machinery, introduce fluorophores with minimal disruption to the cell, and target probes to cell types and subcellular regions of interest [65].

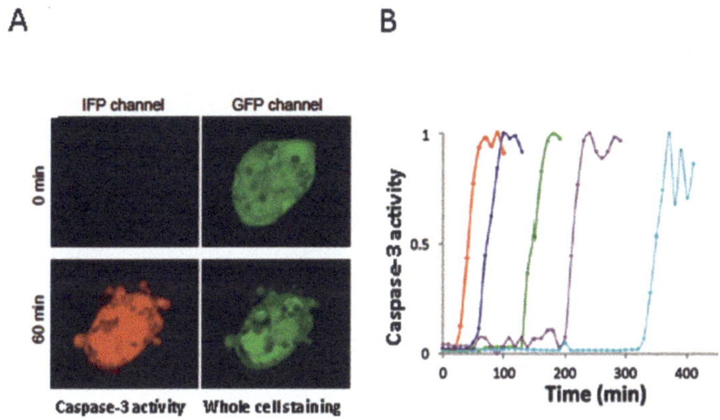

Figure 4.1 ***Morphological change and caspase activation during apoptotic cell death.*** (A) Confocal images of human glioblastoma cells LN229 stably expressing iCasper before and after treatment with 1 μM staurosporin (Magnification: 40x). Blebbing of plasma membrane is observed in cells undergoing apoptosis. (B) Caspase-3 activation as determined by infrared fluorescence of LN229 cells stably expressing iCasper. Time-lapse microscopy reveals cell-to-cell heterogeneity of caspase activation upon addition of 1 μM staurosporin. *(Adapted from T.L. To, et al., Rationally designed fluorogenic protease reporter visualizes spatiotemporal dynamics of apoptosis in vivo, Proc. Natl. Acad. Sci. USA 112 (11) (2015) 3338–3343.)*

6.1 Genetically encoded reporters of caspase-3 activation

Previous genetically encoded caspase reporters are not fluorogenic, which limits their in vivo applications. For example, the FRET-based reporters suffer from weak signals (change of fluorescence within tens of percentage points) and require tedious image processing to obtain the ratiometric signal [66]. More recently, a fluorophore translocation-based reporter has been developed, in which the fluorophore is anchored to plasma membrane and upon caspase cleavage it moves into nucleus. However, such a reporter has limited detection capacity, especially since cell shape change is significant during apoptosis [67]. In general, it is very challenging to use these non-fluorogenic reporters in detecting spatiotemporal dynamics of apoptosis in living animals because of cell heterogeneity, rapid cell shape change, and tissue movement, as well as cellular and tissue autofluorescence.

6.2 Infrared and green fluorogenic caspase-3 reporters (iCasper and ZipGFP)

We have developed iCasper [68] by redesigning a monomeric infrared fluorescent protein (mIFP) [69] such that its chromophore incorporation is regulated by caspase-3 activity (Fig. 4.2A). iCasper is fluorogenic, permitting straightforward detection of caspase activity, and requires no image processing. iCasper directly reports caspase activity and visualizes all phases of apoptosis from the initial stage, when the apoptotic cells start to shrink and round up to the later stage when the apoptotic cells are disintegrated into small apoptotic bodies and even after they are phagocytosed by other cells (Fig. 4.2B). When expressed in transgenic *Drosophila*, iCasper reveals the spatiotemporal coordination between apoptosis and embryonic morphogenesis, as well as the dynamics of apoptosis during tumorigenesis in *Drosophila*. As such, iCasper is a robust fluorogenic caspase-3 reporter that visualizes spatiotemporal dynamics of apoptosis in vivo.

However, the bacterial phytochrome-based iCasper requires the cofactor biliverdin, which seems to be limiting in *Drosophila* and zebrafish [69]. Although expression of heme oxygenase, which converts heme into biliverdin, overcomes this problem, a GFP-based fluorogenic reporter would be preferred since its fluorescence requires no cofactor in animals. Two GFP-based fluorogenic reporters, CA-GFP [70] and VC3AI [71], have been used to image apoptosis in cultured cells. However, neither has been shown to detect apoptosis in animals such as zebrafish. To provide a better solution for in vivo imaging needs, we have developed a GFP-based fluorogenic caspase reporter ZipGFP [72] by "zipping up" each fragment of split GFP in a manner that prevents their association and fluorophore formation until release by caspase-3 cleavage (Fig. 4.2C) [73]. We have expressed ZipGFP in zebrafish embryo to demonstrate its utility in live imaging of apoptosis with spatiotemporal resolution. ZipGFP reports the single-cell dynamics of cycloheximide-induced apoptosis in early zebrafish embryo over the gastrulation stage, as well as the general dynamics of apoptosis during normal embryo development after gastrulation (Fig. 4.2D). The spatial pattern of ZipGFP fluorescence that we observe in the zebrafish embryo is consistent with a previous study based on TUNEL staining. While TUNEL staining lacks temporal resolution, the temporal pattern revealed by ZipGFP illustrates the spatial progression of apoptosis.

Genetically encoded tools for the detection of apoptosis in vivo can be applied to a broad range of organisms. The intrinsic pathway of apoptosis appears to be relevant in all metazoans analyzed to date [74]. As such, many experimental methods for inducing apoptosis in mammalian cells should also be effective in modal organisms such as *C. elegans*, *Drosophila* [75], and zebrafish [76].

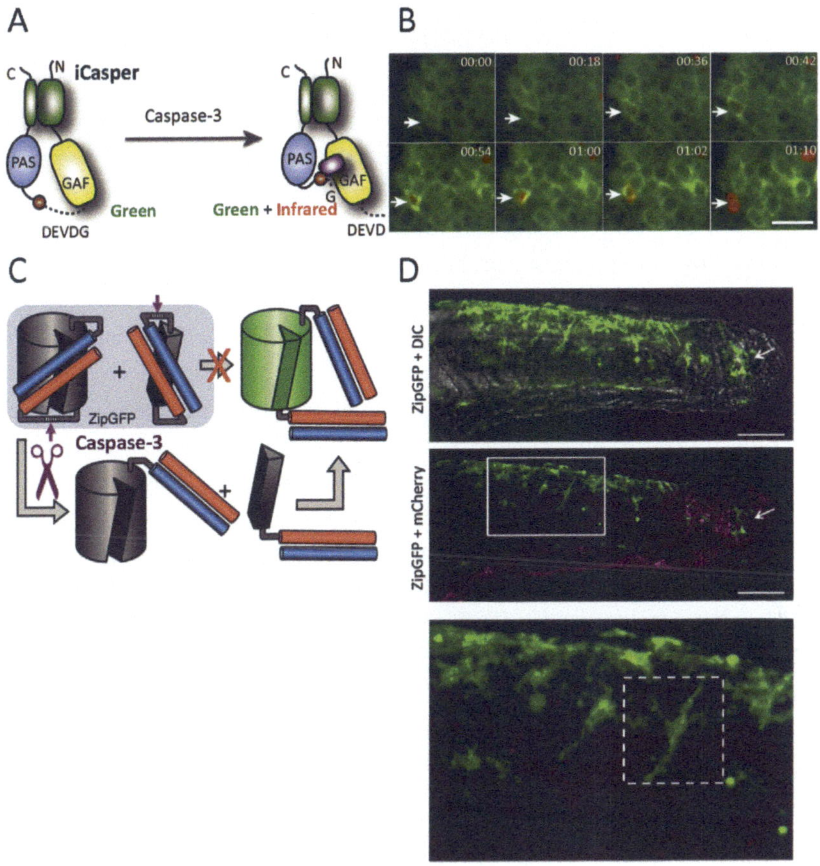

Figure 4.2 ***iCasper and ZipGFP enable the visualization of the spatiotemporal dynamics of apoptosis in live animals.*** (A) Design of an Infrared fluorescent executioner caspase reporter (iCasper). Cartoon showing the cleavage and activation of iCasper by caspase-3 activation during apoptosis. (B) iCasper enables the visualization of the spatiotemporal dynamics of apoptosis in vivo. Time-lapse images of neurons in the ventral nerve cord in *Drosophila* reveal caspase activation and apoptotic cell shape change during central nervous system (CNS) development. The neuron undergoing apoptosis is indicated with the arrow. The number at the top-right corner of each panel refers to time (hh:mm). Scale bar = 10 μm. (C) Design of an caspase-3 activated ZipGFP. GFP can be split into two nonfluorescent fragments (b1−10 and b11) that spontaneously associate and reform the fluorescent GFP. ZipGFP is a pair of modified b1−10 and b11 fragments that are both genetically fused to coiled-coil domains that block the fragments from associating until they are "uncaged" through caspase-3 cleavage (cut sites indicated with purple arrows). (D) Images around the trunk of the zebrafish embryo expressing ZipGFP caspase reporter and mCherry at 48 h post fertilization (hpf). Arrow points to the tail bud. Scale bar = 80 μm. *(A and B) Adapted from T.L. To, et al., Rationally designed fluorogenic protease reporter visualizes spatiotemporal dynamics of apoptosis in vivo, Proc. Natl. Acad. Sci. USA 112 (11) (2015) 3338−3343. (C) Adapted from M.D. Wiens, X. Lu, R.E. Campbell, Spying on cells: toward a perfect sleeper agent, Cell Chem. Biol. 23 (7) (2016) 756−758. (D) Adapted from T.L. To, et al., Rational design of a GFP-based fluorogenic caspase reporter for imaging apoptosis in vivo. Cell Chem. Biol. 23 (7) (2016) 875−882.*

6.3 The future of cell death studies

A more dynamic view on the apoptotic events, enabled by genetically encoded fluorogenic reporters, can perhaps reveal the precise temporal order among key molecular players in apoptosis. High-resolution fluorogenic reporter-based imaging can reveal the intracellular sites for apoptosis signaling. For instance, one can determine whether or not locations such as endoplasmic reticulum and lysosomes are important. Multiplex, high-resolution and dynamic imaging with other fluorescent markers and reporters may allow us to probe the roles of intracellular pH changes and endoplasmic reticulum-mitochondria connection during apoptosis. Genetic screening using the fluorogenic reporters can help identify novel signaling proteins for apoptosis. Moreover, a modular fluorogenic reporter that can be engineered with defined substrate peptide sequences can help us determine which caspase substrates are important for cell death.

Many interesting scientific questions can be investigated with a holistic view on the interconnection between apoptosis and cell signaling events [77]. What about a role for kinases in the cell death pathways? The fact that staurosporine, a broad-spectrum kinase inhibitor, is frequently used to induce apoptosis suggests a strong connection between kinase activity and cell death [78]. One may wonder how apoptotic cells signal to the neighboring cells to induce compensatory cell proliferation, which is likely important in maintaining tissue homeostasis following cell death. Looking forward, we envision the development of additional fluorogenic reporters of cell signaling events (e.g., kinase activation [79]) that can be used together with our caspase-3 reporters, to enable an integral and dynamic view of cell death.

Finally, the therapeutic opportunities enabled by robust detection of cell death cannot be overstated [80]. For instance, assays for apoptotic cell death are highly valuable for cancer drug discovery and development [81], as drug candidates incapable of inducing apoptosis are less likely to be effective. Conversely, drugs that suppress excessive apoptosis may prevent pathological cell death in a large variety of human diseases. Although apoptosis has been the focus of drug discovery and development for many years, recent research has identified the therapeutic promises for other types of cell death, such as necroptosis [51] and autophagy [82]. Much effort has been made on the development to pharmacologically intervene in these novel cell death pathways. With the current and future advances in experimental methods that allow us to dive deep into the mechanisms of cell death, our understanding of diseases and treatments can only improve.

References

[1] G. Kroemer, et al., Classification of cell death: recommendations of the nomenclature committee on cell death 2009, Cell Death Differ. 16 (1) (2009) 3–11.

[2] M. Suzanne, H. Steller, Shaping organisms with apoptosis, Cell Death Differ. 20 (5) (2013) 669–675.

[3] R. Weinlich, et al., Necroptosis in development, inflammation and disease, Nat. Rev. Mol. Cell Biol. 18 (2) (2017) 127–136.

[4] E. Laane, et al., Cell death induced by dexamethasone in lymphoid leukemia is mediated through initiation of autophagy, Cell Death Differ. 16 (7) (2009) 1018–1029.

[5] D.V. Krysko, et al., Apoptosis and necrosis: detection, discrimination and phagocytosis, Methods 44 (3) (2008) 205–221.

[6] C.D. Gregory, A. Devitt, The macrophage and the apoptotic cell: an innate immune interaction viewed simplistically? Immunology 113 (1) (2004) 1–14.

[7] D.J. Klionsky, et al., Guidelines for the use and interpretation of assays for monitoring autophagy (3rd edition), Autophagy 12 (1) (2016) 1–222.

[8] B.E. Fitzwalter, A. Thorburn, Recent insights into cell death and autophagy, FEBS J. 282 (22) (2015) 4279–4288.

[9] T. Vanden Berghe, et al., Molecular crosstalk between apoptosis, necroptosis, and survival signaling, Mol. Cell. Oncol. 2 (4) (2015) e975093.

[10] C. Dive, et al., Analysis and discrimination of necrosis and apoptosis (programmed cell death) by multiparameter flow cytometry, Biochim. Biophys. Acta 1133 (3) (1992) 275–285.

[11] D.V. Krysko, et al., Macrophages use different internalization mechanisms to clear apoptotic and necrotic cells, Cell Death Differ. 13 (12) (2006) 2011–2022.

[12] C. Grimsley, Cues for apoptotic cell engulfment: eat-me, don't eat-me and come-get-me signals, Trends Cell Biol. 13 (12) (2003) 648–656.

[13] K.I. Hulkower, R.L. Herber, Cell migration and invasion assays as tools for drug discovery, Pharmaceutics 3 (1) (2011) 107–124.

[14] K. Lauber, et al., Apoptotic cells induce migration of phagocytes via caspase-3-mediated release of a lipid attraction signal, Cell 113 (6) (2003) 717–730.

[15] D.R. Green, Apoptotic pathways: paper wraps stone blunts scissors, Cell 102 (1) (2000) 1–4.

[16] P.D. Bhola, A. Letai, Mitochondria-judges and executioners of cell death sentences, Mol. Cell 61 (5) (2016) 695–704.

[17] D.G. Nicholls, M.W. Ward, Mitochondrial membrane potential and neuronal glutamate excitotoxicity: mortality and millivolts, Trends Neurosci. 23 (4) (2000) 166–174.

[18] S.W. Perry, et al., Mitochondrial membrane potential probes and the proton gradient: a practical usage guide, Biotechniques 50 (2) (2011) 98–115.

[19] S.T. Smiley, et al., Intracellular heterogeneity in mitochondrial membrane potentials revealed by a J-aggregate-forming lipophilic cation JC-1, Proc. Natl. Acad. Sci. USA 88 (9) (1991) 3671–3675.

[20] R.C. Scaduto Jr., L.W. Grotyohann, Measurement of mitochondrial membrane potential using fluorescent rhodamine derivatives, Biophys. J. 76 (1 Pt 1) (1999) 469–477.

[21] H. Rottenberg, S. Wu, Quantitative assay by flow cytometry of the mitochondrial membrane potential in intact cells, Biochim. Biophys. Acta Mol. Cell Res. 1404 (3) (1998) 393–404.

[22] V. Petronilli, et al., Transient and long-lasting openings of the mitochondrial permeability transition pore can be monitored directly in intact cells by changes in mitochondrial calcein fluorescence, Biophys. J. 76 (2) (1999) 725–734.

[23] G. Denecker, et al., Apoptotic and necrotic cell death induced by death domain receptors, Cell. Mol. Life Sci. 58 (3) (2001) 356–370.

[24] J.C. Goldstein, et al., Cytochrome c is released in a single step during apoptosis, Cell Death Differ. 12 (5) (2005) 453–462.
[25] M. Poreba, et al., Caspase substrates and inhibitors, Cold Spring Harb. Perspect. Biol. 5 (8) (2013) a008680.
[26] P.G. Ekert, J. Silke, D.L. Vaux, Caspase inhibitors, Cell Death Differ. 6 (11) (1999) 1081–1086.
[27] G.V. Chaitanya, A.J. Steven, P.P. Babu, PARP-1 cleavage fragments: signatures of cell-death proteases in neurodegeneration, Cell Commun. Signal. 8 (2010) 31.
[28] M.A. Lagarkova, O.V. Iarovaia, S.V. Razin, Large-scale fragmentation of mammalian DNA in the course of apoptosis proceeds via excision of chromosomal DNA loops and their oligomers, J. Biol. Chem. 270 (35) (1995) 20239–20241.
[29] C.D. Georgiou, I. Papapostolou, K. Grintzalis, Protocol for the quantitative assessment of DNA concentration and damage (fragmentation and nicks), Nat. Protoc. 4 (2) (2009) 125–131.
[30] T. Idziorek, et al., YOPRO-1 permits cytofluorometric analysis of programmed cell death (apoptosis) without interfering with cell viability, J. Immunol. Methods 185 (2) (1995) 249–258.
[31] N.P. Singh, et al., A simple technique for quantitation of low levels of DNA damage in individual cells, Exp. Cell Res. 175 (1) (1988) 184–191.
[32] Z. Darzynkiewicz, D. Galkowski, H. Zhao, Analysis of apoptosis by cytometry using TUNEL assay, Methods 44 (3) (2008) 250–254.
[33] F. Amblard, J.H. Cho, R.F. Schinazi, Cu(I)-catalyzed Huisgen azide-alkyne 1,3-dipolar cycloaddition reaction in nucleoside, nucleotide, and oligonucleotide chemistry, Chem. Rev. 109 (9) (2009) 4207–4220.
[34] M.D. Best, Click chemistry and bioorthogonal reactions: unprecedented selectivity in the labeling of biological molecules, Biochemistry 48 (28) (2009) 6571–6584.
[35] T.M. Marti, et al., H2AX phosphorylation within the G1 phase after UV irradiation depends on nucleotide excision repair and not DNA double-strand breaks, Proc. Natl. Acad. Sci. USA 103 (26) (2006) 9891–9896.
[36] V.A. Fadok, et al., The role of phosphatidylserine in recognition of apoptotic cells by phagocytes, Cell Death Differ. 5 (7) (1998) 551–562.
[37] J. Suzuki, et al., Xk-related protein 8 and CED-8 promote phosphatidylserine exposure in apoptotic cells, Science 341 (6144) (2013) 403–406.
[38] M. van Engeland, et al., Annexin V-affinity assay: a review on an apoptosis detection system based on phosphatidylserine exposure, Cytometry 31 (1) (1998) 1–9.
[39] M.Y. Yang, et al., Reversible phosphatidylserine expression on blood granulocytes related to membrane perturbation but not DNA strand breaks, J. Leukoc. Biol. 71 (2) (2002) 231–237.
[40] I.J. Furlong, et al., Induction of apoptosis by valinomycin: mitochondrial permeability transition causes intracellular acidification, Cell Death Differ. 5 (3) (1998) 214–221.
[41] D. Lagadic-Gossmann, L. Huc, V. Lecureur, Alterations of intracellular pH homeostasis in apoptosis: origins and roles, Cell Death Differ. 11 (9) (2004) 953–961.
[42] M.C. Gendron, et al., Oxidation of pyridine nucleotides during Fas- and ceramide-induced apoptosis in Jurkat cells: correlation with changes in mitochondria, glutathione depletion, intracellular acidification and caspase 3 activation, Biochem. J. 353 (Pt 2) (2001) 357–367.
[43] R.A. Gottlieb, et al., Apoptosis induced in Jurkat cells by several agents is preceded by intracellular acidification, Proc. Natl. Acad. Sci. USA 93 (2) (1996) 654–658.
[44] S. Roy, et al., Maintenance of caspase-3 proenzyme dormancy by an intrinsic "safety catch" regulatory tripeptide, Proc. Natl. Acad. Sci. USA 98 (11) (2001) 6132–6137.
[45] T.J. Rink, R.Y. Tsien, T. Pozzan, Cytoplasmic pH and free Mg2+ in lymphocytes, J. Cell Biol. 95 (1) (1982) 189–196.

[46] J.E. Whitaker, R.P. Haugland, F.G. Prendergast, Spectral and photophysical studies of benzo[c]xanthene dyes: dual emission pH sensors, Anal. Biochem. 194 (2) (1991) 330–344.

[47] G. Miesenbock, D.A. De Angelis, J.E. Rothman, Visualizing secretion and synaptic transmission with pH-sensitive green fluorescent proteins, Nature 394 (6689) (1998) 192–195.

[48] J. Llopis, et al., Measurement of cytosolic, mitochondrial, and Golgi pH in single living cells with green fluorescent proteins, Proc. Natl. Acad. Sci. USA 95 (12) (1998) 6803–6808.

[49] M. Artal-Sanz, et al., Lysosomal biogenesis and function is critical for necrotic cell death in *Caenorhabditis elegans*, J. Cell Biol. 173 (2) (2006) 231–239.

[50] V. Vandevoorde, G. Haegeman, W. Fiers, Induced expression of trimerized intracellular domains of the human tumor necrosis factor (TNF) p55 receptor elicits TNF effects, J. Cell Biol. 137 (7) (1997) 1627–1638.

[51] M. Conrad, et al., Regulated necrosis: disease relevance and therapeutic opportunities, Nat. Rev. Drug Discov. 15 (5) (2016) 348–366.

[52] K. Schulze-Osthoff, et al., Depletion of the mitochondrial electron transport abrogates the cytotoxic and gene-inductive effects of TNF, EMBO J. 12 (8) (1993) 3095–3104.

[53] B.G. Rosser, S.P. Powers, G.J. Gores, Calpain activity increases in hepatocytes following addition of ATP. Demonstration by a novel fluorescent approach, J. Biol. Chem. 268 (31) (1993) 23593–23600.

[54] B.S. Cummings, R.G. Schnellmann, Measurement of cell death in mammalian cells, Curr. Protoc. Pharmacol. (2004) (Chapter 12): Unit 12.8.

[55] F.K. Chan, K. Moriwaki, M.J. De Rosa, Detection of necrosis by release of lactate dehydrogenase activity, Methods Mol. Biol. 979 (2013) 65–70.

[56] G. Kramer, et al., Differentiation between cell death modes using measurements of different soluble forms of extracellular cytokeratin 18, Cancer Res. 64 (5) (2004) 1751–1756.

[57] P. Kuballa, et al., Autophagy and the immune system, Annu. Rev. Immunol. 30 (2012) 611–646.

[58] L. Du, et al., Starving neurons show sex difference in autophagy, J. Biol. Chem. 284 (4) (2009) 2383–2396.

[59] B. Loos, A. du Toit, J.H. Hofmeyr, Defining and measuring autophagosome flux-concept and reality, Autophagy 10 (11) (2014) 2087–2096.

[60] R.J. Youle, D.P. Narendra, Mechanisms of mitophagy, Nat. Rev. Mol. Cell Biol. 12 (1) (2011) 9–14.

[61] B. Chazotte, Labeling lysosomes in live cells with LysoTracker, Cold Spring Harb. Protoc. 2011 (2) (2011) p. pdb prot5571.

[62] C.Y. Lim, R. Zoncu, The lysosome as a command-and-control center for cellular metabolism, J. Cell Biol. 214 (6) (2016) 653–664.

[63] R.C.M. Reis, M.H.F. Sorgine, T. Coelho-Sampaio, A novel methodology for the investigation of intracellular proteolytic processing in intact cells, Eur. J. Cell Biol. 75 (2) (1998) 192–197.

[64] H. Katayama, et al., A sensitive and quantitative technique for detecting autophagic events based on lysosomal delivery, Chem. Biol. 18 (8) (2011) 1042–1052.

[65] R.Y. Tsien, Imagining imaging's future, Nat. Rev. Mol. Cell Biol. (Suppl.) (2003) SS16–SS21.

[66] K. Takemoto, et al., Local initiation of caspase activation in *Drosophila* salivary gland programmed cell death in vivo, Proc. Natl. Acad. Sci. USA 104 (33) (2007) 13367–13372.

[67] P.L. Bardet, et al., A fluorescent reporter of caspase activity for live imaging, Proc. Natl. Acad. Sci. USA 105 (37) (2008) 13901–13905.

[68] T.L. To, et al., Rationally designed fluorogenic protease reporter visualizes spatiotemporal dynamics of apoptosis in vivo, Proc. Natl. Acad. Sci. USA 112 (11) (2015) 3338–3343.
[69] D. Yu, et al., A naturally monomeric infrared fluorescent protein for protein labeling in vivo, Nat. Methods 12 (8) (2015) 763–765.
[70] S.B. Nicholls, et al., Mechanism of a genetically encoded dark-to-bright reporter for caspase activity, J. Biol. Chem. 286 (28) (2011) 24977–24986.
[71] J. Zhang, et al., Visualization of caspase-3-like activity in cells using a genetically encoded fluorescent biosensor activated by protein cleavage, Nat. Commun. 4 (2013) 2157.
[72] T.L. To, et al., Rational design of a GFP-based fluorogenic caspase reporter for imaging apoptosis in vivo, Cell Chem. Biol. 23 (7) (2016) 875–882.
[73] M.D. Wiens, X. Lu, R.E. Campbell, Spying on cells: toward a perfect sleeper agent, Cell Chem. Biol. 23 (7) (2016) 756–758.
[74] R.J. Youle, A. Strasser, The BCL-2 protein family: opposing activities that mediate cell death, Nat. Rev. Mol. Cell Biol. 9 (1) (2008) 47–59.
[75] Y. Fuchs, H. Steller, Programmed cell death in animal development and disease, Cell 147 (4) (2011) 742–758.
[76] P.M. Eimon, A. Ashkenazi, The zebrafish as a model organism for the study of apoptosis, Apoptosis 15 (3) (2010) 331–349.
[77] D.E. Johnson, Programmed cell death regulation: basic mechanisms and therapeutic opportunities, Leukemia 14 (8) (2000) 1340–1344.
[78] H.J. Chae, et al., Molecular mechanism of staurosporine-induced apoptosis in osteoblasts, Pharmacol. Res. 42 (4) (2000) 373–381.
[79] Q. Zhang, et al., Visualizing dynamics of cell signaling in vivo with a phase separation-based kinase reporter, Mol. Cell 69 (2) (2018) 334–346 e4.
[80] L.F. Diaz, et al., Mechanisms of cell death: molecular insights and therapeutic perspectives, Cell Death Differ. 12 (11) (2005) 1449–1456.
[81] G. Wang, et al., Small-molecule activation of the TRAIL receptor DR5 in human cancer cells, Nat. Chem. Biol. 9 (2) (2013) 84–89.
[82] D.C. Rubinsztein, P. Codogno, B. Levine, Autophagy modulation as a potential therapeutic target for diverse diseases, Nat. Rev. Drug Discov. 11 (9) (2012) 709–730.

CHAPTER 5

Proteotoxicity and endoplasmic reticulum stress-mediated cell death

Mengxiong Wang[1], Mary E. Law[2] and Brian K. Law[2,3]
[1]Department of Radiation Oncology, College of Medicine, Stanford University, Palo Alto, CA, United States; [2]Department of Pharmacology & Therapeutics, College of Medicine, University of Florida, Gainesville, FL, United States; [3]UF-Health Cancer Center, University of Florida, Gainesville, FL, United States

1. Introduction to proteotoxicity and UPR pathways

1.1 General ER functions

The Endoplasmic Reticulum (ER) is an essential organelle in eukaryotic cells that regulates multiple cellular processes, such as lipid biogenesis, intracellular calcium homeostasis, and protein folding and secretion. There are two types of ER, Rough ER (RER), and Smooth ER (SER) [1]. As RER is studded with ribosomes, it plays vital roles in assisting and participating in the synthesis and folding of secretory and membrane proteins, while SER is mainly involved in the production and metabolism of fats and steroid hormones [2]. These processes are required for normal cellular functions and cell survival. The ER lumen has a unique cellular environment to perform these processes [3]. For example, calcium ions (Ca^{2+}) are ubiquitous intracellular molecules that mediate various cellular processes and physiological functions. The ER is a major storage organelle for Ca^{2+} and controls its transport through the Ca^{2+}- ATPase [4,5]. Inositol 1,4,5-trisphosphate receptors (IP_3Rs), which regulate the transfer of Ca^{2+} from ER to mitochondria, are also located on the ER [6]. In terms of supporting protein folding, the ER lumen provides a unique oxidizing environment for disulfide bond formation and contains a specialized set of molecular chaperones and folding enzymes. Many of these molecular chaperones are calcium-dependent, such as Glucose-Regulated Protein 78 kDa (GRP78), GRP94, and calreticulin, which are responsible for proofreading newly synthesized proteins [7]. Only properly folded proteins exit the ER through coat protein complex II (COPII)−dependent vesicular transport, while damaged/misfolded proteins are trafficked to the

Mechanisms of Cell Death and Opportunities for Therapeutic Development
ISBN 978-0-12-814208-0
https://doi.org/10.1016/B978-0-12-814208-0.00006-3

cytosol for ubiquitination and degradation by the proteasome. A variety of disturbances reduce ER protein folding capacity and increase the accumulation of damaged/misfolded proteins, such as changes in Ca^{2+} concentrations, disturbances of the redox environment by hypoxia, and disruption of disulfide bond formation by oxidants or reducing agents. When the unfolded or misfolded protein load exceeds the capacity of the ER to fold protein, the cells are under a state of ER stress. Along with the initiation of ER stress, a coordinated adaptive program termed the Unfolded Protein Response (UPR) is triggered to release cells from ER stress via increasing the protein folding capacity of the ER and promoting the ER-Associated Degradation (ERAD) pathway for alleviating ER stress and maintaining ER homeostasis. Mild and chronic ER stress can be released by UPR, while persistent and severe ER stress kill cells by inducing their autophagy, apoptosis, or necroptosis.

1.2 UPR pathways

With accumulation of misfolded/unfolded proteins in the ER, the molecular chaperone GRP78 dissociates from ER stress sensors to assist with the degradation of these proteins, which in turn triggers UPR signaling. There are three pathways that orchestrate UPR, including the IRE1-XBP1, PERK-ATF4-CHOP, and ATF6 cassettes Fig. 5.1. Inositol-Requiring Enzyme 1 (IRE1) is a type I transmembrane receptor, which possesses an endonuclease domain and a Ser/Thr kinase domain. Upon activation of UPR, IRE1 dimerizes/oligomerizes, and autophosphorylates, thus triggering its endonuclease activity that cleaves an intron from the mRNA coding for the bZIP transcription factor XBP1 [8,9]. Cleaved exons are ligated by tRNA ligase RtcB to generate the spliced form of XBP1 mRNA (XBP1s) [10]. The protein encoded by the spliced form of XBP1 mRNA (sXBP1) facilitates UPR gene expression that reestablishes ER function. The activation of IRE1 also induces cleavage of a subset of mRNA that is associated with the ER membrane, thus mediating their decay. This event is termed Regulated IRE1-Dependent Decay (RIDD) [11]. Activation of XBP1s and RIDD are two independent events. IRE1 needs to coordinate with a ligase to cleave the XBP1 intron, while IRE1 needs to interact with mRNA degradation enzymes for induction of RIDD. Some studies suggest that XBP1s may promote cell survival, whereas RIDD activation is more likely to trigger cell apoptosis [12]. In addition to its endoribonuclease activity, the kinase domain of IRE1 interacts with Tumor necrosis factor

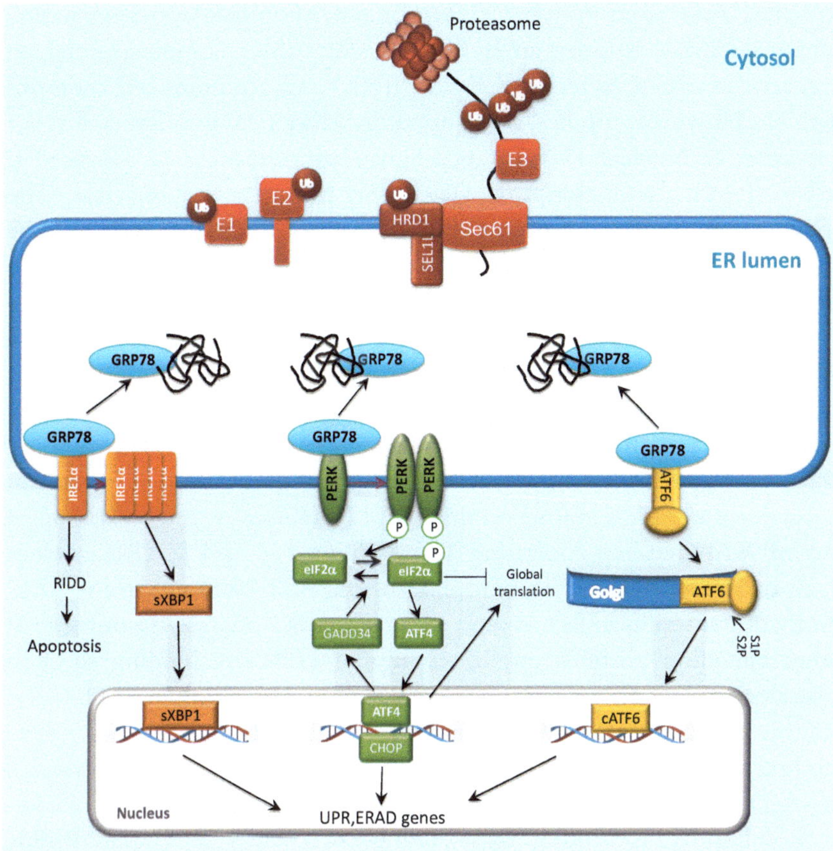

Figure 5.1 The Unfolded Protein Response (UPR).In the absence of unfolded/misfolded proteins, GRP78 maintains the three ER stress sensors, IRE1, PERK, and ATF6, in inactive states. Misfolded proteins bind GRP78, allowing oligomerization and activation of IRE1 and PERK, and translocation of ATF6 to the Golgi, where S1P and S2P sequentially cleave it to yield an active transcription factor that translocates to the nucleus. The activated PERK protein kinase phosphorylates eIF2a to suppress protein translation, facilitating restoration of proteostasis and leading to the upregulation and activation of the ATF4 and CHOP transcription factors. IRE1 possesses both kinase activity and endonuclease activity. IRE1 nuclease activity is required for splicing the mRNA coding for the sXBP1 protein, which is a transcription factor. Collectively, these three arms of UPR function to restore proteostasis, or to trigger cell death if proteostasis cannot be restored. Misfolded proteins may be retro-translocated to the cytoplasm for degradation via the proteasome in a process termed ER-Associated Destruction (ERAD).

Receptor-Associated Factor 2 (TRAF2) and Apoptosis Signal-regulating Kinase 1 (ASK1) to form an IRE1–TRAF2–ASK1 complex, which in turn activates JUN N-terminal Kinase (JNK) and promotes cell apoptosis [13,14]. However, studies also show that IRE1 is important for the activation of Nuclear Factor kappa-light-chain-enhancer of activated B cells (NF–κB), a transcription factor that promotes cell survival. The upregulation of IRE1 during UPR is required for basal IκBα Kinase (IKK), which activates NF–κB by stimulating the degradation of IκBα, an inhibitor of NF–κB [15,16].

The enzyme Protein Kinase R (PKR)-like ER Kinase (PERK) is a Ser/Thr protein kinase, which oligomerizes and autophosphorylates upon dissociation of GRP78. The oligomerization of PERK further leads to phosphorylation and inactivation of the eukaryotic translation initiation factor 2 subunit 1 (eIF2α), which in turn inhibits global protein translation and reduces the protein load on the ER [17]. However, the translation of the mRNA encoding Activating Transcription Factor 4 (ATF4) is selectively activated upon eIF2α phosphorylation. ATF4 belongs to the bZIP family of transcription factors, and it stimulates the transcription of several genes encoding proteins implicated in the UPR by binding to their promoters, such as ER chaperones GRP78 and GRP94. ATF4 also induces transcription of the proapoptotic gene CHOP and forms a heterodimer with CHOP to upregulate the expression of Growth Arrest and DNA-Damage-inducible protein 34 (GADD34), which in turn facilitates dephosphorylation of eIF2α by protein phosphatase 1 (PP1). The PERK pathway plays a critical role in protecting cells from ER stress through translational control. Studies indicate that $PERK^{-/-}$ cells and cells with a non-phosphorylatable eIF2α knock-in are hypersensitive to ER stress. $PERK^{-/-}$ mice bear multiple defects, including small size, bone abnormalities, Type I diabetes, and Wolcott-Rallison syndrome [18–20].

Much like ATF4, Activating Transcription Factor 6 (ATF6), which has two homologs ATF6α and ATF6β, is also a member of the bZIP family of transcription factors. In response to UPR, the dissociation of GRP78 frees ATF6 to translocate to the Golgi apparatus, where ATF6 is cleaved sequentially by proteases S1P and S2P at a juxtamembrane site to generate their active form, cleaved ATF6 (cATF6) [21]. Cytosolic cATF6 then migrates into the nucleus to mediate gene expression. The target genes for ATF6α and ATF6β are different. ATF6α is involved in the regulation of ER-associated genes, such as genes encoding GRP78, PDI, and ER degradation-enhancing α-mannosidase-like protein 1(EDEM1), which promote the degradation of damaged/misfolded proteins [22,23]. Studies also demonstrate that activation of ATF6α promotes gene expression of CHOP and XBP1 [24]. The function of ATF6β in regulating UPR is not clear. Most studies imply that ATF6β acts as a translational repressor for ATF6α (e.g. Ref. [22]).

1.3 Proteotoxicity and ER quality control

Proteotoxicity refers to the irreparable damage to the cell that is caused by an accumulation of impaired and misfolded proteins. At the cellular level, uncontrolled or severe proteotoxicity will ultimately result in cell death. Cells take a series of protective steps, including activation of UPR, ERAD, and autophagy, for ER quality control and reduction of proteotoxicity Fig. 5.2. As discussed above, UPR pathways play vital roles in releasing cells from ER stress. For instance, sXBP1 facilitates the expression of genes responsible for restoring ER folding capacity, and eIF2α phosphorylation induces sustained translational repression. While UPR is a highly efficient and coordinated ER quality control mechanism, it is sometimes unable to repair all damaged/misfolded proteins and fully restore ER function. In such circumstances, these damaged/misfolded proteins need to be degraded by another mechanism. The ERAD machinery is a cellular pathway that targets damaged/misfolded proteins in the ER for ubiquitination and subsequent degradation by the cytosolic proteasome. During ERAD, the proteins in the ER are routed by the cooperation of several enzymes including EDEMs, ER Mannosidase I (ER ManI), and Osteosarcoma amplified-9 (OS-9) [25–28]. EDEMs and ER ManI are responsible for trimming N-glycans, while OS-9 recognizes these trimmed glycans and assembles with the SEL1L-HRD1 complex before directing these substrates to the ubiquitin (Ub)-proteasome system (UPS) in the cytosol. Besides the SEL1L-HRD1 complex, the other retrotranslocon important for routing damaged/misfolded proteins from the ER consists of gp78, UBA domain-containing polytopic protein UBAC2, VCP/p97-binding protein UBXD8, and Derlins-1 and -2 [29]. Once substrates translocate from the ER to the cytosol, they undergo a series of steps, starting with the binding of ubiquitin-activating enzyme E1, along with ATP hydrolysis and ubiquitin activation. Subsequently, ubiquitin-activating enzymes E2 and E3 catalyze the formation of polyubiquitin chains that attach to the lysine residues of damaged/misfolded substrates. Finally, the proteasome recognizes the polyubiquitin chains and degrades the damaged/misfolded substrates.

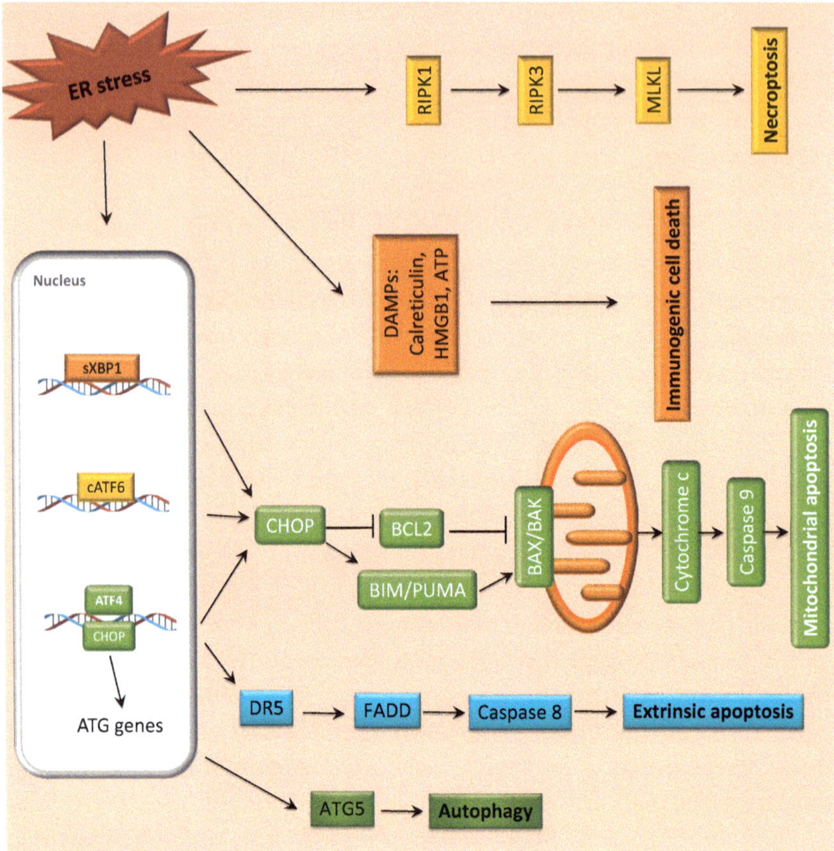

Figure 5.2 Mechanisms of ER Stress-Induced Cell Death.When ER stress is immitigable, cells will die through one of several mechanisms. Activation of the RIPK1-RIPK3-MLKL cascade results in programmed necrosis, or necroptosis. Damage-Associated Molecular Pathogens (DAMPs) that localize to the cell surface during the ER stress response, such as Calreticulin or HMGB1, may trigger immunogenic cell death. Alternatively, some of the autophagy-related genes (ATG) may be upregulated by the UPR pathways, resulting in autophagic cell death. Finally, excessive ER stress may trigger apoptosis in several ways. Various types of ER stress cause transcriptional upregulation of the TRAIL receptor, Death Receptor 5, which can trigger extrinsic apoptosis through Caspase 8/10 activation. UPR-associated CHOP upregulation regulates the expression of proteins that stimulate intrinsic apoptosis through mitochondria-mediated Caspase 9 activation. Although not shown here, extensive cross-talk exists among these cell death pathways.

Damaged/misfolded proteins that are not cleared by ERAD can be transferred to quality control autophagy through the Vps34/beclin-1 initiation machinery [30]. The activation of UPR promotes the recruitment of Hsp104p to the ER and lipid droplet formation at ER aggregates, so that the damaged/misfolded proteins can be transported to the vacuole for degradation. Then the ER itself can be degraded by a selective autophagic machinery termed ER-phagy. During ER-phagy, the ER is recognized by specific ER-phagy receptors. Four ER-phagy receptors have been identified in mammalian cells, including a reticulon-like protein FAM134B, reticulon 3 (RTN3) [31], a component of the ER translocon, SEC62 [32], and a vertebrate-specific protein, CCPG1 [33,34]. These receptors share similar features to LIR/AIM and ER-anchored domains, which serve as a bridge between ER and autophagosomal membranes. Although sharing similar features, they have distinct functions during ER-phagy. FAM134B and RTN3 are, respectively, involved in degradation of different parts of the ER, the sheet ER, and tubular ER. Sec62 is crucial for ER-phagy during recovery from ER stress conditions. CCPG1 was recently identified as a noncanonical receptor, which drives ER-phagy through interaction with Autophagy-related Gene 8 (ATG8) and FAK interacting protein 200 kDa (FIP200). CCPG1-deficient mice have a proteostatic defect, especially in their exocrine pancreata, which suggests that CCPG is more important for controlling proteostasis within pancreatic acinar cells [34].

In summary, this section gives a brief introduction to ER functions, key players in UPR pathways, the definition of proteotoxicities, and means by which the ER quality control machinery maintains ER homeostasis. It is well documented that ER stress is involved in various types of diseases, including neurodegeneration, cancer, and cardiovascular and autoimmune diseases. It is important to understand the role of ER stress pathways in disease initiation and progression, and to identify druggable targets that could be used for disease intervention. In the following sections, we will introduce major players in the ER stress pathways that are crucial for regulating cell death. As deregulation of cellular calcium levels is correlated with multiple cellular processes and physiological functions, disturbances in ER calcium regulation and the induction of cell death will also be discussed. Subsequently, ER stress-related diseases and the potential targets for treating these diseases will be introduced, as well as the drugs that have been developed for targeting these diseases.

2. ER stress pathways and cell death regulation

2.1 Mechanisms of ER stress-mediated mitochondrial apoptosis

Mitochondria are essential organelles that support cell growth and survival as they generate the bulk of the cell's adenosine triphosphate (ATP) supply.

When cells sense irremediable stress or apoptotic signals, signals are sent to mitochondria, triggering the opening of the mitochondrial outer membrane (MOM) pore and the mitochondrial apoptosis pathway [35]. The B-cell lymphoma-2 (BCL-2) family members, BAX and BAK, are crucial mediators for this apoptotic pathway. Under normal conditions, BAX exists as an auto-inhibited monomer in the cytosol with its C-terminal α9 helix occupying its dimerization pocket. BAK, in contrast, is not inhibited by itself, but by the mammalian restricted voltage-dependent anion selective channel (VDAC) isoform VDAC2 that binds to its dimerization pocket, thus obstructing dimerization [36]. In response to apoptotic stimuli, the activator BH3-only molecules (BH3s), including truncated BID (tBID), BIM, Noxa, and PUMA, promote conformational changes of BAX, thus stimulating the targeting and homo-oligomerization of BAX at the MOM. Meanwhile, BH3s trigger the homo-oligomerization of BAK by disrupting its interaction with VDAC2. The activation of BAX and BAK further results in the efflux of apoptogenic factors, cytochrome *c*, and SMAC, followed by the binding of Apaf-1 to cytochrome *c*, and the recruitment of pro-caspase 9. The formation of the apoptosome subsequently activates caspase 9 and downstream enzymes such as caspase 3. In addition, BCL-2 ovarian killer (BOK), sharing 70%–80% sequence homology with BAX and BAK, is functionally redundant with BAX and BAK. However, mice with a double knockout of BAK/BAX exhibit multiple phenotypic abnormalities, suggesting that BOK fails to completely compensate for the deficiency of BAX/BAK [37]. Instead of triggering apoptosis, other BCL-2 family members, such as MCL-1, BCL-2, and BCL-XL, are antiapoptotic, which can prevent cytochrome *c* efflux and preserve the integrity of the mitochondria by sequestering activator BH3s (BID, BIM, Noxa, and PUMA), BAX, BAK, or BOK.

ER stress stimulates mitochondrial apoptotic pathways via selectively activating BH3s, PUMA, Noxa, and BIM [35,38]. Mouse embryonic fibroblasts (MEFs) with BIM or PUMA knockdown become less sensitive to the ER stress inducers, Thapsigargin or Tunicamycin, while the double knockdown of BIM and PUMA in MEFs amplifies the resistance, suggesting that the loss of both proteins impedes the activation of BAX/BAK and thus inhibits ER stress-induced apoptosis. Studies also indicate that ER stress significantly increases the p53-dependent transcription of BH3s [38]. Upon treatment with ER stress inducers, the level of p53 significantly increases, as well as mRNA encoding PUMA and Noxa. In contrast, the expression of PUMA and Noxa is largely suppressed in p53$^{-/-}$ MEFs. BOK, localized at the ER, is also an important mediator for ER stress-induced apoptosis [37,39]. BOK activity is suppressed by an ER-associated Hsp40 family protein JB12 in normal cells, while the inactivation of JB12 during ER stress causes accumulation and elevation of BOK [40]. Studies also show that BOK$^{-/-}$ cells and mice exhibit decreased activation of the mitochondrial apoptotic pathway upon induction of ER stress [37]. However, there is some disagreement about whether the function of BOK is dependent on the activation of BAX/BAK in response to ER stress. One study showed that BOK expression restores the effect of BOK knockout in response to ER stressors, but fails to reverse the reduced mitochondrial apoptotic pathway in BAX$^{-/-}$BAK$^{-/-}$ cells, suggesting that BOK requires activation of BAX/BAK to trigger mitochondrial apoptosis [37]. In contrast, a more recent study indicates that by using proteasome inhibitors to block the degradation of BOK, BOK independently stimulates the mitochondrial apoptotic pathway [40].

Among major elements of ER stress pathways, CHOP transcriptionally regulates expression of BCL-2. Expression of constitutively activated CHOP in cells downregulates BCL-2 expression, thus sensitizing cells to ER stress [41]. Elevated CHOP expression is also correlated with enhanced oxidant injury. Replenishment of BCL-2 in cells restores redox homeostasis and protects cells from ER stress, suggesting that CHOP is a major mediator of mitochondrial apoptosis. Instead of regulating BCL-2, IRE1α directly interacts with BAX and BAK through its BH3 and BH1 domains, and this interaction is enhanced under ER stress. Cells with BAX and BAK double knockout exhibit decreased activation of IRE1α-XBP1 signaling. These results indicate that BAX and BAK regulate cell sensitivity to ER stress through direct interaction with IRE1α and by mediating IRE1α-XBP1 signaling [42]. IRE1α is also important for initiating mitochondrial apoptosis in response to infection-associated ER stress [43]. Upon chemical stress or microbial infection-induced ER stress, IRE1α promotes the translocation of ROS-dependent NLRP3 to mitochondria, which induces the caspase 2-BID mitochondrial damage pathway, resulting in the activation of the inflammasome. Readers are also referred to Chapters 1, 2, 3, and 7 of this volume for topics related to mechanisms of apoptosis.

2.2 Mechanisms of ER stress-mediated extrinsic apoptosis

Activation of the extrinsic apoptotic pathway starts with the ligation of death ligands to tumor necrosis factor-related apoptosis-inducing ligand receptor 1 (TRAIL-R1/death receptor 4 (DR4)) and TRAIL-R2/DR5, which results in the oligomerization of these receptors, thus providing a docking site for the adapter protein Fas-associated protein with death domain (FADD). FADD serves as a bridge between DR4/DR5 and multiple pro-caspase 8 molecules. Pro-caspase 8 molecules bind to FADD through its death effector domain (DED) and bind to each other to nucleate a complex termed death-inducing signaling complex (DISC), which in turn causes the auto-cleavage and activation of caspase 8. Subsequently, activated caspase 8 further cleaves the BCL-2 family protein BID and induces MOMP, or stimulates downstream effector caspases such as caspase 3, 6, and 7 to induce cell death [44].

ER stress activates the extrinsic apoptotic pathway via upregulation of the transcription of mRNAencoding DR4 or DR5. The activation of DR5 by various types of ER stress stimuli (Brefeldin-A, Thapsigargin, etc.) is independent of death ligands. One study showed that either depletion of DR5 ligand TRAIL by siRNA or neutralization of extracellular TRAIL using soluble DR5-Fc fusion proteins does not affect ER stress-induced cell apoptosis [45]. Further investigation demonstrated that the PERK-ATF4-CHOP pathway plays a vital role in activating DR5. By applying the NEDD8 E1 Activating Enzyme (NAE) inhibitor MLN4924 to treat esophageal squamous cell carcinoma, another study found that ATF4 is stabilized by MLN4924, which in turn induces the upregulation of CHOP and DR5 [46]. Activated DR5 further triggers extrinsic apoptosis in these cancer cells, suggesting that ATF4-CHOP-DR5 is a critical pathway for treating cancer. CHOP alone is also important for mediating Thapsigargin-induced cell extrinsic apoptosis. Inhibition of CHOP expression attenuates Thapsigargininduced upregulation of DR5 and thus rescues cells from apoptosis [47]. A more recent study demonstrates that CHOP is bound to the AP-1 binding site in the DR5 gene promoter region and regulates its expression [48]. In contrast to CHOP, IRE1α reverses Thapsigargin-induced apoptosis by mediating DR5 RIDD. A recombinant protein that contains the catalytic domain of IRE1α was found to cleave DR5 mRNA in vitro, and this process was reversed by an IRE1α RNase inhibitor [45].

2.3 Necrosis and ER stress

Compared to apoptosis, necrosis is a caspase-independent, inflammatory type of programmed cell death, which is mediated by the enzymatic activity of receptor-interacting protein kinase 1 (RIPK1), RIPK3, and pseudokinase mixed lineage kinase domain-like protein (MLKL). RIPK1 is a protein Ser/Thr kinase that is a key determinant for cell survival and cell death [49]. $RIPK1^{-/-}$ mice exhibit extensive apoptosis and die 1–3 days after birth, suggesting a pro-survival role for RIPK1 [50]. However, other studies reveal that RIPK1 actually functions as a binary switch between promoting pro-survival and pro-apoptosis signals [51,52]. Upon the activation of DR4/DR5, RIPK1 is upregulated and stabilized through K63-linked polyubiquitination, and subsequently stimulates NF–κB and MAPKs, which in turn promote cell survival. However, if K63-linked ubiquitination of RIPK1 is inhibited or activation of caspase 8 is blocked, deubiquitinated RIPK1 recruits RIPK3, subsequently triggering programmed necrosis. Programmed necrosis (necroptosis) is also discussed in Chapters 1, 2, and 3 of the volume.

One study found that ER stress induces tumor necrosis factor receptor 1 (TNFR1)-mediated necrosis in a murine fibrosarcoma cell line L929 [53]. Knockdown of TNFR1 protects cells from TNF-triggered necrosis but does not affect brefeldin A or Thapsigargin-induced cell death. Further investigation reveals that inhibition of TNFR1 switched the cell death mechanism from RIPK1-dependent necrosis to RIPK1-independent apoptosis in response to activation of ER stress. Similarly, another study illustrated that 2-deoxyglucose (2-DG) induces caspase-dependent apoptosis in rhabdomyosarcoma cells, while deprivation of glucose causes a switch from apoptosis to the necrotic pathway [54]. This study also demonstrates that ATF4 is important for mediating both apoptosis induced by 2-DG, and necrosis triggered by glucose withdrawal.

2.4 ER stress in modulating autophagy

Autophagy refers to a destructive cellular process that involves the rearrangement of subcellular membranes to capture abnormal protein aggregates and damaged organelles, with subsequent targeting to lysosomes for degradation [55]. The degradation products are then released by the lysosome and recycled in metabolic pathways. Autophagy is important for protein and organelle quality control and the maintenance of cellular homeostasis, as it helps digest damaged cellular components and recycle nutrients. Autophagy also prevents toxic waste accumulation in cells and supports survival under conditions of environmental stress and nutrient deprivation. Autophagy starts with the activation of autophagy-related genes (ATGs) by multiple stimuli, including stress, energy shortage, and hypoxia, with modulation from various signaling pathways, including the hypoxia-inducible transcription factors (HIFs), mammalian target of rapamycin (mTOR), and AMP-activated protein (AMPK) cassettes. The assembly of ATG protein complexes facilitates the formation and elongation of double-membrane structures called autophagosomes. Autophagosomes capture damaged proteins or organelles and eventually fuse with late endosomes or lysosomes, where contents are degraded by lysosomal hydrolases. Approximately 33 different core ATGs have been identified. Among these, two ubiquitination-like conjugations, ATG12-ATG5-ATG16L1 and ATG8, are involved in the expansion of the autophagic isolation membrane. The ATG12-ATG5-ATG16L1 complexes function as an E3-like enzyme that catalyzes a ubiquitylation-like reaction. It also induces the conversion of microtubule-associated protein1 light chain 3 (LC3), a homolog of ATG8 in mammalian cells, from the free cytosolic form of LC3 (LC3-I) to the lipid-conjugated form (LC-II), and leads to the closure of autophagosomes and sequestration of their contents. ATG5 and LC3 conjugation systems are required for the canonical autophagy pathway to proceed and LC-I/LC-II is commonly used as an indicator for autophagy induction.

ER stress induces autophagy by increasing autophagosome formation. Treatment of SK-N-SH neuroblastoma cells with ER stress inducers (Thapsigargin and Tunicamycin) for 2–6 h greatly promotes autophagosome formation as seen at the ultrastructural level and indicated by an increased LC3-II/LC3-I ratio [56]. In ATG5-deficient MEFs, ER stress fails to stimulate autophagosome formation and autophagy induction, suggesting that ER stress-induced autophagy is dependent on ATG5 [57]. However, it is unclear which arm of UPR better correlates with autophagy induction. Application of ER stressors to IRE1$\alpha^{-/-}$, ATF6$^{-/-}$, and PERK$^{-/-}$ MEFs results in inhibition of autophagy only in IRE1α-deficient cells, indicating that IRE1α is a key modulator of autophagy induction. Further investigation demonstrated that ER stress-induced autophagy is mediated by the IRE1α-TRAF2-JNK pathway [56]. However, a study of neurodegeneration illustrates that the PERK-elF2α-CHOP pathway is also necessary for ER stress-induced autophagy. Expanded polyglutamine 72 repeat (polyQ72) aggregates cause ATG5-dependent LC3 conversion in MEFs, and also stimulate CHOP upregulation, whereas this process is inhibited in cells expressing dominant-negative PERK or elF2$\alpha^{S51A/S51A}$ mutations [58]. Another study showed that the B-Raf inhibitor PLX4720 triggers ER stress-induced autophagy in melanoma cells. Upon ER stress, only the PERK arm of UPR was significantly upregulated, and inhibition of PERK activity released cells from ER stress-induced autophagy [59]. In the scenario of viral infection, a research group discovered that Japanese encephalitis virus (JEV)-induced autophagy is dependent on the activation of ATF6 and sXBP1 [60]. Inhibition of ATF6 and/or sXBP1 decreases the transcription of genes encoding ER chaperones, ERAD components, and autophagy proteins, thus blocking JEV-induced autophagy. Therefore, different arms or factors in UPR pathways may function differently, depending on various conditions or cell/tissue contexts.

2.5 The regulation of immunogenic cell death by ER stress induction

Compared to most caspase-mediated programmed cell death mechanisms that fail to induce inflammatory or immune responses, immunogenic cell death (ICD) refers to a situation of dying cells expressing a specific antigen that elicits a protective immune response via stimulation of dendritic cells (DCs) and consequent activation of a specific T-cell response. Thus the process of ICD contains two important features: (1) surface exposure or secretion of damage-associated-molecular patterns (DAMPs) and (2) activation of DAMPs-mediated innate and adaptive immunity. Several DAMPs have been identified that mediate immune responses, including surface calreticulin, surface HSP90, secreted ATP, and nonhistone nuclear protein high-mobility group box 1 (HMGB1) [61]. Calreticulin, a 46-kDa Ca^{2+}-binding protein, was originally identified as a molecular chaperone resident in the ER that is crucial for protein quality control and Ca^{2+} homeostasis. Upon induction of ICD, calreticulin translocates to the surface of dying cells and acts as an "eat me" signal by binding to CD91 on DCs and other antigen-presenting cells (APCs) and by facilitating phagocytic activity. Calreticulin surface translocation is a major checkpoint of ICD. Knockdown of calreticulin obliterates immune response activation in multiple tumor models, whereas restoration of calreticulin expression reestablishes immunogenicity [62,63]. Surface HSP90 also releases an "eat me" signal to induce immunity and functions redundantly with calreticulin. The secretion of ATP from dying cells, on the other hand, functions as a "find me" signal by binding to P_2Y_2 purinoceptors or by activating the NLRP3 inflammasome. HMGB1 binds to toll-like receptor 2 (TLR2) and TLR4 on DCs, and triggers the release of proinflammatory cytokines.

Studies illustrate that induction of ER stress is a common characteristic for all ICD stimuli, suggesting an important role for ER stress triggering

ICD [63–65]. For example, photodynamic therapy (PDT) generates ROS-mediated ER stress in cancer cells, which causes the release of immunogenic DAMPs, resulting in the maturation of DCs and activation of the adaptive immune system [64]. Similarly, ER stress induced by Thapsigargin enables the cytotoxic agent cisplatin to elicit translocation of calreticulin to the cancer cell surface and allows restoration of immune function within the tumor microenvironment, resulting in the amplification of therapeutic efficacy [65]. Further investigation revealed that PERK-dependent eIF2α phosphorylation on serine 51 is mandatory for triggering calreticulin surface exposure. Knockout/knockdown of PERK or mutation of the eIF2α phosphorylation site serine 51 abolishes calreticulin surface exposure, whereas deficiency of IRE1 or ATF6 does not affect this process [63]. The mechanism linking PERK and DAMPs secretion is still under investigation. One study suggests that the function of PERK in forming ER-plasma membrane junctions and modulating the dynamics of the actin cytoskeleton is relevant to the trafficking of DAMPs [66]. Although the past studies of ICD mainly focused on cancer therapies, recent research extends its application to wound healing and tissue regeneration [67–69]. Surface calreticulin induces migration and proliferation of keratinocytes that contribute to restoration of the wound surface, and the recruitment of fibroblasts to reestablish the wound defect [69]. Surface calreticulin also activates macrophage-mediated phagocytosis of dead cells [68]. Therefore, it is important to reveal whether ER stress inducers could be used for wound healing and tissue regeneration in future studies.

In summary, this section describes major cell death pathways, including mitochondrial apoptosis, extrinsic apoptosis, autophagy, necrosis, and ICD, and discusses the mechanisms through which ER stress regulates/activates these pathways. Many disagreements and disputes still exist over this topic as researchers perform experiments under different conditions and in different cell/tissue contexts. An emerging direction is to identify cross-talk among these death pathways. For example, one study indicated that autophagy activation is a salvage pathway for T-cell survival and proliferation, whereas stimulation of caspase 8 blocks autophagy by inducing caspase 8-dependent degradation of ATG5-ATG12, thus killing T cells [70]. Other studies report that cross-talk exists between apoptotic and necrotic pathways [71,72]. It is important to investigate how to manipulate cross-talk among these pathways to amplify treatment efficacy for different types of disease.

3. ER calcium regulation and ER stress-mediated cell death

3.1 ER calcium regulation

As discussed previously, the ER is a major storage organelle for Ca^{2+} and controls its release to the cytosol and transport to mitochondria, thus regulating Ca^{2+} homeostasis. Ca^{2+} handling in the ER is orchestrated by multiple mediators, including the sarco/ER Ca^{2+}–ATPase (SERCA), IP_3R, and IP_3-induced Ca^{2+} release, and store-operated calcium entry (SOCE) [73]. SERCA, encoded by three ATP2A1-3 genes located separately on three different chromosomes, has at least 14 isoforms due to alternative splicing at the mRNA level, and these isoforms are species and tissue specific. SERCA is the only active protein pump that regulates uptake of Ca^{2+} from the cytoplasm into the ER lumen. Overactivation of SERCA results in ER Ca^{2+} overload and promotes ER-mitochondria Ca^{2+} transport, thus inducing defective mitochondrial function and cell apoptosis [74]. The transmembrane protein IP_3R, which regulates the trafficking of Ca^{2+} from ER to mitochondria through mitochondria-associated ER membranes (MAMs) also plays a vital role in regulating cell death and survival. Two types of stimuli trigger Ca^{2+} release from the ER to mitochondria, including the binding of IP_3 to IP_3R, and Ca^{2+} induced Ca^{2+} release. Meanwhile, the formation of a Ca^{2+} sensitive chaperone complex of the ER protein sigma-1 receptor (Sig1R) and GRP78 promotes transport of Ca^{2+} from the ER to mitochondria through stabilization of IP_3R [75,76]. Mitochondrial bioenergetics require mobilization of Ca^{2+} from the ER; however, an over-flux of Ca^{2+} to mitochondria disrupts mitochondrial membrane integrity and triggers release of pro-apoptotic factors, such as cytochrome *c*, which in turn triggers cell apoptosis [73]. SOCE, mainly consisting of ER sensor Stromal interaction molecule 1 (STIM1), the transient receptor potential canonical 1 (TRPC1), and pore forming channel Orai, is activated upon depletion of ER Ca^{2+} stores. In response to ER Ca^{2+} depletion, STIM1, an ER-based Ca^{2+} sensing protein, cooperates with Orai to cause Ca^{2+} influx [77].

3.2 Interplay between ER calcium regulation and ER stress

A high level of ER Ca^{2+} is important for maintaining normal ER function, while impairment of the ER Ca^{2+} regulator SERCA inhibits the ability of the cell to pump calcium into the ER, which induces ER stress. The importance of SERCA dysfunction in triggering ER stress-induced cell death has been demonstrated by many studies [73,78–80]. Expression of SERCA in islets and liver of a genetic mouse model of insulin resistance and type II diabetes is decreased. Restoration of SERCA in these mice manifests a reduction of ER stress and ER stress-induced liver cell apoptosis, whereas knockdown of SERCA induces the opposite effect. In addition to obesity/diabetes, many other factors negatively regulate SERCA function and exacerbate ER stress-induced cell death such as ROS, posttranslational modifications (PTMs) of SERCA, nitric oxide, and cytokines [78,79].

ER calcium regulation is correlated with the activity of several major players in UPR, such as ATF6 and CHOP [81–84]. In a cardiac myocyte model system, an ER Ca^{2+} decrease or myocardial infarction is shown to activate the ATF6 branch of UPR, while pharmacological or genetic inhibition of ATF6 reduces cell death caused by this event [82]. Further investigation found that depletion of Ca^{2+} in cardiac myocytes stimulates the translocation of ATF6 from the ER to the nucleus, which in turn promotes the transcription of the gene encoding SERCA2a, suggesting that ATF6 is a major transcriptional regulator of SERCA2a [81]. CHOP, on the other hand, is a mediator for the ER calcium release channel IP_3R1. Studies demonstrated that CHOP transcriptionally increases expression of ERO1α, which in turn triggers the IP3R1-CaMKII pathway to induce cell apoptosis [83,84].

3.3 ER calcium regulation and cell death

ER calcium interacts with the cell death machinery through multiple mechanisms. Most importantly, BCL-2 family proteins play important roles in regulating calcium concentrations in the ER. As discussed previously, BAX and BAK are pro-apoptotic factors that are located at the ER. A study indicated that in BAX and BAK double knockout MEFs, the resting concentration of Ca^{2+} in the ER decreases, which leads to the reduction of Ca^{2+} transport from the ER to mitochondria [85]. Restoration of BAX and BAK expression in these cells recovers Ca^{2+} trafficking, upregulates the Ca^{2+} concentration in mitochondria, and increases cytochrome *c* release, thus promoting cell death. Further investigation found that in BAX and BAK double knockout cells, the interaction between IP_3R and BCL-2 is increased [86], which results in increased leakage of Ca^{2+} from the ER. These findings suggest that the ratio of proapoptotic/antiapoptotic BCL-2 family members can control the ER Ca^{2+} level through IP_3R regulated Ca^{2+} transport.

In addition, there is a strong correlation between ER calcium regulation and cell autophagy. Increases in free Ca^{2+} in the cytosol promote cell autophagy by inhibiting mTOR, a negative regulator of autophagy, and by increasing accumulation of autophagosomes through stimulation of ATG genes and Beclin 1 [87]. ER localized BCL-2, which controls ER calcium transport, inhibits calcium-induced autophagy. The mechanisms by which BCL-2 regulates calcium-dependent autophagy are still under investigation. One explanation is that BCL-2 is an activator of IP_3R. A study indicated that increased IP_3 levels strongly block autophagy that is triggered by nutrient depletion, whereas pharmacologically or genetically inhibiting IP_3R activates autophagy [88]. However, BCL-2 cannot inhibit ER stress-induced autophagy, suggesting that activation of autophagy by inhibiting IP_3R has a different mechanism compared to ER stress-triggered autophagy [88].

In summary, this section introduces ER Ca^{2+} modulation, and the correlation between ER Ca^{2+} handling and cell death regulation. The general notion is that when the ER has lower levels of free Ca^{2+} to release in response to stress signals, the mitochondria would not have sufficient Ca^{2+} to activate/participate in downstream death pathways. Besides Ca^{2+}, other important regulators of ER stress-induced cell death, such as protein tyrosine kinase ABL, SCOTIN (a direct target of p53), and BRUCE (an antiapoptotic protein) are under investigation [4]. Further studies should focus on the interplay among these regulators and how they cooperate to affect ER stress-induced cell death.

4. ER stress-related diseases

4.1 Neurodegenerative diseases

As discussed previously, the ER protein folding and quality control machinery are important for maintaining ER homeostasis. Dysfunction of these processes results in initiation and progression of various types of diseases. Many neurodegenerative diseases are characterized by aggregation and misfolding of specific proteins, including Alzheimer's disease (AD), Parkinson's disease (PD), Huntington's disease (HD), and prion-related disorders (PrDs) [89]. For instance, AD is the cause of 60%–70% of the cases of dementia, which affects approximately 47 million people worldwide [90]. Studies show that ER stress pathways are activated in AD, as manifested by the upregulation of GRP78, PERK, eIF2α, and IRE1α in the brain samples of AD patients [91,92]. The misfolding and aggregation of amyloid-β peptide is the major trigger of ER stress. The amyloid-β peptide, which is a proteolytic product of amyloid-β precursor protein (APP), disrupts the neuronal connections in the brain resulting in neuronal withering [93]. The proteolytic processing of APP is mediated by the integral membrane γ-secretase protease complex, with a major component being Presenilin 1 (PS1) protein [93,94]. Mutation of PS1 promotes the accumulation of amyloid-β plaques, and mutated PS1 interferes with UPR signaling and calcium homeostasis in the ER. Both events contribute to ER stress initiation and trigger ER stress-induced neuronal death. Importantly, ER stress initiation appears mainly in morphologically intact neurons, suggesting that ER stress activation is an early event in AD progression, and indicating that ER stress is a promising target for AD therapeutics [95].

Mounting evidence indicates that targeting the PERK-ATF4-CHOP pathway is an important strategy for preventing neurodegeneration, as ATF4 and CHOP are significantly upregulated in the AD brain compared with the other branches of UPR. ATF4 elevation activates the transcription of glycogen synthase kinase 3 (GSK-3) and protein phosphatase 1 (PP1), which further induces tau protein phosphorylation and promotes neuronal damage [96]. Increased ATF4 expression also activates CHOP, which is an important link between ER stress and neuron apoptosis [97,98]. In AD mouse models, intrahippocampal injections or genetic overproduction of amyloid-β induces upregulation of GRP78 and the PERK-ATF4-CHOP pathway and triggers caspase 12-dependent neuron death [99]. Moreover, in an amyloid-β injection mouse model, impedance of PERK-ATF4-CHOP signaling inhibits neurodegeneration, suggesting that amyloid-β induced neurodegeneration depends on this pathway. Additionally, neuroinflammation is another crucial component responsible for the progression of AD. Three branches of UPR are involved in promoting the secretion of pro-inflammatory mediators. For example, activated IRE1 recruits IκBα kinase, which in turn activates NF−κB by stimulating the degradation of IκBα [15,16]. IRE1 also interacts with TRAF2 to activate ASK1 and JNK, thus stimulating inflammatory signaling [13,14].

Similarly, PD, which results from gradual depletion of dopaminergic neurons in the substantia nigra pars compacta (SNpc) and progressive loss of motor control, is a protein misfolding disorder [100]. ER stress is activated in the PD brain as evidenced by analysis of patients' postmortem brain tissues. Alpha-synuclein is the main component of Lewy bodies and is encoded by the SNCA gene. Mutation of the SNCA gene leads to the accumulation of alpha-synuclein in the brain. Overproduction of alpha-synuclein inhibits vesicular trafficking between the ER and Golgi, thus inducing ER stress [101]. Besides the SNCA gene, ER stress induction is also correlated with other PD genes, such as LRRK2, Parkin, DJ-1, Pael-R, and ATP13A2 [102,103]. Like AD, activation of PERK-ATF4-CHOP signaling contributes to the progression of PD. For instance, overexpression of CHOP exaggerates the loss of dopaminergic neurons in a mouse PD model [104]. Delivery of ATF4 to SNpc by adeno-associated viruses (AAVs) also induces dopaminergic neuron loss [105]. However, unlike AD, the activation of IRE1-XBP1 and ATF6 branches of UPR plays a pro-survival role instead of triggering dopaminergic neuron death [106,107]. Besides AD and PD, other neurodegeneration diseases also involve ER stress activation. The mutant superoxide dismutase (SOD) in amyotrophic lateral sclerosis (ALS) and expanded glutamate homopolymeric sequences in HD lead to accumulation of protein aggregates in the ER and trigger ER stress-induced neuron death [108].

4.2 Stroke

Stroke is a cerebrovascular accident that stems from insufficient blood supply in the brain and subsequent brain cell death. It is a major health problem, which is the second most common cause of death and third most common cause for disability worldwide [109]. There are two types of stroke, including ischemic stroke that results from lack of blood flow, and hemorrhagic stroke that results from bleeding. Ischemic stroke accounts for approximately 87% of all stroke events, mostly due to arterial occlusion or hypotension. There is a strong correlation between ischemic stroke and ER stress, as the hypoxia and hypoglycemia caused by insufficient blood flow disturbs protein folding, and reperfusion of affected tissue activates production of nitric oxide (NO) that perturbs disulfide bond formation [110]. Perturbation of Ca^{2+} homeostasis by ischemia is also a major reason for triggering ER stress in neuronal cells [111]. Various models have demonstrated ER stress activation in neurons after ischemic stroke [112,113]. One study, using a rat model of 10-minute cardiac arrest followed by reperfusion, observed the upregulation of PERK-eIF2α after 10 min of reperfusion, which disappeared after 4 h. Activation of the other pathways of UPR was not detected in this study [112]. A rat model of cerebral ischemia that utilized occlusion of the middle cerebral artery for 2 h followed by reperfusion at different time points, found a time-dependent upregulation of GRP78, ATF4, CHOP, and sXBP1, and corresponding caspase 12 activation. This study suggests that ER stress triggers neuronal loss and brain cell death [113]. Further investigation proves the importance of CHOP in triggering neuronal cell death [114,115]. One study showed that ischemic stroke induces less neuronal cell death and brain damage in CHOP-knockout mice compared to wild-type mice [115].

ER stress-induced autophagy in ischemic stroke is widely studied. Mounting evidence illustrates that autophagy is mainly triggered by the IRE1-XBP1 branch of UPR in ischemic stroke [56]. Unlike apoptosis, ER stress-induced autophagy functions as a double-edged sword that could be protective or hazardous to neuronal cells [116,117]. One study indicated that inhibition of autophagy by 3-Methyladenine (3-MA) in a rat ischemic model blocks Beclin1 expression and thus increases cell death, whereas induction of autophagy by rapamycin decreases cell death and brain damage, indicating the neuroprotective role of autophagy [118]. This study also argued that the inhibition of autophagy triggers cells to undergo necrosis. However, another study, which utilized cerebral ischemia by permanent middle cerebral artery occlusion (pMCAO), showed evidence for strong autophagy induction in rat brain cells, while autophagy inhibitors, 3-MA, Brefeldin A, and Z-Phe-Ala-Fluoromethyl Ketone, strongly alleviated infarct volume, brain edema, and motor deficits, suggesting that autophagy induction is neurotoxic [119]. The disagreement among studies potentially arises from the different strength of autophagy induced in different animal models. Chronic and mild autophagy could clear protein aggregates and alleviate ER stress, while strong autophagy activation could trigger cell death. Another explanation is that when autophagy flux is unobstructed, it may serve a protective role for neuronal cells, whereas if autophagy flux is hindered it may become harmful [116].

4.3 Bipolar disorder

Bipolar disorder is a mental syndrome that affects approximately 2% of people worldwide [120]. Patients with bipolar disorder suffer from periods of depression and periods of mania. While the causes for bipolar disorder are still under investigation, researchers agree that both genetic and environmental factors are involved in the initiation and progression of this disease [121]. One contributing genetic risk factor for bipolar disorder is a polymorphism (−116 C/G mutation) in the promoter region of XBP1 that disrupts the putative binding site of XBP1 and thus decreases XBP1 expression [122]. One study using lymphoblastoid cells to compare gene expression patterns between discordant twins found that XBP1 polymorphism is significantly more prevalent in Japanese patients with bipolar disorder than the control group (odds ratio = 4.6), indicating that the mutation is associated with the pathophysiology of bipolar disorder [122]. By treating lymphoblastoid cells with the ER stress inducer Thapsigargin, this study observed that people with bipolar disorder exhibit reduced expression of XBP1 and GRP78 mRNA, indicating the −116C/G polymorphism of XBP1 results in an impaired ER stress response, which could contribute to the pathology of the illness. The same group published another study that demonstrated that polymorphism in the GRP78 promoter region is also correlated with bipolar disorder [123]. Consistent with these two studies, treatment of leukocytes from bipolar disorder patients and healthy control subjects with Tunicamycin showed that components from the ER stress pathway, including eIF2α, GRP78, CHOP, and sXBP1, are only upregulated in control samples but not in samples from bipolar disorder patients [124]. Additionally, an impaired ER stress response is related to the treatment efficacy of the mood stabilizers valproate and lithium, as studies demonstrated that valproate/lithium induced upregulation of GRP78 mRNA, which is known to have beneficial effects in patients with bipolar disorder by increasing the adaptive response to ER stress [125−128].

4.4 Cardiac disease

A large body of evidence suggests the importance of ER stress in the initiation and progression of various cardiac diseases, including ischemic heart disease, cardiac arrhythmias, and chronic heart failure. Ischemic heart disease causes severe trauma to cardiomyocytes, which usually results in a high rate of cell death after the event [129−131]. The underlying mechanism by which ER stress regulates ischemic heart disease is similar to that of ischemic stroke. Basically, the ischemic condition leads to hypoxia and hypoglycemia in cardiomyocytes and thus disturbs protein folding, which results in the stimulation of UPR. Activation of UPR helps relieve ER stress in cardiomyocytes, while prolonged and immitigable ER stress causes death of these cells. Studies have demonstrated the important role of ER stress in mediating the progression of ischemic heart disease [129,132]. One study indicated that ischemic conditions trigger all three branches of UPR, as indicated by the upregulation of PERK, ATF6, IRE1, and their downstream pathways, in primary cardiomyocytes [130]. Ischemia-induced cardiomyocyte death is mainly mediated by the PERK-ATF4-CHOP arm of UPR, as prolonged activation of CHOP further triggers processing of pro-caspase 12, thus mediating cell apoptosis [129]. Consistent with this study, another study showed that knockout of CHOP in mice reduces reperfusion injury after a myocardial infarction event, and deficiency of both CHOP and caspase 12 protects cardiomyocytes under conditions of hypoxia [133]. Except for the PERK-ATF4-CHOP arm of UPR, many studies show that activation of the other two arms plays a protective role in ischemic heart disease [132,134,135]. By culturing neonatal rat ventricular myocytes under hypoxic conditions for 16 h, one study detected increased XBP1 mRNA splicing. Spliced XBP1 further activates the GRP78 promoter and causes GRP78 upregulation [134]. In contrast, knockdown of XBP1 in these cells increased hypoxia-induced cell death. Similarly, the protective role of sXBP1 is further verified in a murine model of cardiac ischemia/reperfusion injury [132]. Activation of sXBP1 protects cardiomyocytes from directly activating GFAT1, which is a rate-limiting enzyme for the hexosamine biosynthetic pathway (HBP) [132]. HBP is crucial for O-GlcNAc modification and protein glycosylation, and studies demonstrate that HBP upregulation promotes cell survival under various types of stress. In terms of the ATF6 branch, induction of ATF6 in mice decreases cardiomyocyte death after ischemia/reperfusion injury, while knockout of ATF6 in mice induces an opposite effect [135]. Further investigation proves that ATF6 protects cardiomyocytes by stimulating ERAD, especially the component Derlin-3 [136]. Derlin-3 is a retro-translocation channel, which is crucial for clearing misfolded proteins and decreasing ER stress, and thus attenuating caspase activity. Silencing of Derlin-3 in the heart impairs ERAD, which in turn increases cardiomyocyte apoptosis after ischemia/reperfusion injury [136].

The mechanism by which ER stress regulates arrhythmias is primarily due to its effect on $Na_v1.5$ channels [137,138]. The cardiac isoform $Na_v1.5$ is a glycoprotein that must undergo N-glycosylation processing in the ER before being transported to the plasma membrane. The glycosylation process is important for $Na_v1.5$, because only fully glycosylated forms of $Na_v1.5$ traffic to the plasma membrane through the conventional route, while mutations in the gene encoding $Na_v1.5$, SCN5A disrupt glycosylation and trafficking of this protein [138]. One study showed that the PERK-ATF4-CHOP arm of UPR is involved in regulating expression of SCN5A [137]. Treatment of human-induced pluripotent stem cell-derived cardiomyocytes with activators of SCN5A variants, angiotensin II (AngII), and hypoxia, upregulates PERK in these cells, suggesting that PERK is activated by SCN5A splice variants. PERK activation subsequently induces the reduction of Na^+ currents by mediating the degradation of full-length SCN5A mRNA [137]. Changes in Na^+ currents are responsible for the induction of heart arrhythmias. Although this study failed to provide an arrhythmia model to prove the correlation, as splice variants of SCN5A are only present in humans, it suggests that PERK is a potential target for treating cardiac arrhythmias.

ER stress also plays an important role in the development and progression of chronic heart failure. The Lys-Asp-Glu-Leu (KDEL) receptor, which is a retrieval receptor for ER chaperones, is responsible for protein quality control in the ER. Mutant KDEL receptors interrupt the transport of unfolded/misfolded proteins from the ER to the Golgi apparatus and cause accumulation of these proteins in the ER, which thus triggers ER stress. Mice with mutant KDEL receptors develop cardiomyopathy and heart failure [139]. The hearts from these transgenic mice display ER stress as indicated by elevated expression of CHOP. Furthermore, apoptosis is increased in the cardiomyocytes of these mice, suggesting that ER stress promotes the progression of heart failure [139]. The activation of ER stress in failing hearts is also shown by the upregulation of GRP78 and extensive splicing of XBP1 in heart samples from patients with heart failure [140,141]. Further investigation suggests that sXBP1 serves as a cardioprotective factor in the progression of heart failure. Spliced XBP1 is upregulated in the early adaptive phase of heart failure, but it is inhibited by microRNAs, miR-30, and miR-214, in the maladaptive phase [142]. As sXBP1 promotes transcription of VEGFA in cardiomyocytes in response to ER stress, downregulation of sXBP1 blocks VEGFA signaling, which in turn exacerbates heart failure [143]. Besides the XBP1 arm, the PERK-ATF4-CHOP branch is critical in modulating heart failure through multiple mechanisms [144,145]. In cardiac-specific PERK knockout mice, the activity of the ATPase SERCA2a is significantly decreased after thoracic aorta constriction (TAC) [144]. Reduction of SERCA2a activity induces calcium imbalance and aggravates ER stress in cardiomyocytes, which in turn promotes heart failure. Unlike the protective role of PERK, enhanced CHOP expression is correlated with increased cardiomyocyte apoptosis. Mice with CHOP knockout manifest less severe cardiac dysfunction in comparison with wild-type mice after TAC [145]. The mechanism by which CHOP deletion protects cardiomyocytes is still under investigation.

One potential explanation would be the negative feedback loop between CHOP and eIF2α phosphorylation [146]. The deletion of CHOP results in increased phosphorylation of eIF2α and decreased global translation, which may reduce proteostatic burden on cardiomyocytes.

4.5 Diabetes and obesity

Diabetes mellitus (DM), including type I DM and type II DM, is a metabolic disorder in which patients exhibit high blood sugar levels over prolonged time. Type I DM results from excessive loss of insulin-producing β cells, which leads to a deficiency of insulin secretion. The cause of type I DM is still unknown, but recent research suggests ER stress is one of the pathogenic factors that contribute to the progression of type I DM [4,147]. Studies confirm that the ER is extremely well developed both structurally and functionally, and that IRE1, ATF6, and PERK are abundant in pancreatic β cells, indicating their significant roles in promoting the production of insulin and various other glycoproteins [4,148]. PERK knockout mice develop a normal exocrine and endocrine pancreas at birth; however, they experience a significant deterioration in glycemic control before 4 weeks of age, and become diabetic after 4 weeks [149]. These mice are classified as having type I DM, because they exhibit decreased islet size and increased apoptosis of insulin-producing cells. PERK gene mutations in humans lead to Wolcott-Rallison syndrome, which is an autosomal recessive disease. These patients, similar to PERK knockout mice, experience advanced loss of β cells and develop severe type I DM at an early age [150,151]. Moreover, mutation of the phosphorylation site of eIF2α (Ser51Ala) in mice causes a more advanced and progressive β cell deficiency [20]. Phosphorylation of eIF2α, which is partially regulated by PERK, inhibits global protein translation and reduces the protein load on the ER. Mutation of Ser51Ala, on the other hand, exacerbates the protein burden in the ER, thus causing exhaustion of the protein folding capacity of β cells and induces β cell death. Additionally, multiple other pathways are involved in inducing β cell apoptosis, including the IRE1-ASK1-JNK and CHOP-caspase 12 pathways [148,152]. For example, the Akita mouse model is a spontaneously diabetic model that is caused by a mutation in the insulin 2 gene. The Akita mice exhibit progressive hyperglycemia coupled with elevation of CHOP and CHOP-induced β cell death. Inhibition of CHOP gene expression delays the onset of DM in heterozygous Akita mice, suggesting that CHOP is a potential target for treating diabetes before patients develop irreversible loss of β cells [152]. However, CHOP inhibition fails to rescue homozygous Akita mice, suggesting that CHOP is not the only factor that is involved in β cell apoptosis and that other mechanisms remain to be discovered.

Type II DM and its long-term complications have resulted in severe public health problems, which are major causes for morbidity and mortality worldwide. Unlike type I DM, type II DM results from insulin resistance, which is mainly induced by obesity. Insulin resistance coupled with impaired inhibition of glucagon signaling leads to an incongruous elevation of hepatic glucose production, which in turn results in hyperinsulinemia and increased lipid synthesis and storage, thus aggravating obesity. Activation of chronic ER stress is one major mechanism that contributes to the pathogenesis of obesity-induced insulin resistance. Hepatic ER stress is triggered by high fat diet and obesity in mice [153]. Moreover, ER stress inhibits the phosphorylation of insulin receptor substrate-1 (IRS1) by activating the IRE1-TRAF2-JNK pathway, which in turn causes insulin resistance [153]. Besides the IRE1 pathway, eIF2α, which is abundant in liver cells, is another important factor. Homozygous $eIF2\alpha^{S51A}$ mutation inhibits hepatic glucose production in neonatal mice, while heterozygous $eIF2\alpha^{S51A}$ mutation induces adult mice to develop obesity, insulin resistance, and glucose intolerance [20,154], because eIF2α phosphorylation promotes translation of C/EBP, which causes adipogenesis in liver [155].

ATF6 and sXBP1 are highly expressed in pancreatic β cells and play protective roles in the progression of type II DM. Studies found that in dietary and genetically obese mice, expression of ATF6 and sXBP1 gradually decreases with the progression of type II DM [153,156]. High fat diet induces continuous hyperinsulinemia and more significant glucose intolerance in XBP1+/− mice compared to their wild-type controls [153]. Similar effects can be detected in ATF6 knockout mice [156]. Patients with type II DM also express lower levels of ATF6 and sXBP1 in their pancreatic β cells [156]. Decreased ATF6 could result from the activation of the CaMKII-MK2 pathway that is triggered by aberrant glucagon action in hepatocytes [157]. More specifically, abnormal glucagon action in obesity promotes excessive calcium release into the cytoplasm through hyperactivation of the ER calcium channel, which in turn activates CaMKII-MK2. CaMKII further induces phosphorylation of histone deacetylase 4

(HDAC4) and decreases SUMOylation of DACH1. DACH1 suppresses transcription of ATF6 by interacting with nuclear receptor corepressor 1 (NCoR1) [158–160].

4.6 Muscle atrophy

The muscle is the most important organ for the storage of protein and amino acids, which is a major source of energy production during catabolic periods. Under normal conditions, protein synthesis and degradation in muscle is in a balanced state for muscle mass maintenance. Any disturbances that cause muscle protein breakdown to exceed protein synthesis will result in decreased muscle mass, termed muscle atrophy [161]. Muscle atrophy is induced by multiple diseases, such as cancer, AIDS, congestive heart failure, chronic obstructive pulmonary disease (COPD) and renal failure, and also is a consequence of long-term glucocorticoid treatment and aging [162]. Progressive and excessive degradation of skeletal muscle is a devastating complication that aggravates primary diseases and increases morbidity and mortality. There are currently no effective treatments for muscle atrophy.

The Ubiquitin-proteasome machinery and autophagy-lysosome machinery are two of the most critical systems that control the balance between protein synthesis and protein degradation in muscle. The ATP-dependent ubiquitin-proteasome system is the primary pathway responsible for muscle proteolysis, and it involves multiple E3 ubiquitin ligases. One study utilizing transcript profiling identified two genes that encode E3 ligases, namely Muscle RING Finger 1 (MuRF1) and Muscle Atrophy F-box (MAFbx), which are upregulated in in vivo models of muscle atrophy [163]. Knockout of MuRF1 and MAFbx in mice significantly inhibits the progression of muscle atrophy upon denervation. Moreover, by analyzing rat and human tissues, they found that MuRF1 and MAFbx mRNAs are specifically expressed in cardiac and skeletal muscle, suggesting their potential role as druggable targets for treating muscle atrophy. However, the underlying mechanisms by which these two E3 ligases regulate muscle atrophy are still under investigation. One study demonstrated that MuRF1 specifically binds to and ubiquitylates a subset of muscle proteins, including MyBP-C, MyLC1, MyLC2, and myosin heavy chain (MYHC), which play selective roles in stabilizing thick but not thin filament components [164]. The two most widely known targets of MFAbx are MyoD and eukaryotic translation initiation factor 3 subunit f (eIF3-f). Unlike MuRF1 that controls protein degradation, MFAbx may be responsible for protein synthesis, as overexpression of eIF3-f induces hypertrophy in myotubes [161]. However, genetic knockout of MAFbx does not result in larger than normal myofibers in mice [165]. Besides MuRF1 and MFAbx, TRAF6 is another ER ligase that regulates muscle atrophy. Genetic ablation of TRAF6 in mice prevents skeletal muscle wasting in response to starvation, denervation, and cancer cachexia [166,167]. Furthermore, TRAF6 is an upstream inducer of ER stress-related gene expression, including genes encoding ATF4 and CHOP [166].

The autophagy-lysosome machinery is another system that regulates the degradation of unfolded/misfolded proteins. Deregulation of autophagy, either excessive or defective, can cause myopathies [162]. For example, inhibition of autophagy specifically in skeletal muscle increases mitochondrial dysfunction and oxidative stress in autophagy-deficient ($ATG7^{-/-}$) mice, which leads to neuromuscular junction instability and decreased force generation [168]. On the other hand, hyperactivation of autophagy has been detected in multiple muscle atrophy models, including cancer cachexia, long-term glucocorticoid treatment, and aging [169]. There is a crosstalk between ER stress pathways and autophagy in the pathogenesis of muscle atrophy. ER stress is activated in both patient samples and mouse models of muscle atrophy, while deletion of the ER stress marker CHOP in AR113Q mice aggravates muscle atrophy through enhanced autophagy, suggesting that CHOP may be an upstream target for treating muscle atrophy via modulation of autophagy [170].

4.7 Cancer

Emerging evidence suggests that both ER stress and UPR activation are associated with the development of various types of cancer [171]. As cancer cells divide rapidly, consume excessive amounts of glucose, and release lactic acid waste, the cancer microenvironment is typically lacking in oxygen and nutrients, and low in pH. These alterations in metabolism and the tumor microenvironment trigger ER stress through various mechanisms. For example, low oxygen conditions induce the production of cytosolic ROS and disruption of ER chaperone functions through ROS-dependent lipid peroxidation and covalent adduct formation [172,173]. Since oxygen is required for protein folding in the ER, especially disulfide bond formation, hypoxia disturbs this process, which in turn triggers ER stress. In terms of nutrient deprivation, glucose and glutamine shortage interfere with the hexosamine biosynthetic pathway that is essential for N-linked glycosylation, while proline starvation enhances the production of ROS [174,175]. Chronic and mild ER stress initiation in cancer cells can be alleviated by activation of UPR, which increases the ER protein folding capacity via induction of ER chaperone synthesis and promotion of ERAD. Tolerable ER stress also contributes to the development of chemoresistance and facilitates cancer progression by supporting cancer cell viability under oxygen and nutrient deprivation, triggering angiogenesis, and prolonging the cell dormancy period [171]. However, persistent and irrecoverable ER stress induces cancer cell death through multiple death pathways as discussed in the previous section. Thus, targeting ER stress has been proposed as a novel treatment strategy for cancer.

Studies indicate that HER2+ and/or EGFR+ breast cancer cells present a severe proteotoxic burden, since HER-family proteins are rich in cysteine residues that form a large number of disulfide bonds during the protein folding process within the ER [176]. Disruption of the rate-limiting disulfide bond formation step aggravates the proteotoxic burden on the ER and induces irremediable ER stress, which in turn causes cancer cell death [176]. Moreover, genes encoding ERAD pathways are activated in HER2+ breast cancer, indicating that ERAD promotes cancer progression by inhibiting toxic ER stress. Blockade of ERAD in HER2+ breast cancer strongly exacerbates ER stress and stimulates ER stress-induced cell death [177]. Various drugs targeting different arms of UPR have been developed for cancer treatments, but there are disagreements on which ER stress pathway is more responsible for triggering cancer cell death. For instance, one study showed that a small molecule compound, DDP-23, increases production of ROS and thus triggers ER stress [178,179]. Inhibition of the IRE1 arm blocks DDA-23 induced caspase activation and decreases cancer cell apoptosis, suggesting that the IRE1 pathway is important for anticancer actions. Another study demonstrated that the anticancer effects of an ER stress inducer 11-dehydrosinulariolide is reversed by pretreatment of melanoma cells with Salubrinal, indicating that the PERK/eIF2α/CHOP pathway contributes to apoptotic events [180]. Incongruities also exist as to whether ER stress-induced autophagy promotes or suppresses cancer progression. Knockout of ATG5 in cancer cells renders cells significantly more sensitive to ER stress, indicating that impedance of autophagy enhances ER stress-triggered cell death [59,181]. However, a small molecule compound named HA15 that targets GRP78 and inhibits melanoma growth by activating concomitant ER stress induces autophagy and apoptosis both in vivo and in vitro [182]. The treatment efficacy of HA15 is reduced by either inhibiting autophagy or apoptosis, suggesting that autophagy contributes to cancer cell death. All of these uncertainties may be the result of these studies being performed under different cell and tissue contexts, or that there are undiscovered actions of these drugs. Thus, in the next section, we will discuss the underlying mechanisms by which these drugs kill cancer cells and will point out opportunities for further drug development.

In summary, this chapter discusses ER stress-related diseases, including neurodegeneration, stroke, bipolar disorder, cardiac disease, diabetes and obesity, muscle atrophy, and cancer, and the mechanisms by which ER stress contributes to the pathogenesis and progression of these diseases. By uncovering the relationship between ER stress activation and disease progression, potential druggable targets for developing therapeutics are also proposed. FDA-approved ER stress-related drugs and drugs in clinical trials for treating these disorders will be introduced in the next section, as well as their specific mechanisms of action.

5. Disease intervention for ER stress-related disorders

5.1 Neuroprotective drugs

5.1.1 ER chaperone modulators

Many ER chaperone modulators, including the mood stabilizers lithium and Valproate, GRP78-inducible factor x (BIX), and protein disulfide isomerase (PDI) inhibitors, have shown neuroprotective effects in the treatment of neurodegenerative diseases, stroke, bipolar disorder, and muscle atrophy [4,183] Table 5.1. Both Lithium and Valproate are FDA-approved drugs for bipolar disorder and are in clinical trials for multiple neurodegenerative diseases, such as AD, ALS, and spinal muscular atrophy. Chronic lithium and Valproate administration increases GRP78 expression without activating ER stress in cerebral cortical and cerebellar granule cells and have neuroprotective effects [184,185]. Such beneficial effects result from the ability of GRP78 to bind overloaded calcium, thus stabilizing the mitochondrial membrane potential and preventing caspase activation. Buffering excessive intracellular free calcium also protects cells from glutamate toxicity [185]. BIX, which selectively promotes GRP78 expression, protects neural cells from ER stress-induced apoptosis [186]. Posttreatment with BIX after ischemic stroke in animal models significantly decreases infarct volume and reduces neuronal damage [186,187].

Table 5.1 FDA-approved drugs and drugs in clinical trials that target ER stress-related diseases.

Strategy	Drugs	Potential indications	Diseases	References
ER chaperone modulators	Lithium Valproate	Mood stabilizers that increase GRP78 expression; neuroprotection	FDA-approved for bipolar disorders; Phase 1/2 for neurodegenerative diseases; Phase 2/3 for spinal muscular atrophy	[184,185]
Autophagy inducers	Rapamycin	mTOR inhibitor that promotes degradation of aggregate-prone proteins; neuroprotection	Phase 2 for neurodegenerative diseases; Phase 1 for sarcopenia	[194,195]
	Trehalose	mTOR-independent autophagy inducer that increases transcription of ATG genes in neurons; neuroprotection	Phase 3 for bipolar disorder	[196–199]
Chemical ER chaperone	PBA TUDCA	Mimic ER chaperones and decrease ER stress response; cardioprotection; neuroprotection; increase insulin sensitivity	FDA-approved for clinical use; Phase 1/2 for neurodegenerative diseases; Phase 4 for type II DM/obesity	[210–215]
Agents that block CHOP expression	Angiotensin AT_1 receptor antagonists (Telmisartan and olmesartan)	Suppress CHOP expression and caspase activation; cardioprotection	FDA-approved for hypertension; Phase 2/3/4 for cardiovascular diseases	[218]
	Vitamin D receptor agonists (calcitriol and paricalcitol)		FDA-approved for clinical use; Phase 3/4 for cardiovascular diseases	[219]
	Statin (atorvastatin)		FDA-approved for dyslipidemia and prevention for cardiovascular diseases	[220]
Agents that induce	Antidiabetic biguanides (Metformin)	Selectively activates ATF4; cardioprotection	FDA-approved for type II DM; Phase 1/2/3 for solid tumors and haematopoietic malignancies	[223]

ATF4 activation	GLP-1R agonists (GLP-1 and exendin-4)	Increases ATF4 expression; Protect pancreatic β cells	Phase 1/2 for type I DM; phase 4 for type II DM	[222]
Proteasome inhibitors	Bortezomib	Inhibits NF−κB pathway; selectively blocks activity of the 26S proteasome	FDA-approved for relapsed MM and mantle cell lymphoma; Phase 1/2 for solid tumors	[235,255]
	Carfilzomib	Inhibits NF−κB pathway; selectively blocks activity of the 20S proteasome	FDA-approved for relapsed MM; Phase 1/2 for solid tumors	
	Marizomib	Inhibits NF−κB pathway; covalently modifies the active threonine residues of the 20S proteasome	Phase 1/2 for solid tumors and haematopoietic malignancies	[236]
	MLN9708	Inhibits NF−κB pathway; selectively blocks the subunit beta type-5 of the proteasome	Phase 1/2/3 for solid tumors, haematopoietic malignancies, and MM	[237]
Grp78 inhibitors	DHA	Inhibits total and surface GRP78; disrupts binding between GRP78 and PERK	Phase 2/3 for solid tumors and metastatic cancer	[230]
	PAT-SM6	A monoclonal IgM antibody that targets a tumor-specific variant of cancer cell surface GRP78	Phase 1 for MM	[229]
IRE1 inhibitors	Sunitinib	Type I IRE1 kinase inhibitor; activates XBP1s; cardiotoxicity	Phase 1/2 for solid tumors and metastatic cancers	[245]

The rationale for developing PDI inhibitors is that PDIs are important foldases that catalyze formation and isomerization of disulfide bonds in some disease-related proteins, and impedance of PDIs would repress the toxic effects of these proteins [183]. For example, the polyglutamine-expanded huntingtin exon one encoded huntingtin protein contributes to the progression of AD by triggering neuronal death. By screening thousands of synthetic compounds and natural products, one study identified a PDI inhibitor that suppresses the expression of the huntingtin protein and thus rescues neurons [188].

5.1.2 Autophagy inducers

Autophagy is critical for mediating unfolded/misfolded protein degradation. The association between autophagy impairment and neurodegenerative diseases is suggested by studies showing that genetic deletion of ATG5/ATG7 in the central nervous system of mice results in spontaneous neurodegeneration, increased protein aggregate accumulation and pervasive neuronal loss [189−191]. Defects in the autophagy system, such as global alteration in the proteolytic capacity of lysosomes and mutations in p62/SQSTM1, are widely seen in neurodegenerative diseases; therefore, enhancing autophagy is a potential strategy for treating these diseases [192,193]. Preclinical studies showed that rapamycin, an mTOR inhibitor, and multiple rapamycin analogs promote degradation of the most relevant aggregate-prone proteins that are involved in neurodegeneration, such as mutant tau, polyglutamine- and polyalanine-containing proteins, and α-synuclein [194,195]. These autophagy inducers are in clinical trials for treating ALS and Tuberous Sclerosis Complex (TSC). Many mTOR-independent autophagy inducers are also effective for neurodegenerative diseases. One representative is Trehalose, a chemical chaperone that prevents protein aggregation in HD, AD, and PD and enhances degradation of mutant SOD1 in ALS [196−199]. Administration of Trehalose significantly increases transcription of ATG genes in neurons, presumably though FOXO1 activation [199,200]. Besides rapamycin and Trehalose, a series of small molecule autophagy inducers have been identified that exhibit beneficial effects for AD and HD [201,202].

5.1.3 Others

Many preclinical studies have provided other therapeutic opportunities for neuroprotection through modulating ER stress. For example, Salubrinal is a small molecule inhibitor that suppresses eIF2α dephosphorylation via impeding ERO1α [203,204]. Salubrinal reduces neuronal death after stroke by arresting global protein translation and thus decreasing the unfolded protein load [113,204]. Additionally, drugs that activate antioxidant pathways, such as NOS inhibitors, are also neuroprotective and are useful as a posttreatment for stroke, because oxidative stress exacerbates ER stress and ER stress-induced cell death [205]. Taurine, the most abundant amino acid in skeletal and cardiac muscle and brain, protects neurons against glutamate toxicity and oxidative stress-induced cell death [206]. Taurine treatment leads to the downregulation of ATF6 and IRE1, which protects cells after hypoxia/reoxygenation or glutamate treatment [207].

5.2 Treatments for cardiac diseases

5.2.1 Chemical ER chaperones

Chemical chaperones, which mimic ER chaperones, alleviate ER stress by stabilizing proteins in their native conformation, improving ER luminal folding capacity, and increasing trafficking of aberrant proteins [208,209]. Sodium phenylbutyrate (PBA) and tauroursodeoxycholic acid (TUDCA) are two small chemical chaperones that are in clinical trials for neurodegenerative diseases and type II DM. Both drugs alleviate Tunicamycin-induced ER stress in vitro and reduce ER stress levels in the liver of ob/ob mice [210]. Pretreatment of ob/ob mice with these drugs significantly restore insulin sensitivity and increase glucose tolerance [210]. Mice with genetically impaired ER capacity exhibit leptin resistance and more frequently develop obesity on a high fat diet, while PBA and TUDCA significantly reinstate leptin sensitivity [211]. PBA and TUDCA also protect neurons by reducing ER stress in stroke and HD animal models [212,213], and PBA and TUDCA exhibit cardioprotective functions in preclinical studies [214,215]. Oral administration of PBA in mice with TAC- induced cardiac hypertrophy reduces the ER stress response and ER stress-induced apoptosis in the heart [214]. TUDCA reverses oxidative stress-mediated contractile malfunction in mouse cardiomyocytes by relieving ER stress [215].

5.2.2 Agents inhibiting CHOP

Upregulation of CHOP triggers cell death during the progression of cardiac diseases. There are no pharmacological strategies to directly inhibit CHOP, but some agents protect cardiomyocytes by indirectly suppressing CHOP activity. For example, pretreatment with a JNK inhibitor SP600125 significantly reduces cyclic stretch-induced CHOP activation [216]. A p38

MAPK inhibitor, SB203580, abrogates stress-induced CHOP phosphorylation in vivo [217]. Angiotensin AT_1 receptor antagonists (Telmisartan and olmesartan) and vitamin D receptor agonists (calcitriol and paricalcitol) decrease the cardiomyocyte apoptosis that is triggered by TAC or MI/R injury by impeding CHOP expression [140,218,219]. The statin atorvastatin, a lipid-lowering agent that is used for cardiovascular disease prevention, inactivates CHOP and caspase 12, thus reducing cardiomyocyte apoptosis after MI-mediated heart failure [220].

5.2.3 Others

Other drugs, including NOS inhibitors, GRP78 inducers, and Salubrinal, also play cardioprotective roles utilizing the mechanisms discussed above [209].

5.3 Diabetes and obesity therapeutics

5.3.1 Drugs that mildly activate UPR

One strategy to protect pancreatic β cells is to alleviate ER stress to a tolerable state and modulate UPR to selectively activate the survival instead of the death pathway. GLP-1R agonists, such as exendin-4 and GLP-1, are promising candidates, as activation of GLP-1R protects pancreatic β cells through mild stimulation of UPR and induction of ATF4 [221,222]. The antidiabetic biguanides are FDA-approved drugs for type I/II DM, which can increase insulin sensitivity, decrease plasma glucose concentration and gluconeogenesis; however, the underlying mechanisms for their actions are not fully understood. One study demonstrated that a biguanide Metformin protects cardiomyocytes in diabetic patients with heart failure by selectively activating the PERK-ATF4 arm of UPR, and activation of CHOP does not induce cell apoptosis [223]. This study proposed a promising mechanism for understanding the protective function of biguanides. Whether biguanides can protect β cells in the same way as it protects cardiomyocytes is still under investigation.

5.4 Cancer therapeutics

5.4.1 GRP78 inhibitors

As discussed previously, GRP78 is a molecular chaperone that assists with the degradation of misfolded/unfolded proteins. Dissociation of GRP78 from ER stress sensors results in the activation of three arms of UPR. Overexpression of GRP78 is correlated with poor prognosis of various types of cancers [224,225]; therefore, GRP78 inhibitors have been developed as cancer therapeutics. These compounds target GRP78 via different mechanisms, including (1) binding to the ATP-binding site of GRP78 thus interrupting its ATPase activity (Epigallocatechin gallate [226], HA15 [227]); (2) cleaving GRP78 at a specific dileucine motif (L416-417) (SubAB [228]); and (3) targeting a tumor-specific variant of GRP78 (PAT-SM6 [229]). Of the GRP78 inhibitors, DHA and PAT-SM6 are in clinical trials to treat cancer. DHA, which is in clinical trials for treating many solid tumors and metastatic cancers, inhibits total and surface GRP78 expression and specifically disrupts binding between PERK and GRP78 [230,231]. The mechanisms by which DHA reduces GRP78 expression are still under investigation, but studies propose that this action of DHA is correlated with its anticancer effects on both primary and metastatic tumors [230,231]. PAT-SM6, which is a monoclonal antibody generated from human tumors, is being tested for treating multiple myeloma (MM) in a Phase I clinical trial. The anticancer effects of PAT-SM6 result from selectively targeting a tumor-specific variant of surface GRP78 on cancer cells [229,232]. PAT-SM6 efficacy against MM is thought to result from triggering of complement dependent cytotoxicity and antibody-dependent cellular cytotoxicity. Although evidence shows that both DHA and PAT-SM6 inhibit GRP78, they have multiple other actions contributing to their anticancer effects. Further studies are necessary to establish the role of GRP78 inhibition in killing cancer cells.

5.4.2 Proteasome inhibitors

Inhibiting the chymotrypsin-like subunits in the proteasome blocks ERAD, which leads to excessive accumulation of ubiquitylated proteins in the ER and induces ER stress-mediated cell death. Two proteasome inhibitors, Bortezomib and carfilzomib, have been approved for treating patients with relapsed MM [233,234]. The initial idea for applying Bortezomib to MM is based on its effect on inhibiting NF−κB, as IκBα is a substrate of the proteasome. Recent studies suggest that Bortezomib-induced ER stress also assists in its anticancer activities, because cancers cells that display prolonged and severe ER stress in response to Bortezomib are more sensitive to it [235]. Moreover, the preferential therapeutic effect of Bortezomib for MM can be explained by the high synthesis rate of immunoglobulins in MM cells compared to other cancer cells, which is associated with increased protein burden in the ER. Besides Bortezomib and carfilzomib, there are several other proteasome inhibitors that are in clinical trials for targeting hematopoietic malignancies and various types of solid tumors, including Marizomib and MLN9708. Marizomib covalently modifies the active threonine residues of the 20S proteasome thus inhibiting proteasome activity [236], while MLN970 selectively and reversibly inhibits the subunit beta type-5 of the proteasome [237].

5.4.3 Others

Many compounds that target different components of UPR pathways have been tested in preclinical studies, including drugs that block PERK, IRE1, and GADD34/PP1c. Interruption of PERK activity in cancer cells under hypoxic conditions disturbs their adaptation to chronic and mild ER stress [238]. A series of ATP-competitive PERK kinase inhibitors have been developed by GlaxoSmithKline including GSK2656157 and GSK2606414, which exhibit prominent anticancer efficacy in vitro and in vivo [239,240]. However, PERK inhibitors induce significant pancreatic toxicity that limits their application in cancer. Compounds that target the GADD34/PP1c complex to promote eIF2α phosphorylation, such as Salubrinal and Guanabenz, on the other hand, stimulate expression of CHOP and thus increase caspase activation and ER stress-induced cell death [241]. The rationale for developing IRE1 inhibitors is to reduce the adaptive signal from XBP1s in the response to ER stress. Many IRE1 inhibitors have been developed that either target the kinase or RNase domain of IRE1, and are used as cancer treatments [242–244]. MKC-3946 is a representative RNAse domain inhibitor, which significantly suppresses cell viability under hypoxia-triggered ER stress [242]. Sunitinib is a type I IRE1 kinase domain inhibitor that activates XBP1s, but little evidence exists that the anticancer activity of Sunitinib results from its role in modulating UPR [245].

In summary, this section reviews FDA-approved drugs and drugs in clinical trials that treat diseases by modulating UPR pathways. Besides regulating UPR pathways, however, these drugs have multiple other actions that may be responsible for their efficacy. Further studies should focus on evaluating the importance of UPR modulation in contributing to treatment effects. This section also discusses promising new drugs that are in preclinical studies. While these drugs may be limited in their application, these studies provide novel mechanisms and potential strategies for drug development.

5.4.4 Future directions

Successful proteotoxic therapy is complicated by the fact that ER stress can be either pro-apoptotic or pro-survival. In neurodegenerative diseases where proteotoxic stress leads to neuronal death, therapeutic strategies are needed to preserve neuronal viability. In the context of cancer therapy, adaptation to ER stress can lead to drug resistance and increased tumor aggressiveness in some instances. In other cases, transformation by specific oncogenes can produce a synthetic lethal interaction with ER stress. For example, cancer cells transformed by MYC are highly sensitive to activation of the ATF4/CHOP/DR5 death pathway stimulated by the TRAIL death ligand [246–248]. A variety of determinants control the outcome of proteotoxic therapies, including the ability to regulate ER stress in specific subpopulations of cells in the body, the specific mechanisms by which ER stress is induced, and the duration and amplitude of the proteotoxic insult.

5.4.5 Selecting therapeutic modes of ER stress initiation and control of the off-target effects of ER stress induction

Many approaches that initiate ER stress perturb-specific protein folding processes including disulfide bond formation, glycosylation, and proline isomerization. Other strategies include perturbing calcium levels in the ER and suppressing the disposal of misfolded and aggregated proteins by inhibiting autophagy, or proteasomal or lysosomal activity. Different disease states involve different cell populations, each with a distinctive proteome and secretory pathways with capabilities matched to their respective cellular functions.

In addition to a diversity of triggers of ER stress, irreparable disruption of proteostasis can induce cell death through a variety of mechanisms that can include necroptosis, autophagy, immunogenic cell death, and the intrinsic and extrinsic apoptotic pathways. There is extensive cross-talk between these death pathways. The multitude of inducers of and responses to proteotoxicity provides opportunities for finely tuning pharmacological interventions that combat diseases by positively or negatively impacting the cellular proteostasis profile.

In the context of cancer therapy, detailed knowledge of the tumor genetics and status of UPR activation is required to determine whether specific tumors would be killed by proteotoxic insults, or if ER stress would further enhance their viability and aggressiveness. For example, HER2-overexpressing breast cancers rely on ERAD so ERAD inhibitors may provide an effective therapy for these tumors [177,249]. Similarly, overexpression of EGFR or HER2 sensitizes breast tumors to DDAs by triggering irremediable ER stress [176,250]. In general terms, cancers that rely on oncogenic drivers that are integral membrane proteins or tumors that are driven by autocrine loops of secreted growth factors may be particularly vulnerable to proteotoxic therapy.

5.4.6 Potential resistance mechanisms to proteotoxic therapy

Human tumors often acquire resistance to targeted therapies, radiotherapy, and cytotoxic chemotherapy. It is thus likely that over time tumors will also become refractory to proteotoxic therapies. Many routes to resistance could be envisaged. Cancer cells might be able to mitigate proteotoxicity by reducing their overall rates of protein synthesis or downregulate the expression of specific proteotoxic drivers by relying on alternate drivers with a lesser impact on proteostasis. Alternatively, since disruption of proteostasis triggers adaptation through the UPR, therapy-driven cell death may be transitory. Over time, cancer cells are likely to acquire genetic mutations and epigenetic reprogramming that confers further resistance.

Combinations of multiple different classes of therapeutic agents may be necessary to conquer treatment refractory cancers, and in some cases such combinations may produce synergistic antitumor efficacy. For instance, the ability of ER stress to render cells immunogenic through the release of "eat me" signals suggests that proteotoxic therapies may enhance the efficacy of immunotherapies. Further support for this concept comes from the observations that ER stress can lead to upregulation of DR5 as mentioned above. Since one mechanism by which cytotoxic immune cells kill cancer cells is through the release of the DR5 ligand TRAIL [251,252], it is expected that induction of cancer-specific proteotoxicity will not only render cancer cells more visible to immune cells but will also enhance cancer cell killing by cytotoxic immune cells.

5.4.7 Programming cell type-specific induction of ER stress

Several neurodegenerative diseases result from the inability of neurons to appropriately process specific proteins as described above. In these situations, knowledge of the specific mechanisms that interfere with proper protein folding and removal of aggregates is required in order to enhance the processing of the toxic proteins. For example, if rate-limiting steps in protein folding such as disulfide bond formation or proline isomerization are the bottlenecks then enhancing expression of Protein Disulfide Isomerases (PDIs) or Prolyl Isomerases (PIs), respectively, may limit proteotoxicity. Likewise, elevating expression of chaperones such as GRP78 and GRP94 could reduce the formation of protein aggregates and facilitate folding. Since multiple PDIs, PIs, and chaperones operate in the secretory pathway and each acts on different subsets of client proteins, detailed knowledge of the folding pathway of individual proteins will be required to develop rational strategies to combat proteotoxicity. Alternatively, in cases involving specific pathogenic proteins that are not required for neuronal function, selectively suppressing their expression in a variety of ways including RNAi-mediated silencing, genome editing (e.g., CRISPR/Cas9), or epigenetic approaches to reduce gene expression could restore proteostasis. Such genetic approaches could be engineered to target specific cell types by using viruses with the desired tropism combined with cell type-specific promoter elements. In instances where protein toxins essential for cell function or viability are generated by genetic mutation, it may be possible to correct such mutations using genome editing (e.g., Ref. [253]). Another strategy under investigation is to generate induced Pluripotent Stem Cells (iPSCs) using genome editing to produce isogenic iPSC lines with and without a proteotoxic mutation of interest. The mutant and wild-type iPSCs are then differentiated to the target cell type, and drug screens can be performed to identify agents that suppress cell death caused by the proteotoxic mutant (e.g., Ref. [254]).

References

[1] M. Wang, R.J. Kaufman, Protein misfolding in the endoplasmic reticulum as a conduit to human disease, Nature 529 (2016) 326–335.

[2] J.R. Gaut, L.M. Hendershot, The modification and assembly of proteins in the endoplasmic reticulum, Curr. Opin. Cell Biol. 5 (1993) 589–595.

[3] X. Chen, A. Karnovsky, M.D. Sans, P.C. Andrews, J.A. Williams, Molecular characterization of the endoplasmic reticulum: insights from proteomic studies, Proteomics 10 (2010) 4040–4052.

[4] I. Kim, W. Xu, J.C. Reed, Cell death and endoplasmic reticulum stress: disease relevance and therapeutic opportunities, Nat. Rev. Drug Discov. 7 (2008) 1013–1030.

[5] R. Rizzuto, M.R. Duchen, T. Pozzan, Flirting in little space: the ER/mitochondria Ca^{2+} liaison, Sci. STKE 2004 (2004) re1.

[6] H. Ando, K. Kawaai, B. Bonneau, K. Mikoshiba, Remodeling of Ca(2+) signaling in cancer: regulation of inositol 1,4,5-trisphosphate receptors through oncogenes and tumor suppressors, Adv. Biol. Regul. (2017).

[7] Y. Ma, L.M. Hendershot, ER chaperone functions during normal and stress conditions, J. Chem. Neuroanat. 28 (2004) 51–65.

[8] M. Calfon, H. Zeng, F. Urano, J.H. Till, S.R. Hubbard, H.P. Harding, S.G. Clark, D. Ron, IRE1 couples endoplasmic reticulum load to secretory capacity by processing the XBP-1 mRNA, Nature 415 (2002) 92–96.

[9] T.N. Gonzalez, C. Sidrauski, S. Dorfler, P. Walter, Mechanism of non-spliceosomal mRNA splicing in the unfolded protein response pathway, EMBO J. 18 (1999) 3119–3132.
[10] Y. Lu, F.X. Liang, X. Wang, A synthetic biology approach identifies the mammalian UPR RNA ligase RtcB, Mol. Cell 55 (2014) 758–770.
[11] J. Hollien, J.H. Lin, H. Li, N. Stevens, P. Walter, J.S. Weissman, Regulated Ire1-dependent decay of messenger RNAs in mammalian cells, J. Cell Biol. 186 (2009) 323–331.
[12] A.B. Tam, A.C. Koong, M. Niwa, Ire1 has distinct catalytic mechanisms for XBP1/HAC1 splicing and RIDD, Cell Rep. 9 (2014) 850–858.
[13] F. Urano, X. Wang, A. Bertolotti, Y. Zhang, P. Chung, H.P. Harding, D. Ron, Coupling of stress in the ER to activation of JNK protein kinases by transmembrane protein kinase IRE1, Science 287 (2000) 664–666.
[14] H. Nishitoh, A. Matsuzawa, K. Tobiume, K. Saegusa, K. Takeda, K. Inoue, S. Hori, A. Kakizuka, H. Ichijo, ASK1 is essential for endoplasmic reticulum stress-induced neuronal cell death triggered by expanded polyglutamine repeats, Genes Dev. 16 (2002) 1345–1355.
[15] A.B. Tam, E.L. Mercado, A. Hoffmann, M. Niwa, ER stress activates NF-kappaB by integrating functions of basal IKK activity, IRE1 and PERK, PLoS One 7 (2012) e45078.
[16] J. Deng, P.D. Lu, Y. Zhang, D. Scheuner, R.J. Kaufman, N. Sonenberg, H.P. Harding, D. Ron, Translational repression mediates activation of nuclear factor kappa B by phosphorylated translation initiation factor 2, Mol. Cell Biol. 24 (2004) 10161–10168.
[17] S.B. Cullinan, J.A. Diehl, Coordination of ER and oxidative stress signaling: the PERK/Nrf2 signaling pathway, Int. J. Biochem. Cell Biol. 38 (2006) 317–332.
[18] P.C. Zhang, B. McGrath, S.A. Li, A. Frank, F. Zambito, J. Reinert, M. Gannon, K. Ma, K. McNaughton, D.R. Cavener, The PERK eukaryotic initiation factor 2 alpha kinase is required for the development of the skeletal system, postnatal growth, and the function and viability of the pancreas, Mol. Cell Biol. 22 (2002) 3864–3874.
[19] H.P. Harding, Y. Zhang, A. Bertolotti, H. Zeng, D. Ron, Perk is essential for translational regulation and cell survival during the unfolded protein response, Mol. Cell 5 (2000) 897–904.
[20] D. Scheuner, B. Song, E. McEwen, C. Liu, R. Laybutt, P. Gillespie, T. Saunders, S. Bonner-Weir, R.J. Kaufman, Translational control is required for the unfolded protein response and in vivo glucose homeostasis, Mol. Cell 7 (2001) 1165–1176.
[21] J. Ye, R.B. Rawson, R. Komuro, X. Chen, U.P. Dave, R. Prywes, M.S. Brown, J.L. Goldstein, ER stress induces cleavage of membrane-bound ATF6 by the same proteases that process SREBPs, Mol. Cell 6 (2000) 1355–1364.
[22] D.Y. Guan, H. Wang, V.E. Li, Y.Y. Xu, M. Yang, Z.H. Shen, N-glycosylation of ATF6 beta is essential for its proteolytic cleavage and transcriptional repressor function to ATF6 alpha, J. Cell. Biochem. 108 (2009) 825–831.
[23] K. Yamamoto, T. Sato, T. Matsui, M. Sato, T. Okada, H. Yoshida, A. Harada, K. Mori, Transcriptional induction of mammalian ER quality control proteins is mediated by single or combined action of ATF6 alpha and XBP1, Dev. Cell 13 (2007) 365–376.
[24] I. Hirsch, M. Weiwad, E. Prell, D.M. Ferrari, ERp29 deficiency affects sensitivity to apoptosis via impairment of the ATF6-CHOP pathway of stress response, Apoptosis 19 (2014) 801–815.
[25] J. Roth, C. Zuber, Quality control of glycoprotein folding and ERAD: the role of N-glycan handling, EDEM1 and OS-9, Histochem. Cell Biol. 147 (2017) 269–284.

[26] K. Hirao, Y. Natsuka, T. Tamura, I. Wada, D. Morito, S. Natsuka, P. Romero, B. Sleno, L.O. Tremblay, A. Herscovics, K. Nagata, N. Hosokawa, EDEM3, a soluble EDEM homolog, enhances glycoprotein endoplasmic reticulum-associated degradation and mannose trimming, J. Biol. Chem. 281 (2006) 9650–9658.
[27] N. Hosokawa, Y. Kamiya, D. Kamiya, K. Kato, K. Nagata, Human OS-9, a lectin required for glycoprotein endoplasmic reticulum-associated degradation, recognizes mannose-trimmed N-glycans, J. Biol. Chem. 284 (2009) 17061–17068.
[28] N. Hosokawa, L.O. Tremblay, B. Sleno, Y. Kamiya, I. Wada, K. Nagata, K. Kato, A. Herscovics, EDEM1 accelerates the trimming of alpha1,2-linked mannose on the C branch of N-glycans, Glycobiology 20 (2010) 567–575.
[29] B. Mueller, E.J. Klemm, E. Spooner, J.H. Claessen, H.L. Ploegh, SEL1L nucleates a protein complex required for dislocation of misfolded glycoproteins, Proc. Natl. Acad. Sci. USA 105 (2008) 12325–12330.
[30] S.A. Houck, H.Y. Ren, V.J. Madden, J.N. Bonner, M.P. Conlin, J.A. Janovick, P.M. Conn, D.M. Cyr, Quality control autophagy degrades soluble ERAD-resistant conformers of the misfolded membrane protein GnRHR, Mol. Cell 54 (2014) 166–179.
[31] P. Grumati, G. Morozzi, S. Holper, M. Mari, M.I. Harwardt, R. Yan, S. Muller, F. Reggiori, M. Heilemann, I. Dikic, Full length RTN3 regulates turnover of tubular endoplasmic reticulum via selective autophagy, eLife 6 (2017).
[32] A. Khaminets, T. Heinrich, M. Mari, P. Grumati, A.K. Huebner, M. Akutsu, L. Liebmann, A. Stolz, S. Nietzsche, N. Koch, M. Mauthe, I. Katona, B. Qualmann, J. Weis, F. Reggiori, I. Kurth, C.A. Hubner, I. Dikic, Regulation of endoplasmic reticulum turnover by selective autophagy, Nature 522 (2015) 354–358.
[33] N. Mizushima, A dual binding receptor for ER-phagy, Dev. Cell 44 (2018) 133–135.
[34] M.D. Smith, M.E. Harley, A.J. Kemp, J. Wills, M. Lee, M. Arends, A. von Kriegsheim, C. Behrends, S. Wilkinson, CCPG1 is a non-canonical autophagy cargo receptor essential for ER-phagy and pancreatic ER proteostasis, Dev. Cell 44 (2018) 217.
[35] H. Kim, H.C. Tu, D.R. Ren, O. Takeuchi, J.R. Jeffers, G.P. Zambetti, J.J.D. Hsieh, E.H.Y. Cheng, Stepwise activation of BAX and BAK by tBID, BIM, and PUMA initiates mitochondrial apoptosis, Mol. Cell 36 (2009) 487–499.
[36] D.C. Ren, H. Kim, H.C. Tu, T.D. Westergard, J.K. Fisher, J.A. Rubens, S.J. Korsmeyer, J.J.D. Hsieh, E.H.Y. Cheng, The VDAC2-BAK rheostat controls thymocyte survival, Sci. Signal. 2 (2009).
[37] M.A. Carpio, M. Michaud, W.P. Zhou, J.K. Fisher, L.D. Walensky, S.G. Katz, BCL-2 family member BOK promotes apoptosis in response to endoplasmic reticulum stress, Proc. Natl. Acad. Sci. USA 112 (2015) 7201–7206.
[38] J.Z. Li, B. Lee, A.S. Lee, Endoplasmic reticulum stress-induced apoptosis - multiple pathways and activation of p53-up-regulated modulator of apoptosis (PUMA) and NOXA by p53, J. Biol. Chem. 281 (2006) 7260–7270.
[39] J.M. Rodriguez, M.A. Glozak, Y.H. Ma, W.D. Cress, Bok, Bcl-2-related ovarian killer, is cell cycle-regulated and sensitizes to stress-induced apoptosis, J. Biol. Chem. 281 (2006) 22729–22735.
[40] P. Sopha, H.Y. Ren, D.E. Grove, D.M. Cyr, Endoplasmic reticulum stress-induced degradation of DNAJB12 stimulates BOK accumulation and primes cancer cells for apoptosis, J. Biol. Chem. 292 (2017) 11792–11803.
[41] K.D. McCullough, J.L. Martindale, L.O. Klotz, T.Y. Aw, N.J. Holbrook, Gadd153 sensitizes cells to endoplasmic reticulum stress by down-regulating Bcl2 and perturbing the cellular redox state, Mol. Cell Biol. 21 (2001) 1249–1259.

[42] C. Hetz, P. Bernasconi, J. Fisher, A.H. Lee, M.C. Bassik, B. Antonsson, G.S. Brandt, N.N. Iwakoshi, A. Schinzel, L.H. Glimcher, S.J. Korsmeyer, Proapoptotic BAX and BAK modulate the unfolded protein response by a direct interaction with IRE1alpha, Science 312 (2006) 572–576.
[43] D.N. Bronner, B.H. Abuaita, X. Chen, K.A. Fitzgerald, G. Nunez, Y. He, X.M. Yin, M.X. O'Riordan, Endoplasmic reticulum stress activates the inflammasome via NLRP3- and caspase-2-driven mitochondrial damage, Immunity 43 (2015) 451–462.
[44] S. Fulda, Safety and tolerability of TRAIL receptor agonists in cancer treatment, Eur. J. Clin. Pharmacol. 71 (2015) 525–527.
[45] M. Lu, D.A. Lawrence, S. Marsters, D. Acosta-Alvear, P. Kimmig, A.S. Mendez, A.W. Paton, J.C. Paton, P. Walter, A. Ashkenazi, Opposing unfolded-protein-response signals converge on death receptor 5 to control apoptosis, Science 345 (2014) 98–101.
[46] P. Chen, T. Hu, Y. Liang, P. Li, X. Chen, J. Zhang, Y. Ma, Q. Hao, J. Wang, P. Zhang, Y. Zhang, H. Zhao, S. Yang, J. Yu, L.S. Jeong, H. Qi, M. Yang, R.M. Hoffman, Z. Dong, L. Jia, Neddylation inhibition activates the extrinsic apoptosis pathway through ATF4-CHOP-DR5 axis in human esophageal cancer cells, Clin. Cancer Res. 22 (2016) 4145–4157.
[47] H. Yamaguchi, H.G. Wang, CHOP is involved in endoplasmic reticulum stress-induced apoptosis by enhancing DR5 expression in human carcinoma cells, J. Biol. Chem. 279 (2004) 45495–45502.
[48] T.L. Li, L. Su, Y.J. Lei, X.F. Liu, Y.J. Zhang, X.G. Liu, DDIT3 and KAT2A proteins regulate TNFRSF10A and TNFRSF10B expression in endoplasmic reticulum stress-mediated apoptosis in human lung cancer cells, J. Biol. Chem. 290 (2015) 11108–11118.
[49] A. Linkermann, D.R. Green, Necroptosis, New Engl. J. Med. 370 (2014) 455–465.
[50] N. Festjens, T. Vanden Berghe, S. Cornelis, P. Vandenabeele, RIP1, a kinase on the crossroads of a cell's decision to live or die, Cell Death Differ. 14 (2007) 400–410.
[51] D.E. Christofferson, Y. Li, J. Hitomi, W. Zhou, C. Upperman, H. Zhu, S.A. Gerber, S. Gygi, J. Yuan, A novel role for RIP1 kinase in mediating TNF alpha production, Cell Death Dis. 3 (2012).
[52] Q. Luan, L. Jin, C.C. Jiang, K.H. Tay, F. Lai, X.Y. Liu, Y.L. Liu, S.T. Guo, C.Y. Li, X.G. Yan, H.Y. Tseng, X.D. Zhang, RIPK1 regulates survival of human melanoma cells upon endoplasmic reticulum stress through autophagy, Autophagy 11 (2015) 975–994.
[53] S. Saveljeva, S.L. Mc Laughlin, P. Vandenabeele, A. Samali, M.J.M. Bertrand, Endoplasmic reticulum stress induces ligand-independent TNFR1-mediated necroptosis in L929 cells, Cell Death Dis. 6 (2015).
[54] C.L. Leon-Annicchiarico, S. Ramirez-Peinado, D. Dominguez-Villanueva, A. Gonsberg, T.J. Lampidis, C. Munoz-Pinedo, ATF4 mediates necrosis induced by glucose deprivation and apoptosis induced by 2-deoxyglucose in the same cells, FEBS J. 282 (2015) 3647–3658.
[55] Z.P. Xie, D.J. Klionsky, Autophagosome formation: core machinery and adaptations, Nat. Cell Biol. 9 (2007) 1102–1109.
[56] M. Ogata, S.I. Hino, A. Saito, K. Morikawa, S. Kondo, S. Kanemoto, T. Murakami, M. Taniguchi, I. Tanii, K. Yoshinaga, S. Shiosaka, J.A. Hammarback, F. Urano, K. Imaizumi, Autophagy is activated for cell survival after endoplasmic reticulum stress, Mol. Cell Biol. 26 (2006) 9220–9231.

[57] S. Erbil, O. Oral, G. Mitou, C. Kig, E. Durmaz-Timucin, E. Guven-Maiorov, F. Gulacti, G. Gokce, J. Dengjel, O.U. Sezerman, D. Gozuacik, RACK1 is an interaction partner of ATG5 and a novel regulator of autophagy, J. Biol. Chem. 291 (2016) 16753–16765.
[58] Y. Kouroku, E. Fujita, I. Tanida, T. Ueno, A. Isoai, H. Kumagai, S. Ogawa, R.J. Kaufman, E. Kominami, T. Momoi, ER stress (PERK/eIF2 alpha phosphorylation) mediates the polyglutamine-induced LC3 conversion, an essential step for autophagy formation, Cell Death Differ. 14 (2007) 230–239.
[59] X.H. Ma, S.F. Piao, S. Dey, Q. Mcafee, G. Karakousis, J. Villanueva, L.S. Hart, S. Levi, J. Hu, G. Zhang, R. Lazova, V. Klump, J.M. Pawelek, X.W. Xu, W. Xu, L.M. Schuchter, M.A. Davies, M. Herlyn, J. Winkler, C. Koumenis, R.K. Amaravadi, Targeting ER stress-induced autophagy overcomes BRAF inhibitor resistance in melanoma, J. Clin. Invest. 124 (2014) 1406–1417.
[60] M. Sharma, S. Bhattacharyya, K.B. Sharma, S. Chauhan, S. Asthana, M.Z. Abdin, S. Vrati, M. Kalia, Japanese encephalitis virus activates autophagy through XBP1 and ATF6 ER stress sensors in neuronal cells, J. Gen. Virol. 98 (2017) 1027–1039.
[61] N. Rufo, A.D. Garg, P. Agostinis, The unfolded protein response in immunogenic cell death and cancer immunotherapy, Trends Cancer 3 (2017) 643–658.
[62] T. Panaretakis, N. Joza, N. Modjtahedi, A. Tesniere, I. Vitale, M. Durchschlag, G.M. Durchschlag, G.M. Fimia, O. Kepp, M. Piacentini, K.U. Froehlich, P. van Endert, L. Zitvogel, F. Madeo, G. Kroemer, The co-translocation of ERp57 and calreticulin determines the immunogenicity of cell death, Cell Death Differ. 15 (2008) 1499–1509.
[63] T. Panaretakis, O. Kepp, U. Brockmeier, A. Tesniere, A.C. Bjorklund, D.C. Chapman, M. Durchschlag, N. Joza, G. Pierron, P. van Endert, J. Yuan, L. Zitvogel, F. Madeo, D.B. Williams, G. Kroemer, Mechanisms of pre-apoptotic calreticulin exposure in immunogenic cell death, EMBO J. 28 (2009) 578–590.
[64] A.D. Garg, D.V. Krysko, T. Verfaillie, A. Kaczmarek, G.B. Ferreira, T. Marysael, N. Rubio, M. Firczuk, C. Mathieu, A.J.M. Roebroek, W. Annaert, J. Golab, P. de Witte, P. Vandenabeele, P. Agostinis, A novel pathway combining calreticulin exposure and ATP secretion in immunogenic cancer cell death, EMBO J. 31 (2012) 1062–1079.
[65] I. Martins, O. Kepp, F. Schlemmer, S. Adjemian, M. Tailler, S. Shen, M. Michaud, L. Menger, A. Gdoura, N. Tajeddine, A. Tesniere, L. Zitvogel, G. Kroemer, Restoration of the immunogenicity of cisplatin-induced cancer cell death by endoplasmic reticulum stress, Oncogene 30 (2011) 1147–1158.
[66] A.R. van Vliet, F. Giordano, S. Gerlo, I. Segura, S. Van Eygen, G. Molenberghs, S. Rocha, A. Houcine, R. Derua, T. Verfaillie, J. Vangindertael, H. De Keersmaecker, E. Waelkens, J. Tavernier, J. Hofkens, W. Annaert, P. Carmeliet, A. Samali, H. Mizuno, P. Agostinis, The ER stress sensor PERK coordinates ER-plasma membrane contact site formation through interaction with filamin-A and F-actin remodeling, Mol. Cell 65 (2017) 885–899.
[67] L.I. Gold, P. Eggleton, M.T. Sweetwyne, L.B. Van Duyn, M.R. Greives, S.M. Naylor, M. Michalak, J.E. Murphy-Ullrich, Calreticulin: non-endoplasmic reticulum functions in physiology and disease, FASEB J. 24 (2010) 665–683.
[68] S.J. Gardai, K.A. McPhillips, S.C. Frasch, W.J. Janssen, A. Starefeldt, J.E. Murphy-Ullrich, D.L. Bratton, P.A. Oldenborg, M. Michalak, P.M. Henson, Cell-surface calreticulin initiates clearance of viable or apoptotic cells through trans-activation of LRP on the phagocyte, Cell 123 (2005) 321–334.

[69] M.R. Greives, F. Samra, S.C. Pavlides, K.M. Blechman, S.M. Naylor, C.D. Woodrell, C. Cadacio, J.P. Levine, T.A. Asoulin, M. Michalak, S.M. Warren, L.I. Gold, Exogenous calreticulin improves diabetic wound healing, Wound Repair Regen. 21 (2013) 337 (vol. 20, p. 715, 2012).
[70] B.D. Bell, S. Leverrier, B.M. Weist, R.H. Newton, A.F. Arechiga, K.A. Luhrs, N.S. Morrissette, C.M. Walsh, FADD and caspase-8 control the outcome of autophagic signaling in proliferating T cells, Proc. Natl. Acad. Sci. USA 105 (2008) 16677–16682.
[71] W.J. Kaiser, J.W. Upton, A.B. Long, D. Livingston-Rosanoff, L.P. Daley-Bauer, R. Hakem, T. Caspary, E.S. Mocarski, RIP3 mediates the embryonic lethality of caspase-8-deficient mice, Nature 471 (2011) 368–372.
[72] M. You, N. Savaraj, M.T. Kuo, M. Wangpaichitr, J. Varona-Santos, C.J. Wu, D.M. Nguyen, L. Feun, TRAIL induces autophagic protein cleavage through caspase activation in melanoma cell lines under arginine deprivation, Mol. Cell. Biochem. 374 (2013) 181–190.
[73] E.R. Chemaly, L. Troncone, D. Lebeche, SERCA control of cell death and survival, Cell Calcium 69 (2018) 46–61.
[74] T.S. Ma, D.L. Mann, J.H. Lee, G.J. Gallinghouse, SR compartment calcium and cell apoptosis in SERCA overexpression, Cell Calcium 26 (1999) 25–36.
[75] A. Carreras-Sureda, P. Pihan, C. Hetz, Calcium signaling at the endoplasmic reticulum: fine-tuning stress responses, Cell Calcium 70 (2018) 24–31.
[76] T. Hayashi, T.P. Su, Sigma-1 receptor chaperones at the ER-mitochondrion interface regulate Ca(2+) signaling and cell survival, Cell 131 (2007) 596–610.
[77] I. Derler, I. Jardin, C. Romanin, Molecular mechanisms of STIM/Orai communication, Am. J. Physiol. Cell Physiol. 310 (2016) C643–C662.
[78] S.W. Park, Y. Zhou, J. Lee, J. Lee, U. Ozcan, Sarco(endo)plasmic reticulum Ca(2+)–ATPase 2b is a major regulator of endoplasmic reticulum stress and glucose homeostasis in obesity, Proc. Natl. Acad. Sci. USA 107 (2010) 19320–19325.
[79] S. Kang, R. Dahl, W. Hsieh, A. Shin, K.M. Zsebo, C. Buettner, R.J. Hajjar, D. Lebeche, Small molecular allosteric activator of the sarco/endoplasmic reticulum Ca(2+)–ATPase (SERCA) attenuates diabetes and metabolic disorders, J. Biol. Chem. 291 (2016) 5185–5198.
[80] S. Fu, L. Yang, P. Li, O. Hofmann, L. Dicker, W. Hide, X. Lin, S.M. Watkins, A.R. Ivanov, G.S. Hotamisligil, Aberrant lipid metabolism disrupts calcium homeostasis causing liver endoplasmic reticulum stress in obesity, Nature 473 (2011) 528–531.
[81] D.J. Thuerauf, H. Hoover, J. Meller, J. Hernandez, L. Su, C. Andrews, W.H. Dillmann, P.M. McDonough, C.C. Glembotski, Sarco/endoplasmic reticulum calcium ATPase-2 expression is regulated by ATF6 during the endoplasmic reticulum stress response: intracellular signaling of calcium stress in a cardiac myocyte model system, J. Biol. Chem. 276 (2001) 48309–48317.
[82] S. Doroudgar, D.J. Thuerauf, M.C. Marcinko, P.J. Belmont, C.C. Glembotski, Ischemia activates the ATF6 branch of the endoplasmic reticulum stress response, J. Biol. Chem. 284 (2009) 29735–29745.
[83] T.A. Seimon, A. Obstfeld, K.J. Moore, D.T. Golenbock, I. Tabas, Combinatorial pattern recognition receptor signaling alters the balance of life and death in macrophages, Proc. Natl. Acad. Sci. USA 103 (2006) 19794–19799.
[84] G. Li, M. Mongillo, K.T. Chin, H. Harding, D. Ron, A.R. Marks, I. Tabas, Role of ERO1-alpha-mediated stimulation of inositol 1,4,5-triphosphate receptor activity in endoplasmic reticulum stress-induced apoptosis, J. Cell Biol. 186 (2009) 783–792.

[85] L. Scorrano, S.A. Oakes, J.T. Opferman, E.H. Cheng, M.D. Sorcinelli, T. Pozzan, S.J. Korsmeyer, BAX and BAK regulation of endoplasmic reticulum Ca^{2+}: a control point for apoptosis, Science 300 (2003) 135–139.
[86] S.A. Oakes, L. Scorrano, J.T. Opferman, M.C. Bassik, M. Nishino, T. Pozzan, S.J. Korsmeyer, Proapoptotic BAX and BAK regulate the type 1 inositol trisphosphate receptor and calcium leak from the endoplasmic reticulum, Proc. Natl. Acad. Sci. USA 102 (2005) 105–110.
[87] M. Hoyer-Hansen, L. Bastholm, P. Szyniarowski, M. Campanella, G. Szabadkai, T. Farkas, K. Bianchi, N. Fehrenbacher, F. Elling, R. Rizzuto, I.S. Mathiasen, M. Jaattela, Control of macroautophagy by calcium, calmodulin-dependent kinase kinase-beta, and Bcl-2, Mol. Cell 25 (2007) 193–205.
[88] A. Criollo, M.C. Maiuri, E. Tasdemir, I. Vitale, A.A. Fiebig, D. Andrews, J. Molgo, J. Diaz, S. Lavandero, F. Harper, G. Pierron, D. di Stefano, R. Rizzuto, G. Szabadkai, G. Kroemer, Regulation of autophagy by the inositol trisphosphate receptor, Cell Death Differ. 14 (2007) 1029–1039.
[89] F. Chiti, C.M. Dobson, Protein misfolding, functional amyloid, and human disease, Annu. Rev. Biochem. 75 (2006) 333–366.
[90] A. Wimo, M. Guerchet, G.C. Ali, Y.T. Wu, A.M. Prina, B. Winblad, L. Jonsson, Z. Liu, M. Prince, The worldwide costs of dementia 2015 and comparisons with 2010, Alzheimer Dement. 13 (2017) 1–7.
[91] U. Unterberger, R. Hoftberger, E. Gelpi, H. Flicker, H. Budka, T. Voigtlander, Endoplasmic reticulum stress features are prominent in Alzheimer disease but not in prion diseases in vivo, J. Neuropathol. Exp. Neurol. 65 (2006) 348–357.
[92] L.D. Stutzbach, S.X. Xie, A.C. Naj, R. Albin, S. Gilman, V.M.Y. Lee, J.Q. Trojanowski, B. Devlin, G.D. Schellenberg, P.G.S. Grp, The unfolded protein response is activated in disease-affected brain regions in progressive supranuclear palsy and Alzheimer's disease, Acta Neuropathol. Commun. 1 (2013).
[93] T. Katayama, K. Imaizumi, N. Sato, K. Miyoshi, T. Kudo, J. Hitomi, T. Morihara, T. Yoneda, F. Gomi, Y. Mori, Y. Nakano, J. Takeda, T. Tsuda, Y. Itoyama, O. Murayama, A. Takashima, P. St George-Hyslop, M. Takeda, M. Tohyama, Presenilin-1 mutations downregulate the signalling pathway of the unfolded-protein response, Nat. Cell Biol. 1 (1999) 479–485.
[94] F. Terro, C. Czech, F. Esclaire, W. Elyaman, C. Yardin, M.C. Baclet, N. Touchet, G. Tremp, L. Pradier, J. Hugon, Neurons overexpressing mutant presenilin-1 are more sensitive to apoptosis induced by endoplasmic reticulum-Golgi stress, J. Neurosci. Res. 69 (2002) 530–539.
[95] J.J.M. Hoozemans, E.S. van Haastert, D.A.T. Nijholt, A.J.M. Rozemuller, W. Scheper, Activation of the unfolded protein response is an early event in Alzheimer's and Parkinson's disease, Neurodegener. Dis. 10 (2012) 212–215.
[96] N. Wei, L.Q. Zhu, D. Liu, ATF4: a novel potential therapeutic target for Alzheimer's disease, Mol. Neurobiol. 52 (2015) 1765–1770.
[97] J.H. Lee, S.M. Won, J. Suh, S.J. Son, G.J. Moon, U.J. Park, B.J. Gwag, Induction of the unfolded protein response and cell death pathway in Alzheimer's disease, but not in aged Tg2576 mice, Exp. Mol. Med. 42 (2010) 386–394.
[98] S.O. Yoon, D.J. Park, J.C. Ryu, H.G. Ozer, C. Tep, Y.J. Shin, T.H. Lim, L. Pastorino, A.J. Kunwar, J.C. Walton, A.H. Nagahara, K.P. Lu, R.J. Nelson, M.H. Tuszynski, K. Huang, JNK3 perpetuates metabolic stress induced by A beta peptides, Neuron 75 (2012) 824–837.

[99] J. Baleriola, C.A. Walker, Y.Y. Jean, J.F. Crary, C.M. Troy, P.L. Nagy, U. Hengst, Axonally synthesized ATF4 transmits a neurodegenerative signal across brain regions, Cell 158 (2014) 1159–1172.
[100] M. Torres, J.M. Matamala, C. Duran-Aniotz, V.H. Cornejo, A. Foley, C. Hetz, ER stress signaling and neurodegeneration: at the intersection between Alzheimer's disease and Prion-related disorders, Virus Res. 207 (2015) 69–75.
[101] N. Thayanidhi, J.R. Helm, D.C. Nycz, M. Bentley, Y. Liang, J.C. Hay, Alpha-synuclein delays endoplasmic reticulum (ER)-to-Golgi transport in mammalian cells by antagonizing ER/Golgi SNAREs, Mol. Biol. Cell 21 (2010) 1850–1863.
[102] G. Mercado, V. Castillo, P. Soto, A. Sidhu, ER stress and Parkinson's disease: pathological inputs that converge into the secretory pathway, Brain Res. 1648 (2016) 626–632.
[103] S. Tsujii, M. Ishisaka, H. Hara, Modulation of endoplasmic reticulum stress in Parkinson's disease, Eur. J. Pharmacol. 765 (2015) 154–156.
[104] R.M. Silva, V. Ries, T.F. Oo, O. Yarygina, V. Jackson-Lewis, E.J. Ryu, P.D. Lu, S.J. Marciniak, D. Ron, S. Przedborski, N. Kholodilov, L.A. Greene, R.E. Burke, CHOP/GADD153 is a mediator of apoptotic death in substantia nigra dopamine neurons in an in vivo neurotoxin model of parkinsonism, J. Neurochem. 95 (2005) 974–986.
[105] J.C. Gully, V.G. Sergeyev, Y. Bhootada, H. Mendez-Gomez, C.A. Meyers, S. Zolotukhin, M.S. Gorbatyuk, O.S. Gorbatyuk, Up-regulation of activating transcription factor 4 induces severe loss of dopamine nigral neurons in a rat model of Parkinson's disease, Neurosci. Lett. 627 (2016) 36–41.
[106] P. Valdes, G. Mercado, R.L. Vidal, C. Molina, G. Parsons, F.A. Court, A. Martinez, D. Galleguillos, D. Armentano, B.L. Schneider, C. Hetz, Control of dopaminergic neuron survival by the unfolded protein response transcription factor XBP1, Proc. Natl. Acad. Sci. USA 111 (2014) 6804–6809.
[107] N. Egawa, K. Yamamoto, H. Inoue, R. Hikawa, K. Nishi, K. Mori, R. Takahashi, The endoplasmic reticulum stress sensor, ATF6 alpha, protects against neurotoxin-induced dopaminergic neuronal death, J. Biol. Chem. 286 (2011) 7947–7957.
[108] H. Nishitoh, H. Kadowaki, A. Nagai, T. Maruyama, T. Yokota, H. Fukutomi, T. Noguchi, A. Matsuzawa, K. Takeda, H. Ichijo, ALS-linked mutant SOD1 induces ER stress- and ASK1-dependent motor neuron death by targeting Derlin-1, Genes Dev. 22 (2008) 1451–1464.
[109] A. Kumar, Aakriti, V. Gupta, A review on animal models of stroke: an update, Brain Res. Bull. 122 (2016) 35–44.
[110] T. Gotoh, M. Mori, Nitric oxide and endoplasmic reticulum stress, Arterioscler. Thromb. Vasc. Biol. 26 (2006) 1439–1446.
[111] G. Begum, D. Kintner, Y. Liu, S.W. Cramer, D.D. Sun, DHA inhibits ER Ca^{2+} release and ER stress in astrocytes following in vitro ischemia, J. Neurochem. 120 (2012) 622–630.
[112] R. Kumar, G.S. Krause, H. Yoshida, K. Mori, D.J. DeGracia, Dysfunction of the unfolded protein response during global brain ischemia and reperfusion, J. Cereb. Blood Flow Metab. 23 (2003) 462–471.
[113] V.P. Nakka, A. Gusain, R. Raghubir, Endoplasmic reticulum stress plays critical role in brain damage after cerebral ischemia/reperfusion in rats, Neurotox. Res. 17 (2010) 189–202.
[114] I.L. Llorente, T.C. Burgin, D. Perez-Rodriguez, B. Martinez-Villayandre, C.C. Perez-Garcia, A. Fernandez-Lopez, Unfolded protein response to global ischemia following 48 h or reperfusion in the rat brain: the effect of age and meloxicam, J. Neurochem. 127 (2013) 701–710.

[115] S. Tajiri, S. Oyadomari, S. Yano, M. Morioka, T. Gotoh, J.I. Hamada, Y. Ushio, M. Mori, Ischemia-induced neuronal cell death is mediated by the endoplasmic reticulum stress pathway involving CHOP, Cell Death Differ. 11 (2004) 403–415.
[116] Y. Yin, G. Sun, E. Li, K. Kiselyov, D.D. Sun, ER stress and impaired autophagy flux in neuronal degeneration and brain injury, Ageing Res. Rev. 34 (2017) 3–14.
[117] M.M. Niu, X.H. Dai, W. Zou, X.P. Yu, W. Teng, Q.X. Chen, X.W. Sun, W.W. Yu, H.H. Ma, P. Liu, Autophagy, endoplasmic reticulum stress and the unfolded protein response in intracerebral hemorrhage, Transl. Neurosci. 8 (2017) 37–48.
[118] S. Carloni, S. Girelli, C. Scopa, G. Buonocore, M. Longini, W. Balduini, Activation of autophagy and Akt/CREB signaling play an equivalent role in the neuroprotective effect of rapamycin in neonatal hypoxia-ischemia, Autophagy 6 (2010) 366–377.
[119] Y.D. Wen, R. Sheng, L.S. Zhang, R. Han, X. Zhang, X.D. Zhang, F. Han, K. Fukunaga, Z.H. Qin, Neuronal injury in rat model of permanent focal cerebral ischemia is associated with activation of autophagic and lysosomal pathways, Autophagy 4 (2008) 762–769.
[120] K.R. Merikangas, R. Jin, J.P. He, R.C. Kessler, S. Lee, N.A. Sampson, M.C. Viana, L.H. Andrade, C. Hu, E.G. Karam, M. Ladea, M.E. Medina-Mora, Y. Ono, J. Posada-Villa, R. Sagar, J.E. Wells, Z. Zarkov, Prevalence and correlates of bipolar spectrum disorder in the world mental health survey initiative, Arch. Gen. Psychiatr. 68 (2011) 241–251.
[121] Y. Kim, R. Santos, F.H. Gage, M.C. Marchetto, Molecular mechanisms of bipolar disorder: progress made and future challenges, Front. Cell. Neurosci. 11 (2017) 30.
[122] C. Kakiuchi, K. Iwamoto, M. Ishiwata, M. Bundo, T. Kasahara, I. Kusumi, T. Tsujita, Y. Okazaki, S. Nanko, H. Kunugi, T. Sasaki, T. Kato, Impaired feedback regulation of XBP1 as a genetic risk factor for bipolar disorder, Nat. Genet. 35 (2003) 171–175.
[123] C. Kakiuchi, M. Ishiwata, S. Nanko, H. Kunugi, Y. Minabe, K. Nakamura, N. Mori, K. Fujii, T. Umekage, M. Tochigi, K. Kohda, T. Sasaki, K. Yamada, T. Yoshikawa, T. Kato, Functional polymorphisms of HSPA5: possible association with bipolar disorder, Biochem. Biophys. Res. Commun. 336 (2005) 1136–1143.
[124] B. Pfaffenseller, B. Wollenhaupt-Aguiar, G.R. Fries, G.D. Colpo, R.K. Burque, G. Bristot, P. Ferrari, K.M. Cereser, A.R. Rosa, F. Klamt, F. Kapczinski, Impaired endoplasmic reticulum stress response in bipolar disorder: cellular evidence of illness progression, Int. J. Neuropsychopharmacol. 17 (2014) 1453–1463.
[125] M.S. Breen, C.H. White, T. Shekhtman, K. Lin, D. Looney, C.H. Woelk, J.R. Kelsoe, Lithium-responsive genes and gene networks in bipolar disorder patient-derived lymphoblastoid cell lines, Pharmacogenom. J. 16 (2016) 446–453.
[126] C.D. Bown, J.F. Wang, L.T. Young, Increased expression of endoplasmic reticulum stress proteins following chronic valproate treatment of rat C6 glioma cells, Neuropharmacology 39 (2000) 2162–2169.
[127] A.J. Kim, Y. Shi, R.C. Austin, G.H. Werstuck, Valproate protects cells from ER stress-induced lipid accumulation and apoptosis by inhibiting glycogen synthase kinase-3, J. Cell Sci. 118 (2005) 89–99.
[128] T. Hiroi, H. Wei, C. Hough, P. Leeds, D.M. Chuang, Protracted lithium treatment protects against the ER stress elicited by thapsigargin in rat PC12 cells: roles of intracellular calcium, GRP78 and Bcl-2, Pharmacogenom. J. 5 (2005) 102–111.
[129] E. Szegezdi, A. Duffy, M.E. O'Mahoney, S.E. Logue, L.A. Mylotte, T. O'Brien, A. Samali, ER stress contributes to ischemia-induced cardiomyocyte apoptosis, Biochem. Biophys. Res. Commun. 349 (2006) 1406–1411.

[130] A. Azfer, J. Niu, L.M. Rogers, F.M. Adamski, P.E. Kolattukudy, Activation of endoplasmic reticulum stress response during the development of ischemic heart disease, Am. J. Physiol. Heart Circ. Physiol. 291 (2006) H1411–H1420.
[131] S. Wang, P. Binder, Q. Fang, Z. Wang, W. Xiao, W. Liu, X. Wang, Endoplasmic reticulum stress in the heart: insights into mechanisms and drug targets, Br. J. Pharmacol. (2017).
[132] Z.V. Wang, Y. Deng, N. Gao, Z. Pedrozo, D.L. Li, C.R. Morales, A. Criollo, X. Luo, W. Tan, N. Jiang, M.A. Lehrman, B.A. Rothermel, A.H. Lee, S. Lavandero, P.P.A. Mammen, A. Ferdous, T.G. Gillette, P.E. Scherer, J.A. Hill, Spliced X-box binding protein 1 couples the unfolded protein response to hexosamine biosynthetic pathway, Cell 156 (2014) 1179–1192.
[133] K. Terai, Y. Hiramoto, M. Masaki, S. Sugiyama, T. Kuroda, M. Hori, I. Kawase, H. Hirota, AMP-activated protein kinase protects cardiomyocytes against hypoxic injury through attenuation of endoplasmic reticulum stress, Mol. Cell Biol. 25 (2005) 9554–9575.
[134] D.J. Thuerauf, M. Marcinko, N. Gude, M. Rubio, M.A. Sussman, C.C. Glembotski, Activation of the unfolded protein response in infarcted mouse heart and hypoxic cultured cardiac myocytes, Circ. Res. 99 (2006) 275–282.
[135] J.J. Martindale, R. Fernandez, D. Thuerauf, R. Whittaker, N. Gude, M.A. Sussman, C.C. Glembotski, Endoplasmic reticulum stress gene induction and protection from ischemia/reperfusion injury in the hearts of transgenic mice with a tamoxifen-regulated form of ATF6, Circ. Res. 98 (2006) 1186–1193.
[136] P.J. Belmont, W.J. Chen, M.N. San Pedro, D.J. Thuerauf, N. Gellings Lowe, N. Gude, B. Hilton, R. Wolkowicz, M.A. Sussman, C.C. Glembotski, Roles for endoplasmic reticulum-associated degradation and the novel endoplasmic reticulum stress response gene Derlin-3 in the ischemic heart, Circ. Res. 106 (2010) 307–316.
[137] G. Gao, A. Xie, J. Zhang, A.M. Herman, E.M. Jeong, L. Gu, M. Liu, K.C. Yang, T.J. Kamp, S.C. Dudley, Unfolded protein response regulates cardiac sodium current in systolic human heart failure, Circ. Arrhythm Electrophysiol. 6 (2013) 1018–1024.
[138] A. Mercier, R. Clement, T. Harnois, N. Bourmeyster, P. Bois, A. Chatelier, Na(v)1.5 channels can reach the plasma membrane through distinct N-glycosylation states, Biochim. Biophys. Acta 1850 (2015) 1215–1223.
[139] H. Hamada, M. Suzuki, S. Yuasa, N. Mimura, N. Shinozuka, Y. Takada, M. Suzuki, T. Nishino, H. Nakaya, H. Koseki, T. Aoe, Dilated cardiomyopathy caused by aberrant endoplasmic reticulum quality control in mutant KDEL receptor transgenic mice, Mol. Cell Biol. 24 (2004) 8007–8017.
[140] K. Okada, T. Minamino, Y. Tsukamoto, Y. Liao, O. Tsukamoto, S. Takashima, A. Hirata, M. Fujita, Y. Nagamachi, T. Nakatani, C. Yutani, K. Ozawa, S. Ogawa, H. Tomoike, M. Hori, M. Kitakaze, Prolonged endoplasmic reticulum stress in hypertrophic and failing heart after aortic constriction: possible contribution of endoplasmic reticulum stress to cardiac myocyte apoptosis, Circulation 110 (2004) 705–712.
[141] T. Sawada, T. Minamino, H.Y. Fu, M. Asai, K. Okuda, T. Isomura, S. Yamazaki, Y. Asano, K. Okada, O. Tsukamoto, S. Sanada, H. Asanuma, M. Asakura, S. Takashima, M. Kitakaze, I. Komuro, X-box binding protein 1 regulates brain natriuretic peptide through a novel AP1/CRE-like element in cardiomyocytes, J. Mol. Cell. Cardiol. 48 (2010) 1280–1289.
[142] Q. Duan, C. Chen, L. Yang, N. Li, W. Gong, S. Li, D.W. Wang, MicroRNA regulation of unfolded protein response transcription factor XBP1 in the progression of cardiac hypertrophy and heart failure in vivo, J. Transl. Med. 13 (2015) 363.

[143] R. Ghosh, K.L. Lipson, K.E. Sargent, A.M. Mercurio, J.S. Hunt, D. Ron, F. Urano, Transcriptional regulation of VEGF-A by the unfolded protein response pathway, PLoS One 5 (2010) e9575.

[144] D. Mekahli, G. Bultynck, J.B. Parys, H. De Smedt, L. Missiaen, Endoplasmic-reticulum calcium depletion and disease, Cold Spring Harb. Perspect. Biol. 3 (2011).

[145] H.Y. Fu, T. Minamino, O. Tsukamoto, Y. Asano, T. Isomura, M. Kitakaze, I. Komuro, Ablation of C/EBP homologous protein attenuates ER-mediated apoptosis and cardiac dysfunction induced by pressure overload, J. Card. Fail. 16 (2010) S145–S146.

[146] H.P. Harding, I. Novoa, Y.H. Zhang, H.Q. Zeng, R. Wek, M. Schapira, D. Ron, Regulated translation initiation controls stress-induced gene expression in mammalian cells, Mol. Cell 6 (2000) 1099–1108.

[147] E. Sozen, N.K. Ozer, Impact of high cholesterol and endoplasmic reticulum stress on metabolic diseases: an updated mini-review, Redox Biol. 12 (2017) 456–461.

[148] E. Araki, S. Oyadomari, M. Mori, Endoplasmic reticulum stress and diabetes mellitus, Intern. Med. 42 (2003) 7–14.

[149] H.P. Harding, H. Zeng, Y. Zhang, R. Jungries, P. Chung, H. Plesken, D.D. Sabatini, D. Ron, Diabetes mellitus and exocrine pancreatic dysfunction in perk−/− mice reveals a role for translational control in secretory cell survival, Mol. Cell 7 (2001) 1153–1163.

[150] M. Delepine, M. Nicolino, T. Barrett, M. Golamaully, G.M. Lathrop, C. Julier, EIF2AK3, encoding translation initiation factor 2-alpha kinase 3, is mutated in patients with Wolcott-Rallison syndrome, Nat. Genet. 25 (2000) 406–409.

[151] C.M. Thornton, D.J. Carson, F.J. Stewart, Autopsy findings in the Wolcott-Rallison syndrome, Pediatr. Pathol. Lab. Med. 17 (1997) 487–496.

[152] S. Oyadomari, A. Koizumi, K. Takeda, T. Gotoh, S. Akira, E. Araki, M. Mori, Targeted disruption of the Chop gene delays endoplasmic reticulum stress-mediated diabetes, J. Clin. Invest. 109 (2002) 525–532.

[153] U. Ozcan, Q. Cao, E. Yilmaz, A.H. Lee, N.N. Iwakoshi, E. Ozdelen, G. Tuncman, C. Gorgun, L.H. Glimcher, G.S. Hotamisligil, Endoplasmic reticulum stress links obesity, insulin action, and type 2 diabetes, Science 306 (2004) 457–461.

[154] D. Scheuner, D.V. Mierde, B. Song, D. Flamez, J.W.M. Creemers, K. Tsukamoto, M. Ribick, F.C. Schuit, R.J. Kaufman, Control of mRNA translation preserves endoplasmic reticulum function in beta cells and maintains glucose homeostasis, Nat. Med. 11 (2005) 757–764.

[155] S. Oyadomari, H.P. Harding, Y.H. Zhang, M. Oyadomari, D. Ron, Dephosphorylation of translation initiation factor 2 alpha enhances glucose tolerance and attenuates hepatosteatosis in mice, Cell Metabol. 7 (2008) 520–532.

[156] F. Engin, T. Nguyen, A. Yermalovich, G.S. Hotamisligil, Aberrant islet unfolded protein response in type 2 diabetes, Sci. Rep. 4 (2014) 4054.

[157] Y. Wang, G. Li, J. Goode, J.C. Paz, K. Ouyang, R. Screaton, W.H. Fischer, J. Chen, I. Tabas, M. Montminy, Inositol-1,4,5-trisphosphate receptor regulates hepatic gluconeogenesis in fasting and diabetes, Nature 485 (2012) 128–132.

[158] L. Ozcan, D.S. Ghorpade, Z. Zheng, J.C. de Souza, K. Chen, M. Bessler, M. Bagloo, B. Schrope, R. Pestell, I. Tabas, Hepatocyte DACH1 is increased in obesity via nuclear exclusion of HDAC4 and promotes hepatic insulin resistance, Cell Rep. 15 (2016) 2214–2225.

[159] M. Abu-Farha, A. Tiss, J. Abubaker, A. Khadir, F. Al-Ghimlas, I. Al-Khairi, E. Baturcam, P. Cherian, N. Elkum, M. Hammad, J. John, S. Kavalakatt, S. Warsame, K. Behbehani, S. Dermime, M. Dehbi, Proteomics analysis of human obesity reveals the epigenetic factor HDAC4 as a potential target for obesity, PLoS One 8 (2013) e75342.

[160] J. Backs, K. Song, S. Bezprozvannaya, S. Chang, E.N. Olson, CaM kinase II selectively signals to histone deacetylase 4 during cardiomyocyte hypertrophy, J. Clin. Invest. 116 (2006) 1853–1864.
[161] S.C. Bodine, L.M. Baehr, Skeletal muscle atrophy and the E3 ubiquitin ligases MuRF1 and MAFbx/atrogin-1, Am. J. Physiol. Endocrinol. Metab. 307 (2014) E469–E484.
[162] K.R. Bohnert, J.D. McMillan, A. Kumar, Emerging roles of ER stress and unfolded protein response pathways in skeletal muscle health and disease, J. Cell. Physiol. 233 (2018) 67–78.
[163] S.C. Bodine, E. Latres, S. Baumhueter, V.K. Lai, L. Nunez, B.A. Clarke, W.T. Poueymirou, F.J. Panaro, E. Na, K. Dharmarajan, Z.Q. Pan, D.M. Valenzuela, T.M. DeChiara, T.N. Stitt, G.D. Yancopoulos, D.J. Glass, Identification of ubiquitin ligases required for skeletal muscle atrophy, Science 294 (2001) 1704–1708.
[164] S. Cohen, J.J. Brault, S.P. Gygi, D.J. Glass, D.M. Valenzuela, C. Gartner, E. Latres, A.L. Goldberg, During muscle atrophy, thick, but not thin, filament components are degraded by MuRF1-dependent ubiquitylation, J. Cell Biol. 185 (2009) 1083–1095.
[165] L.M. Baehr, M. Tunzi, S.C. Bodine, Muscle hypertrophy is associated with increases in proteasome activity that is independent of MuRF1 and MAFbx expression, Front. Physiol. 5 (2014).
[166] P.K. Paul, S. Bhatnagar, V. Mishra, S. Srivastava, B.G. Darnay, Y. Choi, A. Kumar, The E3 ubiquitin ligase TRAF6 intercedes in starvation-induced skeletal muscle atrophy through multiple mechanisms, Mol. Cell Biol. 32 (2012) 1248–1259.
[167] P.K. Paul, S.K. Gupta, S. Bhatnagar, S.K. Panguluri, B.G. Darnay, Y. Choi, A. Kumar, Targeted ablation of TRAF6 inhibits skeletal muscle wasting in mice, J. Cell Biol. 191 (2010) 1395–1411.
[168] S. Carnio, F. LoVerso, M.A. Baraibar, E. Longa, M.M. Khan, M. Maffei, M. Reischl, M. Canepari, S. Loefler, H. Kern, B. Blaauw, B. Friguet, R. Bottinelli, R. Rudolf, M. Sandri, Autophagy impairment in muscle induces neuromuscular junction degeneration and precocious aging, Cell Rep. 8 (2014) 1509–1521.
[169] F. Penna, D. Costamagna, F. Pin, A. Camperi, A. Fanzani, E.M. Chiarpotto, G. Cavallini, G. Bonelli, F.M. Baccino, P. Costelli, Autophagic degradation contributes to muscle wasting in cancer cachexia, Am. J. Pathol. 182 (2013) 1367–1378.
[170] Z.G. Yu, A.M. Wang, H. Adachi, M. Katsuno, G. Sobue, Z.Y. Yue, D.M. Robins, A.P. Lieberman, Macroautophagy is regulated by the UPR-mediator CHOP and accentuates the phenotype of SBMA mice, PLoS Genet. 7 (2011).
[171] J.R. Cubillos-Ruiz, S.E. Bettigole, L.H. Glimcher, Tumorigenic and immunosuppressive effects of endoplasmic reticulum stress in cancer, Cell 168 (2017) 692–706.
[172] R.D. Guzy, B. Hoyos, E. Robin, H. Chen, L.P. Liu, K.D. Mansfield, M.C. Simon, U. Hammerling, P.T. Schumacker, Mitochondrial complex III is required for hypoxia-induced ROS production and cellular oxygen sensing, Cell Metabol. 1 (2005) 401–408.
[173] J.R. Cubillos-Ruiz, P.C. Silberman, M.R. Rutkowski, S. Chopra, A. Perales-Puchalt, M. Song, S. Zhang, S.E. Bettigole, D. Gupta, K. Holcomb, L.H. Ellenson, T. Caputo, A.H. Lee, J.R. Conejo-Garcia, L.H. Glimcher, ER stress sensor XBP1 controls anti-tumor immunity by disrupting dendritic cell homeostasis, Cell 161 (2015) 1527–1538.
[174] A.L. Huber, J. Lebeau, P. Guillaumot, V. Petrilli, M. Malek, J. Chilloux, F. Fauvet, L. Payen, A. Kfoury, T. Renno, E. Chevet, S.N. Manie, p58(IPK)-mediated attenuation of the proapoptotic PERK-CHOP pathway allows malignant progression upon low glucose, Mol. Cell 49 (2013) 1049–1059.

[175] N. Sahu, D. Dela Cruz, M. Gao, W. Sandoval, P.M. Haverty, J.F. Liu, J.P. Stephan, B. Haley, M. Classon, G. Hatzivassiliou, J. Settleman, Proline starvation induces unresolved ER stress and hinders mTORC1-dependent tumorigenesis, Cell Metabol. 24 (2016) 753–761.
[176] R.B. Ferreira, M. Wang, M.E. Law, B.J. Davis, A.N. Bartley, P.J. Higgins, M.S. Kilberg, K.E. Santostefano, N. Terada, C.D. Heldermon, R.K. Castellano, B.K. Law, Disulfide bond disrupting agents activate the unfolded protein response in EGFR- and HER2-positive breast tumor cells, Oncotarget 8 (2017) 28971–28989.
[177] N. Singh, R. Joshi, K. Komurov, HER2-mTOR signaling-driven breast cancer cells require ER-associated degradation to survive, Sci. Signal. 8 (2015) ra52.
[178] J.J. Ventura, A. Hubner, C. Zhang, R.A. Flavell, K.M. Shokat, R.J. Davis, Chemical genetic analysis of the time course of signal transduction by JNK, Mol. Cell 21 (2006) 701–710.
[179] S.Y. Shin, J.M. Lee, M.S. Lee, D. Koh, H. Jung, Y. Lim, Y.H. Lee, Targeting cancer cells via the reactive oxygen species-mediated unfolded protein response with a novel synthetic polyphenol conjugate, Clin. Cancer Res. 20 (2014) 4302–4313.
[180] T.R. Su, F.J. Tsai, J.J. Lin, H.H. Huang, C.C. Chiu, J.H. Su, Y.T. Yang, J.Y. Chen, B.S. Wong, Y.J. Wu, Induction of apoptosis by 11-dehydrosinulariolide via mitochondrial dysregulation and ER stress pathways in human melanoma cells, Mar. Drugs 10 (2012) 1883–1898.
[181] K. Imaizumi, Autophagy is activated for cell survival after ER stress, J. Pharmacol. Sci. 103 (2007) 45.
[182] M. Cerezo, A. Lehraiki, A. Millet, F. Rouaud, M. Plaisant, E. Jaune, T. Botton, C. Ronco, P. Abbe, H. Amdouni, T. Passeron, V. Hofman, B. Mograbi, A.S. Dabert-Gay, D. Debayle, D. Alcor, N. Rabhi, J.S. Annicotte, L. Heliot, M. Gonzalez-Pisfil, C. Robert, S. Morera, A. Vigouroux, P. Gual, M.M.U. Ali, C. Bertolotto, P. Hofman, R. Ballotti, R. Benhida, S. Rocchi, Compounds triggering ER stress exert anti-melanoma effects and overcome BRAF inhibitor resistance, Cancer Cell 30 (2016) 183.
[183] A. Rivas, R.L. Vidal, C. Hetz, Targeting the unfolded protein response for disease intervention, Expert Opin. Ther. Targets 19 (2015) 1203–1218.
[184] C.D. Bown, J.F. Wang, B. Chen, L.T. Young, Regulation of ER stress proteins by valproate: therapeutic implications, Bipolar Disord. 4 (2002) 145–151.
[185] L. Shao, X. Sun, L. Xu, L.T. Young, J.F. Wang, Mood stabilizing drug lithium increases expression of endoplasmic reticulum stress proteins in primary cultured rat cerebral cortical cells, Life Sci. 78 (2006) 1317–1323.
[186] Y. Oida, J. Hamanaka, K. Hyakkoku, M. Shimazawa, T. Kudo, K. Imaizumi, T. Yasuda, H. Hara, Post-treatment of a BiP inducer prevents cell death after middle cerebral artery occlusion in mice, Neurosci. Lett. 484 (2010) 43–46.
[187] T. Kudo, S. Kanemoto, H. Hara, N. Morimoto, T. Morihara, R. Kimura, T. Tabira, K. Imaizumi, M. Takeda, A molecular chaperone inducer protects neurons from ER stress, Cell Death Differ. 15 (2008) 364–375.
[188] B.G. Hoffstrom, A. Kaplan, R. Letso, R.S. Schmid, G.J. Turmel, D.C. Lo, B.R. Stockwell, Inhibitors of protein disulfide isomerase suppress apoptosis induced by misfolded proteins, Nat. Chem. Biol. 6 (2010) 900–906.
[189] R.L. Vidal, S. Matus, L. Bargsted, C. Hetz, Targeting autophagy in neurodegenerative diseases, Trends Pharmacol. Sci. 35 (2014) 583–591.
[190] T. Hara, K. Nakamura, M. Matsui, A. Yamamoto, Y. Nakahara, R. Suzuki-Migishima, M. Yokoyama, K. Mishima, I. Saito, H. Okano, N. Mizushima, Suppression of basal autophagy in neural cells causes neurodegenerative disease in mice, Nature 441 (2006) 885–889.

[191] M. Komatsu, S. Waguri, T. Chiba, S. Murata, J. Iwata, I. Tanida, T. Ueno, M. Koike, Y. Uchiyama, E. Kominami, K. Tanaka, Loss of autophagy in the central nervous system causes neurodegeneration in mice, Nature 441 (2006) 880–884.
[192] F. Fecto, J. Yan, S.P. Vemula, E. Liu, Y. Yang, W. Chen, J.G. Zheng, Y. Shi, N. Siddique, H. Arrat, S. Donkervoort, S. Ajroud-Driss, R.L. Sufit, S.L. Heller, H.X. Deng, T. Siddique, SQSTM1 mutations in familial and sporadic amyotrophic lateral sclerosis, Arch. Neurol. 68 (2011) 1440–1446.
[193] J.H. Lee, W.H. Yu, A. Kumar, S. Lee, P.S. Mohan, C.M. Peterhoff, D.M. Wolfe, M. Martinez-Vicente, A.C. Massey, G. Sovak, Y. Uchiyama, D. Westaway, A.M. Cuervo, R.A. Nixon, Lysosomal proteolysis and autophagy require presenilin 1 and are disrupted by Alzheimer-related PS1 mutations, Cell 141 (2010) 1146–1158.
[194] F.M. Menzies, K. Moreau, D.C. Rubinsztein, Protein misfolding disorders and macroautophagy, Curr. Opin. Cell Biol. 23 (2011) 190–197.
[195] V. Schaeffer, I. Lavenir, S. Ozcelik, M. Tolnay, D.T. Winkler, M. Goedert, Stimulation of autophagy reduces neurodegeneration in a mouse model of human tauopathy, Brain 135 (2012) 2169–2177.
[196] M. Tanaka, Y. Machida, S.Y. Niu, T. Ikeda, N.R. Jana, H. Doi, M. Kurosawa, M. Nekooki, N. Nukina, Trehalose alleviates polyglutamine-mediated pathology in a mouse model of Huntington disease, Nat. Med. 10 (2004) 148–154.
[197] J. Perucho, M.J. Casarejos, A. Gomez, R.M. Solano, J.G. de Yebenes, M.A. Mena, Trehalose protects from aggravation of amyloid pathology induced by isoflurane anesthesia in APP(swe) mutant mice, Curr. Alzheimer Res. 9 (2012) 334–343.
[198] J.A. Rodriguez-Navarro, L. Rodriguez, M.J. Casarejos, R.M. Solano, A. Gomez, J. Perucho, A.M. Cuervo, J. de Yebenes, M.A. Mena, Trehalose ameliorates dopaminergic and tau pathology in parkin deleted/tau overexpressing mice through autophagy activation, Neurobiol. Dis. 39 (2010) 423–438.
[199] K. Castillo, M. Nassif, V. Valenzuela, F. Rojas, S. Matus, G. Mercado, F.A. Court, B. van Zundert, C. Hetz, Trehalose delays the progression of amyotrophic lateral sclerosis by enhancing autophagy in motoneurons, Autophagy 9 (2013) 1308–1320.
[200] R.L. Vidal, C. Hetz, Unspliced XBP1 controls autophagy through FoxO1, Cell Res. 23 (2013) 463–464.
[201] S. Sarkar, E.O. Perlstein, S. Imarisio, S. Pineau, A. Cordenier, R.L. Maglathlin, J.A. Webster, T.A. Lewis, C.J. O'Kane, S.L. Schreiber, D.C. Rubinsztein, Small molecules enhance autophagy and reduce toxicity in Huntington's disease models, Nat. Chem. Biol. 3 (2007) 331–338.
[202] L.H. Zhang, J. Yu, H.L. Pan, P. Hu, Y. Hao, W.Q. Cai, H. Zhu, A.D. Yu, X. Xie, D.W. Ma, J.Y. Yuan, Small molecule regulators of autophagy identified by an image-based high-throughput screen, Proc. Natl. Acad. Sci. USA 104 (2007) 19023–19028.
[203] M. Boyce, K.F. Bryant, C. Jousse, K. Long, H.P. Harding, D. Scheuner, R.J. Kaufman, D.W. Ma, D.M. Coen, D. Ron, J.Y. Yuan, A selective inhibitor-of eIF2 alpha dephosphorylation protects cells from ER stress, Science 307 (2005) 935–939.
[204] A.L. Sokka, N. Putkonen, G. Mudo, E. Pryazhnikov, S. Reijonen, L. Khiroug, N. Belluardo, D. Lindholm, L. Korhonen, Endoplasmic reticulum stress inhibition protects against excitotoxic neuronal injury in the rat brain, J. Neurosci. 27 (2007) 901–908.
[205] K. Ding, H.D. Wang, J.G. Xu, T. Li, L. Zhang, Y. Ding, L. Zhu, J. He, M.L. Zhou, Melatonin stimulates antioxidant enzymes and reduces oxidative stress in experimental traumatic brain injury: the Nrf2-ARE signaling pathway as a potential mechanism, Free Radic. Biol. Med. 73 (2014) 1–11.
[206] J.Y. Wu, H. Prentice, Role of taurine in the central nervous system, J. Biomed. Sci. 17 (2010).

[207] C.L. Pan, H. Prentice, A.L. Price, J.Y. Wu, Beneficial effect of taurine on hypoxia- and glutamate-induced endoplasmic reticulum stress pathways in primary neuronal culture, Amino Acids 43 (2012) 845–855.
[208] D.H. Perlmutter, Chemical chaperones: a pharmacological strategy for disorders of protein folding and trafficking, Pediatr. Res. 52 (2002) 832–836.
[209] T. Minamino, I. Komuro, M. Kitakaze, Endoplasmic reticulum stress as a therapeutic target in cardiovascular disease, Circ. Res. 107 (2010) 1071–1082.
[210] U. Ozcan, E. Yilmaz, L. Ozcan, M. Furuhashi, E. Vaillancourt, R.O. Smith, C.Z. Gorgun, G.S. Hotamisligil, Chemical chaperones reduce ER stress and restore glucose homeostasis in a mouse model of type 2 diabetes, Science 313 (2006) 1137–1140.
[211] L. Ozcan, A.S. Ergin, A. Lu, J. Chung, S. Sarkar, D. Nie, M.G. Myers Jr., U. Ozcan, Endoplasmic reticulum stress plays a central role in development of leptin resistance, Cell Metabol. 9 (2009) 35–51.
[212] X. Qi, T. Hosoi, Y. Okuma, M. Kaneko, Y. Nomura, Sodium 4-phenylbutyrate protects against cerebral ischemic injury, Mol. Pharmacol. 66 (2004) 899–908.
[213] C.D. Keene, C.M.P. Rodrigues, T. Eich, M.S. Chhabra, C.J. Steer, W.C. Low, Tauroursodeoxycholic acid, a bile acid, is neuroprotective in a transgenic model of Huntington's disease, Proc. Natl. Acad. Sci. USA 99 (2002) 10671–10676.
[214] C.S. Park, H. Cha, E.J. Kwon, P.K. Sreenivasaiah, D.H. Kim, The chemical chaperone 4-phenylbutyric acid attenuates pressure-overload cardiac hypertrophy by alleviating endoplasmic reticulum stress, Biochem. Biophys. Res. Commun. 421 (2012) 578–584.
[215] R. Guo, H. Ma, F. Gao, L. Zhong, J. Ren, Metallothionein alleviates oxidative stress-induced endoplasmic reticulum stress and myocardial dysfunction, J. Mol. Cell. Cardiol. 47 (2009) 228–237.
[216] W.P. Cheng, B.W. Wang, K.G. Shyu, Regulation of GADD153 induced by mechanical stress in cardiomyocytes, Eur. J. Clin. Invest. 39 (2009) 960–971.
[217] X.Z. Wang, D. Ron, Stress-induced phosphorylation and activation of the transcription factor CHOP (GADD153) by p38 MAP kinase, Science 272 (1996) 1347–1349.
[218] V. Sukumaran, K. Watanabe, P.T. Veeraveedu, N. Gurusamy, M.L. Ma, R.A. Thandavarayan, A.P. Lakshmanan, K. Yamaguchi, K. Suzuki, M. Kodama, Olmesartan, an AT(1) antagonist, attenuates oxidative stress, endoplasmic reticulum stress and cardiac inflammatory mediators in rats with heart failure induced by experimental autoimmune myocarditis, Int. J. Biol. Sci. 7 (2011) 154–167.
[219] T.B. Yao, X.Y. Ying, Y.C. Zhao, A.C. Yuan, Q. He, H. Tong, S. Ding, J.L. Liu, X. Peng, E.H. Gao, J. Pu, B. He, Vitamin D receptor activation protects against myocardial reperfusion injury through inhibition of apoptosis and modulation of autophagy, Antioxid. Redox Signal. 22 (2015) 633–650.
[220] X.J. Song, C.Y. Yang, B. Liu, Q. Wei, M.T. Korkor, J.Y. Liu, P. Yang, Atorvastatin inhibits myocardial cell apoptosis in a rat model with post-myocardial infarction heart failure by downregulating ER stress response, Int. J. Med. Sci. 8 (2011) 564–572.
[221] S.G. Fonseca, J. Gromada, F. Urano, Endoplasmic reticulum stress and pancreatic beta-cell death, Trends Endocrinol. Metabol. 22 (2011) 266–274.
[222] B. Yusta, L.L. Baggio, J.L. Estall, J.A. Koehler, D.P. Holland, H. Li, D. Pipeleers, Z. Ling, D.J. Drucker, GLP-1 receptor activation improves beta cell function and survival following induction of endoplasmic reticulum stress, Cell Metabol. 4 (2006) 391–406.
[223] T. Quentin, M. Steinmetz, A. Poppe, S. Thoms, Metformin differentially activates ER stress signaling pathways without inducing apoptosis, Dis. Model. Mech. 5 (2012) 259–269.

[224] C.C. Chiu, C.Y. Lin, L.Y. Lee, Y.J. Chen, T.F. Kuo, J.T.C. Chang, C.T. Liao, H.M. Wang, T.C. Yen, C.R. Shen, S.K. Liao, A.J. Cheng, Glucose-regulated protein 78 regulates multiple malignant phenotypes in head and neck cancer and may serve as a molecular target of therapeutic intervention, Mol. Cancer Ther. 7 (2008) 2788–2797.

[225] L.H. Zhang, X.A. Zhang, Roles of GRP78 in physiology and cancer, J. Cell. Biochem. 110 (2010) 1299–1305.

[226] S.P. Ermakova, B.S. Kang, B.Y. Choi, H.S. Choi, T.F. Schuster, W.Y. Ma, A.M. Bode, Z.G. Dong, (-)-Epigallocatechin gallate overcomes resistance to etoposide-induced cell death by targeting the molecular chaperone glucose-regulated protein 78, Cancer Res. 66 (2006) 9260–9269.

[227] M. Cerezo, A. Lehraiki, A. Millet, F. Rouaud, M. Plaisant, E. Jaune, T. Botton, C. Ronco, P. Abbe, H. Amdouni, T. Passeron, V. Hofman, B. Mograbi, A.S. Dabert-Gay, D. Debayle, D. Alcor, N. Rabhi, J.S. Annicotte, L. Heliot, M. Gonzalez-Pisfil, C. Robert, S. Morera, A. Vigouroux, P. Gual, M.M.U. Ali, C. Bertolotto, P. Hofman, R. Ballotti, R. Benhida, S. Rocchi, Compounds triggering ER stress exert anti-melanoma effects and overcome BRAF inhibitor resistance, Cancer Cell 30 (2016) 183 (vol. 29, p. 805, 2016).

[228] J.J. Wolfson, K.L. May, C.M. Thorpe, D.M. Jandhyala, J.C. Paton, A.W. Paton, Subtilase cytotoxin activates PERK, IRE1 and ATF6 endoplasmic reticulum stress-signalling pathways, Cell Microbiol. 10 (2008) 1775–1786.

[229] Z. Rosenes, T.D. Mulhern, D.M. Hatters, L.L. Ilag, B.E. Power, C. Hosking, F. Hensel, G.J. Howlett, Y.F. Mok, The anti-cancer IgM monoclonal antibody PAT-SM6 binds with high avidity to the unfolded protein response regulator GRP78, PLoS One 7 (2012) e44927.

[230] E. Fasano, S. Serini, E. Piccioni, A. Toesca, G. Monego, A.R. Cittadini, F.O. Ranelletti, G. Calviello, DHA induces apoptosis by altering the expression and cellular location of GRP78 in colon cancer cell lines, Biochim. Biophys. Acta 1822 (2012) 1762–1772.

[231] C.Y. Lee, W.H. Sit, S.T. Fan, K. Man, I.W. Jor, L.L. Wong, M.L. Wan, K.C. Tan-Un, J.M. Wan, The cell cycle effects of docosahexaenoic acid on human metastatic hepatocellular carcinoma proliferation, Int. J. Oncol. 36 (2010) 991–998.

[232] N. Rauschert, S. Braendlein, E. Holzinger, F. Hensel, H.K. Muller-Hermelink, H.P. Vollmers, A new tumor-specific variant of GRP78 as target for antibody-based therapy, Lab. Invest. 88 (2008) 375–386.

[233] R. Alexanian, K. Delasalle, M. Wang, S. Thomas, D. Weber, Curability of multiple myeloma, Bone Marrow Res. 2012 (2012) 916479.

[234] P.G. Richardson, P. Sonneveld, M.W. Schuster, D. Irwin, E.A. Stadtmauer, T. Facon, J.L. Harousseau, D. Ben-Yehuda, S. Lonial, H. Goldschmidt, D. Reece, J.F. San-Miguel, J. Blade, M. Boccadoro, J. Cavenagh, W.S. Dalton, A.L. Boral, D.L. Esseltine, J.B. Porter, D. Schenkein, K.C. Anderson, I. Assessment of proteasome inhibition for extending remissions, bortezomib or high-dose dexamethasone for relapsed multiple myeloma, N. Engl. J. Med. 352 (2005) 2487–2498.

[235] T. Kawaguchi, K. Miyazawa, S. Moriya, T. Ohtomo, X.F. Che, M. Naito, M. Itoh, A. Tomoda, Combined treatment with bortezomib plus bafilomycin A1 enhances the cytocidal effect and induces endoplasmic reticulum stress in U266 myeloma cells: crosstalk among proteasome, autophagy-lysosome and ER stress, Int. J. Oncol. 38 (2011) 643–654.

[236] T. Mujtaba, Q.P. Dou, Advances in the understanding of mechanisms and therapeutic use of bortezomib, Discov. Med. 12 (2011) 471–480.
[237] G. Augello, M. Modica, A. Azzolina, R. Puleio, G. Cassata, M.R. Emma, C. Di Sano, A. Cusimano, G. Montalto, M. Cervello, Preclinical evaluation of antitumor activity of the proteasome inhibitor MLN2238 (ixazomib) in hepatocellular carcinoma cells, Cell Death Dis. 9 (2018) 28.
[238] M.X. Bi, C. Naczki, M. Koritzinsky, D. Fels, J. Blais, N.P. Hu, H. Harding, I. Novoa, M. Varia, J. Raleigh, D. Scheuner, R.J. Kaufman, J. Bell, D. Ron, B.G. Wouters, C. Koumenis, ER stress-regulated translation increases tolerance to extreme hypoxia and promotes tumor growth, EMBO J. 24 (2005) 3470–3481.
[239] C. Atkins, Q. Liu, E. Minthorn, S.Y. Zhang, D.J. Figueroa, K. Moss, T.B. Stanley, B. Sanders, A. Goetz, N. Gaul, A.E. Choudhry, H. Alsaid, B.M. Jucker, J.M. Axten, R. Kumar, Characterization of a novel PERK kinase inhibitor with antitumor and antiangiogenic activity, Cancer Res. 73 (2013) 1993–2002.
[240] J.M. Axten, J.R. Medina, Y.H. Feng, A. Shu, S.P. Romeril, S.W. Grant, W.H.H. Li, D.A. Heerding, E. Minthorn, T. Mencken, C. Atkins, Q. Liu, S. Rabindran, R. Kumar, X. Hong, A. Goetz, T. Stanley, J.D. Taylor, S.D. Sigethy, G.H. Tomberlin, A.M. Hassell, K.M. Kahler, L.M. Shewchuk, R.T. Gampe, Discovery of 7-Methyl-5-(1-{[3-(trifluoromethyl)phenyl]acetyl}-2,3-dihydro-1H-indol-5-yl)-7H-pyrrolo[2,3-d]pyrimidin-4-amine (GSK2606414), a potent and selective first-in-class inhibitor of protein kinase R (PKR)-like endoplasmic reticulum kinase (PERK), J. Med. Chem. 55 (2012) 7193–7207.
[241] Y. Teng, M. Gao, J. Wang, Q. Kong, H. Hua, T. Luo, Y. Jiang, Inhibition of eIF2alpha dephosphorylation enhances TRAIL-induced apoptosis in hepatoma cells, Cell Death Dis. 5 (2014) e1060.
[242] N. Mimura, M. Fulciniti, G. Gorgun, T. Tai, D. Cirstea, L. Santo, Y.G. Hu, C. Fabre, J.M.H. Ohguchi, T. Kiziltepe, H. Ikeda, Y. Kawano, M. French, M. Blumenthal, V. Tam, N.L. Kertesz, U.M. Malyankar, M. Hokenson, T. Pham, Q.P. Zeng, J.B. Patterson, P.G. Richardson, N.C. Munshi, K.C. Anderson, Blockade of XBP1 splicing by inhibition of IRE1 alpha is a promising therapeutic option in multiple myeloma, Blood 119 (2012) 5772–5781.
[243] K. Volkmann, J.L. Lucas, D. Vuga, X.P. Wang, D. Brumm, C. Stiles, D. Kriebel, A. Der-Sarkissian, K. Krishnan, C. Schweitzer, Z. Liu, U.M. Malyankar, D. Chiovitti, M. Canny, D. Durocher, F. Sicheri, J.B. Patterson, Potent and selective inhibitors of the inositol-requiring enzyme 1 endoribonuclease, J. Biol. Chem. (2011) 286.
[244] B.C.S. Cross, P.J. Bond, P.G. Sadowski, B.K. Jha, J. Zak, J.M. Goodman, R.H. Silverman, T.A. Neubert, I.R. Baxendale, D. Ron, H.P. Harding, The molecular basis for selective inhibition of unconventional mRNA splicing by an IRE1-binding small molecule, Proc. Natl. Acad. Sci. USA 109 (2012) E869–E878.
[245] B.K. Jha, I. Polyakova, P. Kessler, B. Dong, B. Dickerman, G.C. Sen, R.H. Silverman, Inhibition of RNase L and RNA-dependent protein kinase (PKR) by sunitinib impairs antiviral innate immunity, J. Biol. Chem. 286 (2011) 26319–26326.
[246] S. Rottmann, Y. Wang, M. Nasoff, Q.L. Deveraux, K.C. Quon, A TRAIL receptor-dependent synthetic lethal relationship between MYC activation and GSK3beta/FBW7 loss of function, Proc. Natl. Acad. Sci. USA 102 (2005) 15195–15200.
[247] Y. Wang, I.H. Engels, D.A. Knee, M. Nasoff, Q.L. Deveraux, K.C. Quon, Synthetic lethal targeting of MYC by activation of the DR5 death receptor pathway, Cancer Cell 5 (2004) 501–512.

[248] J. Klefstrom, E.W. Verschuren, G. Evan, c-Myc augments the apoptotic activity of cytosolic death receptor signaling proteins by engaging the mitochondrial apoptotic pathway, J. Biol. Chem. 277 (2002) 43224–43232.
[249] S. Arora, E.A. Golemis, A new strategy to ERADicate HER2-positive breast tumors? Sci. Signal. 8 (2015) fs11.
[250] R.B. Ferreira, M.E. Law, S.C. Jahn, B.J. Davis, C.D. Heldermon, M. Reinhard, R.K. Castellano, B.K. Law, Novel agents that downregulate EGFR, HER2, and HER3 in parallel, Oncotarget 6 (2015) 10445–10459.
[251] E. Jablonska, J. Jablonski, M. Marcinczyk, Z. Grabowska, L. Piotrowski, The release of soluble forms of TRAIL and DR5 by neutrophils of oral cavity cancer patients, Folia Histochem. Cytobiol. 46 (2008) 177–183.
[252] L. Martinez-Lostao, A. Anel, J. Pardo, How do cytotoxic lymphocytes kill cancer cells? Clin. Cancer Res. 21 (2015) 5047–5056.
[253] A. Jain, G. Zode, R.B. Kasetti, F.A. Ran, W. Yan, T.P. Sharma, K. Bugge, C.C. Searby, J.H. Fingert, F. Zhang, A.F. Clark, V.C. Sheffield, CRISPR-Cas9-based treatment of myocilin-associated glaucoma, Proc. Natl. Acad. Sci. USA 114 (2017) 11199–11204.
[254] T. Yoshida, Y. Ozawa, K. Suzuki, K. Yuki, M. Ohyama, W. Akamatsu, Y. Matsuzaki, S. Shimmura, K. Mitani, K. Tsubota, H. Okano, The use of induced pluripotent stem cells to reveal pathogenic gene mutations and explore treatments for retinitis pigmentosa, Mol. Brain 7 (2014) 45.
[255] K.B. Delasalle, L.H. Wang, D. Weber, S.A. Giralt, S. Thomas, J. Shah, B. Handy, R.Z. Orlowski, R. Alexanian, Curability of multiple myeloma, Blood 114 (2009) 1487.

CHAPTER 6

Protein phase separation in cell death and survival

Michael C. Chung[1], Daiqing Liao[2] and Juan Guan[1,2]
[1]Department of Physics, University of Florida, Gainesville, FL, United States; [2]Department of Anatomy and Cell Biology, College of Medicine, University of Florida, Gainesville, FL, United States

1. Phase separation

It has been known early in the study of cells that there exist discrete compartments, well-separated from their environment by lipid membranes. These compartments serve many purposes and form the bulk of our understanding of cellular function. However, in the past decade, there has been a surge in the study of membraneless structures that play important, diverse roles in cellular function. These membraneless structures, or condensates, unlike their membrane-bound counterparts, form via phase separation. Phase separation, a concept well known to those having experimented with the mixing of oil and vinegar, is marked chiefly by the partitioning of components from their surrounding environment, such that there exist distinct, coexisting phases. As we shall see, these condensates perform incredibly varied roles throughout biology and have a hand in cell death. The purpose of this section is to familiarize readers with the basic ideas of phase separation, particularly their biological relevance, hallmarks, and various tests one may perform to probe phase separation in a system of interest.

1.1 Biological relevance

As mentioned previously, over the past decade, there has been mounting evidence that phase separation plays a crucial role in cellular function. Broadly, phase separation may play key roles as diverse as sequestration to prevent reactions, locally concentrating molecular players to incite reactions or nucleate structures, generation of forces to shape membrane-bound structures, and formation of filters such as the nuclear pore complex [1,2]. For example, in transcription regulation, it has been determined that important machinery such as Mediator and RNA polymerase II form

Mechanisms of Cell Death and Opportunities for Therapeutic Development
ISBN 978-0-12-814208-0
https://doi.org/10.1016/B978-0-12-814208-0.00004-X

phase separated condensates on enhancer sites [3]. These phase-separated bodies may serve to concentrate and recruit the appropriate factors to enhance gene activation [4].

Given the diverse function of these phase separated condensates, it should be no surprise that there are many different classes of important membraneless bodies in cells, which span varying length scales and cellular locations [5]. Interestingly, a few of these bodies have been known and studied for nearly a century, with the range of their function recently being revealed. One such membraneless body is the nucleolus, which plays a crucial role in ribosomal assembly. Another increasingly studied class of nuclear condensates, with implications in various pathologies, are PML bodies, thus named due to the presence of promyelocytic leukemia protein. It is suggested that these bodies play important roles in tumor suppression and sequestration of apoptotic factors. In the cytosol, there are also several important classes of phase separated condensates. For example, stress granules (SGs), which form in response to cellular stress signals, serve to sequester mRNA, transcription factors, and components of the small ribosomal subunit. Closely related and often interfacing with SGs are processing bodies, or P-bodies. P-bodies are importantly involved in mRNA processing and modification, containing sets of proteins related to transport and translation.

These condensates also exist in differing physical states that may be of physiological importance. Such states, which importantly form a continuum, are known to be liquid-, gel-, or solid-like [6]. Liquid-like membraneless organelles include a host of familiar cellular bodies such as Cajal Bodies that regulate small nuclear ribronucleoproteins (snRNPs), the nucleolus which plays a key role in ribosome biogenesis, paraspeckles, and much more. This biophysical state permits dynamic exchange of specific macromolecules with its surrounding environment. The gel-like and solid states, which are typified by their less dynamic nature, increasingly appear to be associated with pathological conditions. ALS, amyloid lateral sclerosis, a neurodegenerative disease that attacks nerve cells, is one of many neurodegenerative diseases known to be driven by pathological protein aggregates and associated gel or solid condensates [7,8]. On the other hand, these biophysical states may also play important roles as stable platforms for cellular processes that may be enhanced by such stability, such as macroautophagic degradation [9,10].

Naturally, for phase separation to play a critical role in a biological context, a cell must be able to regulate the composition and other biophysical properties of the condensates. Although this is still an area of active research, it has been shown that multivalency, or the number of interaction possibilities, is a critical determinant of phase separation [11]. Importantly, this would permit a cell to regulate these condensates via posttranslational modifications (PTMs) and/or involving other cellular proteins to modulate the phase separation behavior [12].

1.2 Hallmarks of phase separation

While there lacks comprehensive theory concerning the cell's ability to dynamically regulate condensates in vivo, there has been much work aimed at elucidating the molecular hallmarks of phase separated bodies and how they come into fruition. Among the molecular signatures, the concept of multivalency is of utmost importance. Rosen and others have clearly demonstrated with engineered small ubiquitin-like modifier (SUMO)-SUMO-interacting motif (SIM) constructs that the critical concentration at which one has condensates (the phase boundary) depends strongly on the valency of the constructs [11]. Expanding on this, they have put forth a "client-scaffold" model in which high valency macromolecule "scaffolds" self-associate and recruit lower valency "clients" to available scaffold sites [13]. More recently, it has been proposed that macromolecules classified by their valency may be a powerful tool to understand their function [14].

Another important phase separation motif, closely related to that of valency, is that of intrinsically disordered regions (IDRs), protein domains that do not adopt a folded structure [15,16]. These sequences are typically populated by charged and polar amino acid residues [1]. Additionally, aromatic residues, particularly Tyrosine and Phenylalanine, are often found throughout IDRs and assist in the macromolecular binding via various interactions. One such known example is DEAD Box protein 4 (DDX4), with IDRs at both N and C termini. DDX4's N-terminal IDR was sufficient to induce phase separation in the system studied [17]. However, it should be mentioned that the role of IDRs in the induction of phase separation is still being uncovered. For example, it has recently been implied in the study of SGs that IDRs may instead play a more modulatory role, where for G3BP, IDRs juxtaposed against the RNA-binding domain may prevent RNA binding [14]. It is also implied that IDRs control the transition to gel-like states in phase-separated systems involving multivalent proteins [18].

1.3 Testing phase separation

There are several methods that have become a staple of testing condensates in vivo and in vitro; we will only discuss a few such methods here. Perhaps the most simplistic methods to test the biophysical properties of a condensate include observing its morphology and interactions. Liquid-like assemblies will respond to shear forces, appear spherically shaped, and tend to merge with or wet other liquid condensates. This merging or wetting will depend on the relative surface tension of each droplet [19]. Another commonly used method to probe the biophysical state of phase separated bodies is hexanediol. Hexanediol is an aliphatic alcohol that disrupts hydrophobic bonds that are important for liquid states to form. It has been shown to readily dissolve liquid-like P-bodies, while more solid-like protein aggregates resisted such dissolution [20]. Fluorescence recovery after photobleaching (FRAP), which has traditionally been used to extract quantitative information concerning the dynamics of a cellular region, has been used frequently in assessing the biophysical state of the condensate being studied [21,22]. In FRAP studies, one laser ablates a fluorescent region and monitors its subsequent intensity recovery. In a liquid-like state, one typically observes nearly complete recovery in second to minute timescales, while over the same timescales in solid-like or gel-like states, one sees very low recovery.

2. Cell death pathways

Given the numerous intrinsic or extrinsic death pathways that a cell may take, in this section we discuss several relevant pathways and describe the involvement of phase separation in each. It should be noted that the sections may not be mutually exclusive, as the research in each pathway is ongoing.

2.1 Apoptosis

Apoptosis is a highly regulated, programmed cell death that is marked by blebbing and fragmentation of the cellular body. Most forms of apoptosis rely chiefly on the cysteine protease caspase. In the intrinsic pathway, the caspase cascade is initiated internally by pro-apoptotic factors released from the mitochondria. The extrinsic pathway, on the other hand, is triggered by extracellular receptor binding, notably the FAS receptor [23]. Several specific, known phase separated proteins that feature prominently in the apoptotic pathway will be discussed, with a focus on the involvement of phase separation. As we shall see, PTMs function as a critical regulatory mechanism that serves to sequester these key proteins in important classes of membraneless bodies. Readers are also referred to Chapters 1–3 of this volume for more detailed discussions on apoptosis.

2.1.1 DAXX

DAXX, the death-domain associated protein, has been shown to be tightly associated with the extrinsic pathway, in addition to being involved in gene regulation [24]. In the cytosol, DAXX may induce apoptosis via the ASK1/JNK signaling pathway from the FAS receptor. Alternatively, apoptosis may be triggered through the TGF-β receptor, where DAXX interacts with the cytoplasmic domain of the receptor and leads to JNK activation. It is hypothesized that TGF-β signaling may be suppressed via DAXX repression of SMAD4, an important mediator of the TGF-β pathway [25]. Further, HIPK2 may promote TGF-β-mediated death via colocalization with DAXX in promyelocytic leukemia nuclear bodies (PML-NBs), where it also phosphorylates DAXX. In the nucleus, DAXX serves as a repressor of certain antiapoptotic factors and an activator of other pro-apoptotic factors. The nuclear presence of DAXX is where we will turn our focus. Importantly, DAXX contains two SIM domains, both on the N and C termini which are both flanked by IDRs (Fig. 6.1). The SIM domains are biochemically similar and are subject to serine phosphorylation, dramatically increasing the binding affinity to SUMO [26]. This SIM-SUMO binding is crucial in the localization of DAXX to PML-NBs, which additionally requires SUMOylation of DAXX and PML to recruit DAXX [27,28]. Given that there are many phosphorylation sites in DAXX, particularly in the IDRs, this potential multivalent interaction

Figure 6.1 The domains of DAXX. The N and C-termini are SUMO-Interacting motifs (SIMs). From 55 to 144, we have Daxx helical bundle (4HB), 180–384 is the Histone Binding Domain (HBD), 434–496 is an acidic domain. Also shown are the important IDRs, which are speculated to promote phase separation.

between multiple SIM-SUMO sites may drive phase separation. This PML localization of DAXX has been suggested as a means to sequester DAXX, thus inhibiting its repression of pro-apoptotic factors [28–30].

2.1.2 YAP

YAP, or the yes-associated protein, is a transcriptional coactivator of the Hippo signaling pathway, which is known to regulate organ size and development [31,32]. When the Hippo pathway is active, YAP and related transcriptional coactivator PDZ-binding motif (TAZ) are retained in the cytoplasm. Upon Hippo deactivation, YAP migrates to the nucleus, where it assists in the transcription of genes that control cell proliferation and apoptosis [33,34]. Given the relative importance of YAP, it is surprising that little is known about the mechanisms by which YAP affects gene expression upon nuclear translocation.

There may be two relevant contributions of liquid-liquid phase separation (LLPS) to the regulation of the Hippo pathway. It has been shown in one study that when subject to hyperosmotic stress, YAP forms liquid-like condensates in the cytoplasm [35]. Notably, these condensates, which recruited varied sets of proteins, were found to colocalize with Nemo-like kinase (NLK), which phosphorylates Ser128 on YAP and results in nuclear migration. This phase separated colocalization may serve as a reaction crucible, in which NLK may rapidly phosphorylate the appropriate YAP domains, in addition to retaining important kinases. Under similar hyperosmotic conditions, it was found that TEAD1, TAZ, and other factors colocalize with YAP condensates in the nucleus. This nuclear phase separation may serve to restructure the genomic environment and concentrate important factors for gene activation. In particular, in addition to TEAD1 and TAZ colocalization, RNA polymerase-II and newly transcribed RNA were found to eventually localize to the exterior of these droplets. Importantly, removal of YAP IDRs prevented phase separation and inhibited downstream transcription. It should be noted that YAP has a bivalent role in apoptosis [36]. Many studies indicate that YAP is over-expressed in various cancers [37–39]. Further, it was shown via the use of short hairpin RNA (shRNA) [40–42] and small interfering RNA [43–57] that silencing of the YAP gene induced apoptosis. Taken together, this implies that YAP serves an inhibitory role for the apoptotic pathway. However, it has also been shown that YAP interaction with p73 promotes apoptosis via expression of p53AIP1 [58], Bax [59,60], DR5 [60], and PUMA [61]. Furthermore, a recent study revealed that Hippo signaling inhibition activates YAP, which in turn promotes the expression of pro-ferroptotic genes such as TFRC1 and ACSL4, resulting in ferroptosis [62].

2.1.3 cGAS-STING

The cGAS-STING signaling pathway, whose crucial players are cyclic GMP-AMP synthase (cGAS) and the corresponding receptor, stimulator of interferon genes (STING), serves to detect cytosolic DNA and elicit an innate immune response via the production of type-I interferons [63,64]. Notably, both cGAS and STING have been shown to possess other functions, such as mediating DNA repair and induction of lysosomal cell death [65].

cGAS resides as an inactive protein in the cytosol. Upon binding DNA, it changes conformation, dimerizes, and actively serves to produce cyclic GMP-AMP (cGAMP) from the corresponding ambient, intracellular GTP and ATP. This cGAMP then activates STING, which begins the cascade that eventually produces type-I interferon response. In this pathway, the dominant involvement of condensate formation is in the cGAS-DNA complex. cGAS, which requires dimerization and DNA to begin active signaling, has been shown to require sufficient lengths of DNA to ensure binding [66,67]. Interestingly, once it properly binds, cGAS-DNA dimer complexes tend to form "ladder-like" structures in which DNA fragments serve as the rails and cGAS pairs form the rungs. This structure has been proposed to stabilize the complex through multivalent interactions of the N-terminal positively charged regions and DNA [68]. Through these multivalent interactions and other self-oligomerization aids such as the C-site for DNA binding on cGAS, it has been shown that these complexes lead to phase separated bodies (Fig. 6.2). These condensates, by locally

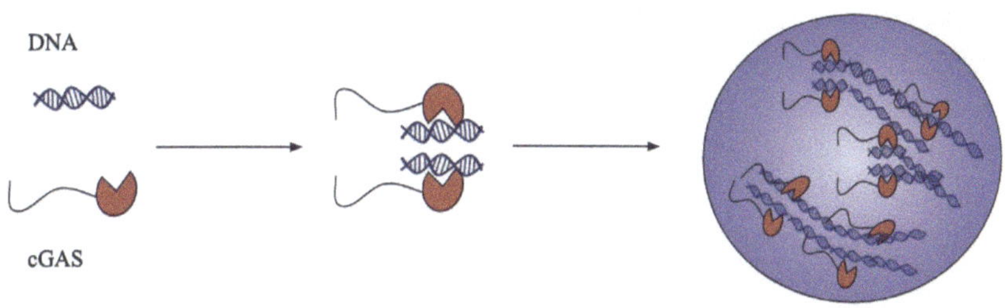

Figure 6.2 cGAS binds to DNA, dimerizes and forms ladder-like complexes that may eventually phase separate.

concentrating cGAS, may serve to further stabilize the cGAS-DNA complex and promote the production of cGAMP. Downstream, STING serves to promote cell death via the apoptotic pathway and also via lysosomal cell death. It has been shown that STING incites T-cell death via the interaction of IRF3 and Bcl2-associated X gene (BAX) [69].

2.2 Autophagic death and aggregates

Although apoptosis may be rightly considered the most well-known cell death pathway available to organisms, there are certainly other routes to cell death that play an important role in a myriad of highly varied cellular contexts. Lysosomal cell death, briefly mentioned with cGAS-STING, is one such cell death mechanism. However, in the context of phase separation, we turn our focus to the less known autophagic cell death and associated formation of pathological aggregates. Thus, in this section we discuss a flow to the eventual autophagic-dependent death of a cell, highlighting the involvement and importance of phase condensate formation (Fig. 6.3).

2.2.1 Aggregates and PTMs

Some of the more well-known instances of pathologies and subsequent cell death result from protein and macromolecular aggregates, which are often associated with gel- or solid-like phase separated condensates. Specifically, misfolded proteins and their aggregates are found in many neurological disorders, including amyotrophic lateral sclerosis (ALS), Huntington's, Alzheimer's, Parkinson's disease, and dementia [70,71]. For example, it has been shown that TAR DNA-binding protein 43 (TDP-43), which typically resides in the nucleus and is associated with frontotemporal dementia and ALS, is tightly associated with neuronal death. Evidence suggests that TDP-43 undergoes LLPS in the cytoplasm, where it accumulates phosphorylated TDP-43 and other nuclear transport machinery [72]. Over time, these cytosolic granules undergo a gel-like transition, eventually depleting the nucleus of its native TDP-43 via sequestration and resulting in cell death. Notably, these TDP-43 granules were shown to be independent of SGs.

As with other cell death pathways, PTMs serve as a crucial regulatory mechanism for such aggregates. In fact, the aggregation of TDP-43 and other RNA-binding proteins such as heterogeneous nuclear ribonucleoprotein A1 (hnRNP A1) in associated granules is strongly influenced by such modifications [73]. One such PTM of note is poly(ADP-ribosyl)ation

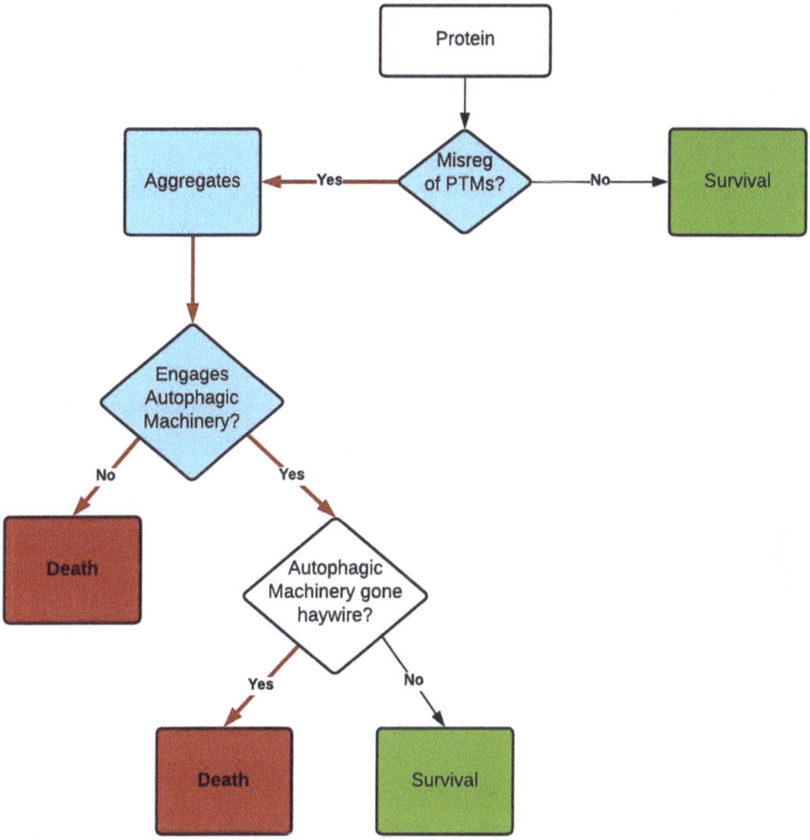

Figure 6.3 A flowchart outlining a possible progression to cell death. Areas in blue denote the possible involvement of phase separation.

(PARylation), which utilizes poly(ADP-ribose) polymerases (PARPs) to add ADP-ribose to specific sets of a protein [74], upon which poly(ADP-ribosyl) glycohydrolases (PARGs) may reversibly remove them. It has been shown that PARylation is a key PTM that regulates the phase separation and SG association of TDP-43 and hnRNP A1 [73], these granules being distinct from the SG-independent TDP-43 granules mentioned previously. Specifically, it was determined that increased levels of cellular PARylation via PARG knockdown inhibit the disassembly of SGs associated with TDP-43 and hnRNP A1. This delay may result in the eventual pathogenic progression of these cellular condensates. It is also worth mentioning that it has been shown that PAR may induce LLPS on DNA break sites [75].

2.2.2 Autophagic death

Autophagy is an important membrane-based mechanism used to recycle cellular materials and circumvent possible pathogenic development derived from various structures, such as damaged organelles or aggregated proteins [76,77]. There are several distinct types of autophagy: microautophagy, macroautophagy, and chaperone-mediated autophagy. In microautophagy, macromolecules to be degraded are directly loaded into lysosomes [78]. Chaperone-mediated autophagy utilizes molecular chaperones that recognize specific motifs to translocate their targets across lysosomal boundaries [79]. Macroautophagy, typically just referred to as autophagy in these circumstances, interestingly makes use of the formation of autophagosomes, membrane-bound bodies that eventually fuse with lysosomes to form autolysosomes. Additionally, within these classes, there is selective and nonselective autophagy. As the name suggests, selective autophagy targets specific cellular structures and is responsible for eliminating protein aggregates and other defunct organelles. By contrast, nonselective autophagy degrades parts of the cytoplasm to recycle molecular components [80].

Autophagic-dependent cell death is a highly context-specific cell death pathway that makes use of the autophagic machinery [81]. It is suggested that autophagic cell death plays an important role in programmed cell death, including that of midgut degradation in drosophila [82]. There are still debates concerning the exact way in which autophagy brings about cellular death, but it is speculated that it may be due to bulk degradation or targeted degradation of critical organelles [77]. Although there is much to still be learned about this mode of cell death, several important molecular players are beginning to be understood. One of particular importance is mTOR and the associated mTORC1 complex. The mechanistic target of rapamycin, mTOR, is a kinase that forms a key component in two protein complexes, mTORC1 and mTORC2, that regulate a host of cellular processes such as cell survival and growth. In the context of autophagy, mTOR has been shown to work in conjunction with SEPA-1 and EPG-2 to regulate the degradation of PGL granules, aggregates formed from P granule components, notably PGL proteins, in *Caenorhabditis elegans* under stress-induced conditions [9]. SEPA-1 and mTORC1 dictate the phase separation of PGL granules. SEPA-1, with its self-oligomerization domain, encourages LLPS and acts as a degradation receptor via binding to LGG-1, the Atg8 homolog in *C. elegans*. Additionally, mTORC1-driven phosphorylation serves as an important PTM, enhancing LLPS of PGL-1 and PGL-3 by lowering the threshold concentration for their formation.

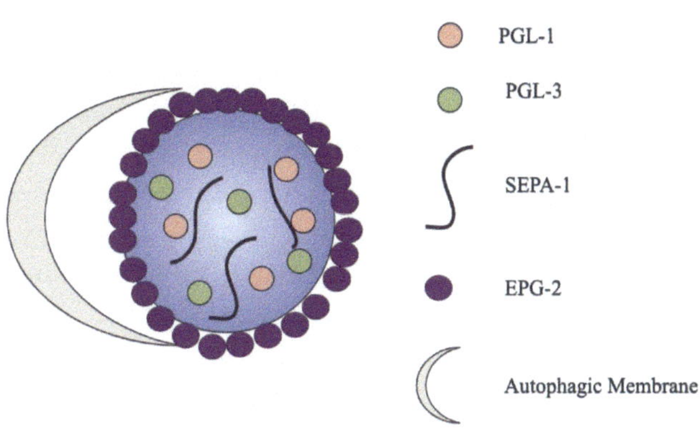

Figure 6.4 EPG-2 aids in the transition of liquid- to gel-like states and the subsequent autophagic envelopment.

EPG-2, an important scaffold protein that appears to coat the surface of these PGL granules, modulates the biophysical properties of such phase separated bodies [9]. EPG-2 may regulate the size of these condensates by inhibiting the fusion of the droplets. It was also shown that EPG-2 induces transition to gel-like states for SEPA-1/PGL1,3 granules. Interestingly, this collective biophysical state, the gel and size, is an important determinant for autophagic degradation. It is conjectured that the gel-like state acts as a stabilizing substrate for the developing autophagic membranes (Fig. 6.4).

3. Cell survival: cancer and the avoidance of cell death

We have thus far considered the involvement of phase separation in a few prominent cell death pathways. The natural complement to death is to consider how a cell may overcome such death mechanisms; that is, how a cell may utilize such mechanisms to survive. Specifically, we consider how phase separation or associated molecular players may be involved in cancer, in both the hijacking of certain cellular machinery, such as the autophagic machinery, and how such machinery may prevent oncogenic expression.

3.1 Autophagy and cancer

The involvement of autophagy in cell survival and death, particularly the interface of this process with cancer, has been intensely debated. Several studies suggest that autophagy serves as a complement to cellular death processes, such as apoptosis. It is proposed that this autophagic death would act as an auxiliary mechanism to control oncogenic development. In contrast, other studies imply that autophagy instead aids cancer, preserving critical cellular functions needed by cancerous growth. Here we will briefly explore the apparently conflicting evidence.

In several studies, it was shown that autophagy promotes cell death in cancer. In one study, it was determined that the autophagy related 5 protein (Atg5), upregulated via the Rac1/MKK7/JNK-signaling pathway, is essential for H-Ras-associated autophagic cell death in fibroblasts [83]. Ras, a critical signal transducer, is noted to be involved in as many as 30% of human cancers [84] and Atg5 is an important protein in the formation of phagophores, the nascent autophagosome membrane, around their cargo. It was observed that overactive H-Ras resulted in autophagic vacuoles, which were caspase independent, inferred from the use of caspase inhibitors. Further, autophagic inhibitors such as Bafilomycin A1 lessened this H-Ras-induced death and vacuole formation. Additionally, targeted knockdown of Atg5 by small interfering RNA (siRNA) and subsequent inhibition of cell death implies the critical importance of Atg5. In another study, it was found that Ras-associated upregulation of Noxa and Beclin-1 promotes autophagic cell death [85]. Noxa, a BH3-only protein, is a well-known p53 response factor and has important apoptotic functionality, in part by binding to Mcl-1 in mitochondria [86]. Beclin-1, which is the mammalian ortholog of yeast Atg6, plays an important role in autophagy, where it is involved in the PI3K complex in early phagosome formation [87]. This system was also determined to be caspase-independent, with both factors attenuating the autophagic death upon their reduced expression. Taken together, these studies imply that autophagy may serve as a mechanism complementary to apoptosis, thus aiding in the circumvention of oncogenic potential.

It was argued in a separate study that autophagy instead promotes cancer survival, given that autophagy functions in part to recycle cellular components and degrade damaged organelles, serving as an important survival mechanism under stressful conditions. In Ras-mutated aggressive cancers, it was found that there is upregulation of basal autophagic activity. This increase in autophagic activity in aggressive cancers was proposed to preserve and support mitochondrial function, which Ras-expressing cells require to maintain themselves, notably through the Warburg Effect [88].

In light of the evidence, how is one to reconcile the seemingly conflicting results? It has been proposed in the previous studies that cancer initiation is prevented in part via autophagic activity, but counterintuitively sustains the aggressive cancers. In other words, the same mechanism that aids in the early prevention of cancerous development via the degradation of damaged tissue may eventually support cancer growth by maintaining mitochondrial metabolic function.

3.2 Cancer and phase separation

Cancer is marked by constitutive activation in signaling pathways. However, the mechanistic understanding for such constitutive activation is often lacking. Several case studies provide some important indications for the possible role of phase separation in cancerous development and its retention [71]. One major function of phase separation in the context of cancer may be signal amplification [89]. In immune cells, the activation of the T-cell receptor pathway eventually leads to the phase separation of downstream effector proteins [90]. This serves as a selective microenvironment that only permits certain factors to enter the condensate, permitting these clusters to amplify signals. In the previously mentioned cGAS system, we see similar amplification effects via the formation of these cGAS-DNA clusters. One can imagine if phase separation can enhance signaling in such cases, then mutations may result in constitutive signal amplification via modification of these signaling clusters, this perhaps resulting in the uncontrolled proliferative processes characteristic of cancerous development.

Given the importance of transcriptional dysregulation in cancer development, the involvement of phase separation in gene regulation may be another important function of phase separation in oncogenic contexts. Particularly, IDR-driven phase separation in which gene fusion products involving IDRs may result in the development of many malignancies. One such example is found in Ewing's sarcoma, in which the fusion of the IDR of EWS and the FLI protein [89]. Additionally, mis-localization of transcriptional condensates can drive oncogene expression [91]. It has been implied that cancerous mutations such as insertions or deletions in the genomes can lead to the development of oncogenic super-enhancer sites. For example, it has been found that insertion or deletion events in the vicinity of the TAL1 gene in T-cell acute lymphoblastic leukemia lead to superenhancer-driven oncogenic expression of TAL1 via the formation of an MYB TF binding site [92].

4. Conclusion

Although the involvement of phase separation in cell death and survival is still being debated and there is much to be uncovered, it is clear that the partitioning of various macromolecules from the surrounding intracellular environment has the potential to play a crucial role in the orchestration of death and survival processes. In cell death and pathogenesis, the development of gel-like or solid-like condensates is becoming an important hallmark, notably among neurodegenerative diseases. However, as we have seen with EPG-2 and the autophagic machinery, these less dynamic states may also play an important role as stable substrates for the development of nascent autophagosomes, allowing a cell to reliably degrade targeted proteins. Functionally, the ability to sequester apoptotic factors or concentrate important signaling molecules has also been noted as a potentially important way in which phase separation can serve to activate or repress cell death pathways. Particularly, the modification of valency via PTMs can serve as an important regulatory mechanism for these condensates in their respective contexts.

In cell survival, the natural complement to cell death, phase separation may serve important functions in oncogenic development. Autophagy, discussed at length in relation to cell death, has a somewhat dual role in cancer and cell survival. Whether or not autophagy serves to aid cancer development is still being elucidated. It is hypothesized that autophagy represses cancer initiation, yet serves to maintain cancerous growth for aggressive cancers via the maintenance of the mitochondrial function. More generally, phase separation may feature prominently in cancerous constitutive signaling by signal amplification and transcriptional dysregulation, perhaps allowing for novel treatment options in the future. Given the current progress in understanding the importance of phase separation in cellular function, it will be exciting to see further developments in the field and what new insights and approaches this organizational principle may offer.

References

[1] Y. Shin, C.P. Brangwynne, Liquid phase condensation in cell physiology and disease, Science 357 (6357) (2017) 1253.

[2] S. Boeynaems, S. Alberti, N.L. Fawzi, T. Mittag, M. Polymenidou, F. Rousseau, J. Schymkowitz, J. Shorter, B. Wolozin, L. Van Den Bosch, P. Tompa, M. Fuxreiter, Protein phase separation: a new phase in cell biology, Trends Cell Biol. 28 (6) (2018) 420–435.

[3] W.-K. Cho, J.-H. Spille, M. Hecht, C. Lee, C. Li, V. Grube, I.I. Cisse, Mediator and RNA polymerase II clusters associate in transcription-dependent condensates, Science 361 (6400) (2018) 412–415.
[4] A.J. Plys, R.E. Kingston, Dynamic condensates activate transcription, Science 361 (6400) (2018) 329–330.
[5] D.M. Mitrea, R.W. Kriwacki, Phase separation in biology; functional organization of a higher order, Cell Commun. Signal. 14 (1) (2016) 1.
[6] A.A. Hyman, C.A. Weber, F. Jülicher, Liquid-liquid phase separation in biology, Annu. Rev. Cell Dev. Biol. 30 (1) (2014) 39–58.
[7] D. Pakravan, G. Orlando, V. Bercier, L. Van Den Bosch, Role and therapeutic potential of liquid–liquid phase separation in amyotrophic lateral sclerosis, J. Mol. Cell Biol. (2020), https://doi.org/10.1093/jmcb/mjaa049.
[8] A. Patel, H.O. Lee, L. Jawerth, S. Maharana, M. Jahnel, M.Y. Hein, S. Stoynov, J. Mahamid, S. Saha, T.M. Franzmann, A. Pozniakovski, I. Poser, N. Maghelli, L.A. Royer, M. Weigert, E.W. Myers, S. Grill, D. Drechsel, A.A. Hyman, S. Alberti, A liquid-to-solid phase transition of the ALS protein FUS accelerated by disease mutation, Cell 162 (5) (2015) 1066–1077.
[9] G. Zhang, Z. Wang, Z. Du, H. Zhang, MTOR regulates phase separation of PGL granules to modulate their autophagic degradation, Cell 174 (6) (2018) 1492–1506. e22.
[10] D. Sun, R. Wu, P. Li, L. Yu, Phase separation in regulation of aggrephagy, J. Mol. Biol. 432 (1) (2020) 160–169 (Molecular Mechanisms of Selective Autophagy).
[11] P. Li, S. Banjade, H.-C. Cheng, S. Kim, B. Chen, L. Guo, M. Llaguno, J.V. Hollingsworth, D.S. King, S.F. Banani, P.S. Russo, Q.-X. Jiang, B.T. Nixon, M.K. Rosen, Phase transitions in the assembly of multivalent signalling proteins, Nature 483 (7389) (2012) 336–340.
[12] S. Alberti, A. Gladfelter, T. Mittag, Considerations and challenges in studying liquid-liquid phase separation and biomolecular condensates, Cell 176 (3) (2019) 419–434.
[13] S.F. Banani, A.M. Rice, W.B. Peeples, Y. Lin, S. Jain, R. Parker, M.K. Rosen, Compositional control of phase-separated cellular bodies, Cell 166 (3) (2016) 651–663.
[14] D.W. Sanders, N. Kedersha, D.S.W. Lee, A.R. Strom, V. Drake, J.A. Riback, D. Bracha, J.M. Eeftens, A. Iwanicki, A. Wang, M.-T. Wei, G. Whitney, S.M. Lyons, P. Anderson, W.M. Jacobs, P. Ivanov, C.P. Brangwynne, Competing protein-RNA interaction networks control multiphase intracellular organization, Cell 181 (2) (2020) 306–324. e28.
[15] C.J. Oldfield, V.N. Uversky, A.K. Dunker, L. Kurgan, Chapter 1 Introduction to intrinsically disordered proteins and regions, in: N. Salvi (Ed.), Intrinsically Disordered Proteins, Academic Press, 2019, pp. 1–34.
[16] V.N. Uversky, Intrinsically disordered proteins in overcrowded mi lieu: membrane-less organelles, phase separation, and intrinsic disorder, Curr. Opin. Struct. Biol. 44 (2017) 18–30.
[17] T.J. Nott, E. Petsalaki, P. Farber, D. Jervis, E. Fussner, A. Plochowietz, T.D. Craggs, D.P. Bazett-Jones, T. Pawson, J.D. Forman-Kay, A.J. Baldwin, Phase transition of a disordered nuage protein generates environmentally responsive membraneless organelles, Mol. Cell 57 (5) (2015) 936–947.
[18] T.S. Harmon, A.S. Holehouse, M.K. Rosen, R.V. Pappu, Intrinsically disordered linkers determine the interplay between phase separation and gelation in multivalent proteins, eLife 6 (2017) e30294.

[19] C.P. Brangwynne, C.R. Eckmann, D.S. Courson, A. Rybarska, C. Hoege, J. Gharakhani, F. Jülicher, A.A. Hyman, Germline P granules are liquid droplets that localize by controlled dissolution/condensation, Science 324 (5935) (2009) 1729–1732.
[20] S. Kroschwald, S. Maharana, A. Simon, Hexanediol: a chemical probe to investigate the material properties of membrane-less compartments, Matters 3 (5) (2017) e201702000010.
[21] B.L. Sprague, J.G. McNally, FRAP analysis of binding: proper and fitting, Trends Cell Biol. 15 (2) (2005) 84–91.
[22] N.O. Taylor, M.-T. Wei, H.A. Stone, C.P. Brangwynne, Quantifying dynamics in phase-separated condensates using fluorescence recovery after photobleaching, Biophys. J. 117 (7) (2019) 1285–1300.
[23] X. Yang, R. Khosravi-Far, H.Y. Chang, D. Baltimore, Daxx, a novel Fas-binding protein that activates JNK and apoptosis, Cell 89 (7) (1997) 1067–1076.
[24] I. Mahmud, D. Liao, DAXX in cancer: phenomena, processes, mechanisms and regulation, Nucleic Acids Res. 47 (15) (2019) 7734–7752.
[25] R. Perlman, W.P. Schiemann, M.W. Brooks, H.F. Lodish, R.A. Weinberg, TGF-β-induced apoptosis is mediated by the adapter protein Daxx that facilitates JNK activation, Nat. Cell Biol. 3 (8) (2001) 708–714.
[26] E. Escobar-Cabrera, M. Okon, D.K. Lau, C.F. Dart, A.M. Bonvin, L.P. McIntosh, Characterizing the N and C-terminal small ubiquitin-like modifier (sumo)-interacting motifs of the scaffold protein DAXX, J. Biol. Chem. 286 (22) (2011) 19816–19829.
[27] A.M. Ishov, A.G. Sotnikov, D. Negorev, O.V. Vladimirova, N. Neff, T. Kamitani, E.T. Yeh, J.F. Strauss III, G.G. Maul, PML is critical for ND10 formation and recruits the PML-interacting protein Daxx to this nuclear structure when modified by SUMO-1, JCB (J. Cell Biol.) 147 (2) (1999) 221–234.
[28] H. Li, C. Leo, J. Zhu, X. Wu, J. O'Neil, E.-J. Park, J.D. Chen, Sequestration and inhibition of Daxx-mediated transcriptional repression by PML, Mol. Cell Biol. 20 (5) (2000) 1784–1796.
[29] F. Lehembre, S. Müller, P.P. Pandolfi, A. Dejean, Regulation of Pax3 transcriptional activity by SUMO-1-modified PML, Oncogene 20 (1) (2001) 1–9.
[30] S.F. Banani, H.O. Lee, A.A. Hyman, M.K. Rosen, Biomolecular condensates: organizers of cellular biochemistry, Nat. Rev. Mol. Cell Biol. 18 (5) (2017) 285–298.
[31] D. Pan, The hippo signaling pathway in development and cancer, Dev. Cell 19 (4) (2010) 491–505.
[32] I. Lian, J. Kim, H. Okazawa, J. Zhao, B. Zhao, J. Yu, A. Chinnaiyan, M.A. Israel, L.S.B. Goldstein, R. Abujarour, S. Ding, K.-L. Guan, The role of YAP transcription coactivator in regulating stem cell self-renewal and differentiation, Genes Dev. 24 (11) (2010) 1106–1118.
[33] X. Varelas, The Hippo pathway effectors TAZ and YAP in development, homeostasis and disease, Development 141 (8) (2014) 1614–1626.
[34] G.T.K. Boopathy, W. Hong, Role of Hippo pathway-YAP/TAZ signaling in angiogenesis, Front. Cell Dev. Biol. 7 (2019) 49.
[35] D. Cai, D. Feliciano, P. Dong, E. Flores, M. Gruebele, N. Porat-Shliom, S. Sukenik, Z. Liu, J. Lippincott-Schwartz, Phase separation of YAP reorganizes genome topology for long-term YAP target gene expression, Nat. Cell Biol. 21 (12) (2019) 1578–1589.
[36] X. Zhang, A. Abdelrahman, B. Vollmar, D. Zechner, The ambivalent function of YAP in apoptosis and cancer, Int. J. Mol. Sci. 19 (12) (2018) 3770.

[37] M.T. Salcedo Allende, J. Zeron-Medina, J. Hernandez, T. Macarulla, J. Balsells, X. Merino, H. Allende, J. Tabernero, S. Ramon, Y. Cajal, Overexpression of Yes associated protein 1, an independent prognostic marker in patients with pancreatic ductal adenocarcinoma, correlated with liver metastasis and poor prognosis, Pancreas 46 (7) (2017) 913–920.
[38] E. Rozengurt, J. Sinnett-Smith, G. Eibl, Yes-associated protein (YAP) in pancreatic cancer: at the epicenter of a targetable signaling network associated with patient survival, Signal Transduct. Target. Ther. 3 (1) (2018) 11.
[39] M. Maugeri-Saccà, R. De Maria, The hippo pathway in normal development and cancer, Pharmacol. Ther. 186 (2018) 60–72.
[40] C. Lin, X. Xu, YAP1-TEAD1-Glut1 axis dictates the oncogenic phenotypes of breast cancer cells by modulating glycolysis, Biomed. Pharmacother. 95 (2017) 789–794.
[41] J. Zhao, X. Li, Y. Yang, D. Zhu, C. Zhang, D. Liu, K. Wu, S. Zhao, Effect of YAP1 silencing on esophageal cancer, Onco Targets Ther. 9 (2016) 3137–3146.
[42] L. Lin, A.J. Sabnis, E. Chan, V. Olivas, L. Cade, E. Pazarentzos, S. Asthana, D. Neel, J.J. Yan, X. Lu, L. Pham, M.M. Wang, N. Karachaliou, M.G. Cao, J.L. Manzano, J.L. Ramirez, J.M.S. Torres, F. Buttitta, C.M. Rudin, E.A. Collisson, A. Algazi, E. Robinson, I. Osman, E. Muñoz-Couselo, J. Cortes, D.T. Frederick, Z.A. Cooper, M. McMahon, A. Marchetti, R. Rosell, K.T. Flaherty, J.A. Wargo, T.G. Bivona, The hippo effector YAP promotes resistance to RAF and MEK-targeted cancer therapies, Nat. Genet. 47 (3) (2015) 250–256.
[43] W. Li, Y. Cao, J. Xu, Y. Wang, W. Li, Q. Wang, Z. Hu, Y. Hao, L. Hu, Y. Sun, G. Xu, G. Ao, YAP transcriptionally regulates COX-2 expression and GCCSysm-4 (G-4), a dual YAP/COX-2 inhibitor, overcomes drug resistance in colorectal cancer, J. Exp. Clin. Cancer Res. 36 (1) (2017) 144.
[44] Z. Zhou, J. Zhu, C. Gao, L. Li, C. Zhou, H. Wang, X. Liu, siRNA targeting YAP gene inhibits gastric carcinoma growth and tumor metastasis in SCID mice, Oncol. Lett. 11 (4) (2016) 2806–2814.
[45] Z. Zhou, J.-S. Zhu, Z.-P. Xu, Q. Zhang, Lentiviral vector-mediated siRNA knockdown of the YAP gene inhibits growth and induces apoptosis in the SGC7901 gastric cancer cell line, Mol. Med. Rep. 4 (6) (2011) 1075–1082.
[46] D.F. Tschaharganeh, X. Chen, P. Latzko, M. Malz, M.M. Gaida, K. Felix, S. Ladu, S. Singer, F. Pinna, N. Gretz, C. Sticht, M.L. Tomasi, S. Delogu, M. Evert, B. Fan, S. Ribback, L. Jiang, S. Brozzetti, F. Bergmann, F. Dombrowski, P. Schirmacher, D.F. Calvisi, K. Breuhahn, Yes-associated protein up-regulates jagged-1 and activates the notch pathway in human hepatocellular carcinoma, Gastroenterology 144 (7) (2013) 1530–1542. e12.
[47] Y. Tian, B. Tang, C. Wang, D. Sun, R. Zhang, N. Luo, Z. Han, R. Liang, Z. Gao, L. Wang, Metformin mediates resensitivity to 5-fluorouracil in hepatocellular carcinoma via the suppression of YAP, Oncotarget 7 (29) (2016) 46230–46241.
[48] X. Wang, B. Wu, Z. Zhong, Downregulation of YAP inhibits proliferation, invasion and increases cisplatin sensitivity in human hepatocellular carcinoma cells, Oncol. Lett. 16 (1) (2018) 585–593.
[49] X. Zhou, J. Su, S. Feng, L. Wang, X. Yin, J. Yan, Z. Wang, Antitumor activity of curcumin is involved in down-regulation of YAP/TAZ expression in pancreatic cancer cells, Oncotarget 7 (48) (2016) 79076–79088.
[50] Z. Jiang, X. Chen, K. Chen, L. Sun, L. Gao, C. Zhou, M. Lei, W. Duan, Z. Wang, Q. Ma, J. Ma, YAP inhibition by resveratrol via activation of AMPK enhances the sensitivity of pancreatic cancer cells to gemcitabine, Nutrients 8 (10) (2016) 546.

[51] C.H. Diep, K.M. Zucker, G. Hostetter, A. Watanabe, C. Hu, R.M. Munoz, D.D. Von Hoff, H. Han, Down-regulation of yes associated protein 1 expression reduces cell proliferation and clonogenicity of pancreatic cancer cells, PLoS One 7 (3) (2012) 1–9.
[52] X. Wang, D. Sun, J. Tai, S. Chen, M. Yu, D. Ren, L. Wang, TFAP2C promotes stemness and chemotherapeutic resistance in colorectal cancer via inactivating hippo signaling pathway, J. Exp. Clin. Cancer Res. 37 (1) (2018) 27.
[53] H. Li, F. He, X. Zhao, Y. Zhang, X. Chu, C. Hua, Y. Qu, Y. Duan, L. Ming, YAP inhibits the apoptosis and migration of human rectal cancer cells via suppression of JNK-Drp1-mitochondrial fission-Htra2/Omi pathways, Cell. Physiol. Biochem. 44 (5) (2017) 2073–2089.
[54] K. Ma, Q. Xu, S. Wang, W. Zhang, M. Liu, S. Liang, H. Zhu, N. Xu, Nuclear accumulation of Yes-Associated Protein (YAP) maintains the survival of doxorubicin-induced senescent cells by promoting survivin expression, Cancer Lett. 375 (1) (2016) 84–91.
[55] Y. Oku, N. Nishiya, S. Sugiyama, H. Sato, Y. Uehara, Sensitisation of cancer cells to MLN8237, an Aurora-A inhibitor, by YAP/TAZ inactivation, Anticancer Res. 38 (6) (2018) 3471–3476.
[56] X. Sheng, W. Li, D. Wang, K. Chen, J. Cao, Z. Luo, J. He, M. Li, W. Liu, C. Yu, YAP is closely correlated with castration-resistant prostate cancer, and downregulation of YAP reduces proliferation and induces apoptosis of PC-3 cells, Mol. Med. Rep. 12 (4) (2015) 4867–4876.
[57] K.K. Slemmons, L.E.S. Crose, E. Rudzinski, R.C. Bentley, C.M. Linardic, Role of the YAP oncoprotein in priming RAS-driven rhabdomyosarcoma, PLoS One 10 (10) (2015) 1–18.
[58] E. Lapi, S. Di Agostino, S. Donzelli, H. Gal, E. Domany, G. Rechavi, P.P. Pandolfi, D. Givol, S. Strano, X. Lu, G. Blandino, PML, YAP, and p73 are components of a proapoptotic autoregulatory feedback loop, Mol. Cell 32 (6) (2008) 803–814.
[59] M. Howell, C. Borchers, S.L. Milgram, Heterogeneous nuclear ribonuclear protein U associates with YAP and regulates its co-activation of Bax transcription, J. Biol. Chem. 279 (25) (2004) 26300–26306.
[60] Q. Xiao, Z. Qian, W. Zhang, J. Liu, E. Hu, J. Zhang, M. Li, J. Wang, F. Kong, Y. Li, R. Wang, X. Tan, D. He, X. Xiao, Depletion of CABYR-a/b sensitizes lung cancer cells to TRAIL-induced apoptosis through YAP/p73mediated DR5 upregulation, Oncotarget 7 (8) (2016) 9513–9524.
[61] D. Matallanas, D. Romano, K. Yee, K. Meissl, L. Kucerova, D. Piazzolla, M. Baccarini, J.K. Vass, W. Kolch, E. O'Neill, RASSF1A elicits apoptosis through an MST2 pathway directing proapoptotic transcription by the p73 tumor suppressor protein, Mol. Cell 27 (6) (2007) 962–975.
[62] J. Wu, A.M. Minikes, M. Gao, H. Bian, Y. Li, B.R. Stockwell, Z.-N. Chen, X. Jiang, Intercellular interaction dictates cancer cell ferroptosis via NF2–YAP signalling, Nature 572 (7769) (2019) 402–406.
[63] L. Sun, J. Wu, F. Du, X. Chen, Z.J. Chen, Cyclic GMP-AMP synthase is a cytosolic DNA sensor that activates the type I interferon pathway, Science 339 (6121) (2013) 786–791.
[64] J. Wu, L. Sun, X. Chen, F. Du, H. Shi, C. Chen, Z.J. Chen, Cyclic GMP-AMP is an endogenous second messenger in innate immune signaling by cytosolic DNA, Science 339 (6121) (2013) 826–830.

[65] M.M. Gaidt, T.S. Ebert, D. Chauhan, K. Ramshorn, F. Pinci, S. Zuber, F. O'Duill, J.L. Schmid-Burgk, F. Hoss, R. Buhmann, G. Wittmann, E. Latz, M. Subklewe, V. Hornung, The DNA inflammasome in human myeloid cells is initiated by a STING-cell death program upstream of NLRP3, Cell 171 (5) (2017) 1110–1124 e18.

[66] L. Andreeva, B. Hiller, D. Kostrewa, C. Lässig, C.C. de Oliveira Mann, D. Jan Drexler, A. Maiser, M. Gaidt, H. Leonhardt, V. Hornung, K.-P. Hopfner, cGAS senses long and HMGB/TFAM-bound U-turn DNA by forming protein–DNA ladders, Nature 549 (7672) (2017) 394–398.

[67] S. Luecke, A. Holleufer, M.H. Christensen, K.L. Jønsson, G.A. Boni, L.K. Sørensen, M. Johannsen, M.R. Jakobsen, R. Hartmann, S.R. Paludan, cGAS is activated by DNA in a length-dependent manner, EMBO Rep. 18 (10) (2017) 1707–1715.

[68] M. Du, Z.J. Chen, DNA-induced liquid phase condensation of cGAS activates innate immune signaling, Science 361 (6403) (2018) 704–709.

[69] S.R. Paludan, L.S. Reinert, V. Hornung, DNA-stimulated cell death: implications for host defence, inflammatory diseases and cancer, Nat. Rev. Immunol. 19 (3) (2019) 141–153.

[70] W.M. Babinchak, W.K. Surewicz, Liquid–liquid phase separation and its mechanistic role in pathological protein aggregation, J. Mol. Biol. 432 (7) (2020) 1910–1925.

[71] S. Alberti, D. Dormann, Liquid–liquid phase separation in disease, Annu. Rev. Genet. 53 (1) (2019) 171–194.

[72] F. Gasset-Rosa, S. Lu, H. Yu, C. Chen, Z. Melamed, L. Guo, J. Shorter, S. Da Cruz, D.W. Cleveland, Cytoplasmic TDP-43 de-mixing independent of stress granules drives inhibition of nuclear import, loss of nuclear TDP-43, and cell death, Neuron 102 (2) (2019) 339–357. e7.

[73] Y. Duan, A. Du, J. Gu, G. Duan, C. Wang, X. Gui, Z. Ma, B. Qian, X. Deng, K. Zhang, L. Sun, K. Tian, Y. Zhang, H. Jiang, C. Liu, Y. Fang, PARylation regulates stress granule dynamics, phase separation, and neurotoxicity of disease-related RNA-binding proteins, Cell Res. 29 (3) (2019) 233–247.

[74] Y. Zhang, J. Wang, M. Ding, Y. Yu, Site-specific characterization of the Aspand Glu-ADP-ribosylated proteome, Nat. Methods 10 (10) (2013) 981–984.

[75] M. Altmeyer, K.J. Neelsen, F. Teloni, I. Pozdnyakova, S. Pellegrino, M. Grøfte, M.-B.D. Rask, W. Streicher, S. Jungmichel, M.L. Nielsen, J. Lukas, Liquid demixing of intrinsically disordered proteins is seeded by poly(ADP-ribose), Nat. Commun. 6 (1) (2015) 8088.

[76] L. Galluzzi, I. Vitale, S.A. Aaronson, J.M. Abrams, D. Adam, P. Agostinis, E.S. Alnemri, L. Altucci, I. Amelio, D.W. Andrews, M. Annicchiarico-Petruzzelli, A.V. Antonov, E. Arama, E.H. Baehrecke, N.A. Barlev, N.G. Bazan, F. Bernassola, M.J.M. Bertrand, K. Bianchi, M.V. Blagosklonny, K. Blomgren, C. Borner, P. Boya, C. Brenner, M. Campanella, E. Candi, D. Carmona-Gutierrez, F. Cecconi, F.K.-M. Chan, N.S. Chandel, E.H. Cheng, J.E. Chipuk, J.A. Cidlowski, A. Ciechanover, G.M. Cohen, M. Conrad, J.R. Cubillos-Ruiz, P.E. Czabotar, V. D'Angiolella, T.M. Dawson, V.L. Dawson, V. De Laurenzi, R. De Maria, K.-M. Debatin, R.J. DeBerardinis, M. Deshmukh, N. Di Daniele, F. Di Virgilio, V.M. Dixit, S.J. Dixon, C.S. Duckett, B.D. Dynlacht, W.S. El-Deiry, J.W. Elrod, G.M. Fimia, S. Fulda, A.J. García-Sáez, A.D. Garg, C. Garrido, E. Gavathiotis, P. Golstein, E. Gottlieb, D.R. Green, L.A. Greene, H. Gronemeyer, A. Gross, G. Hajnoczky, J.M. Hardwick, I.S. Harris, M.O. Hengartner, C. Hetz, H. Ichijo, M. Jäättelä, B. Joseph, P.J. Jost, P.P. Juin, W.J. Kaiser, M. Karin, T. Kaufmann, O. Kepp,

A. Kimchi, R.N. Kitsis, D.J. Klionsky, R.A. Knight, S. Kumar, S.W. Lee, J.J. Lemasters, B. Levine, A. Linkermann, S.A. Lipton, R.A. Lockshin, C. López-Otín, S.W. Lowe, T. Luedde, E. Lugli, M. MacFarlane, F. Madeo, M. Malewicz, W. Malorni, G. Manic, J.-C. Marine, S.J. Martin, J.-C. Martinou, J.P. Medema, P. Mehlen, P. Meier, S. Melino, E.A. Miao, J.D. Molkentin, U.M. Moll, C. Muñoz-Pinedo, S. Nagata, G. Nuñez, A. Oberst, M. Oren, M. Overholtzer, M. Pagano, T. Panaretakis, M. Pasparakis, J.M. Penninger, D.M. Pereira, S. Pervaiz, M.E. Peter, M. Piacentini, P. Pinton, J.H.M. Prehn, H. Puthalakath, G.A. Rabinovich, M. Rehm, R. Rizzuto, C.M.P. Rodrigues, D.C. Rubinsztein, T. Rudel, K.M. Ryan, E. Sayan, L. Scorrano, F. Shao, Y. Shi, J. Silke, H.-U. Simon, A. Sistigu, B.R. Stockwell, A. Strasser, G. Szabadkai, S.W.G. Tait, D. Tang, N. Tavernarakis, A. Thorburn, Y. Tsujimoto, B. Turk, T. Vanden Berghe, P. Vandenabeele, M.G. Vander Heiden, A. Villunger, H.W. Virgin, K.H. Vousden, D. Vucic, E.F. Wagner, H. Walczak, D. Wallach, Y. Wang, J.A. Wells, W. Wood, J. Yuan, Z. Zakeri, B. Zhivotovsky, L. Zitvogel, G. Melino, G. Kroemer, Molecular mechanisms of cell death: recommendations of the nomenclature committee on cell death, Cell Death Differ. 25 (3) (2018) 486–541.

[77] D. Denton, S. Kumar, Autophagy-dependent cell death, Cell Death Differ. 26 (4) (2019) 605–616.

[78] W.-W. Li, J. Li, J.-K. Bao, Microautophagy: lesser-known self-eating, Cell. Mol. Life Sci. 69 (7) (2012) 1125–1136.

[79] S. Kaushik, A.M. Cuervo, Chaperone-mediated autophagy: a unique way to enter the lysosome world, Trends Cell Biol. 22 (8) (2012) 407–417.

[80] P.O. Seglen, P.B. Gordon, I. Holen, Non-selective autophagy, Semin. Cell Biol. 1 (6) (1990) 441–448.

[81] D. Denton, S. Nicolson, S. Kumar, Cell death by autophagy: facts and apparent artefacts, Cell Death Differ. 19 (1) (2012) 87–95.

[82] S.K. Dasari, E. Schejter, S. Bialik, A. Shkedy, V. Levin-Salomon, S. LevinZaidman, A. Kimchi, Death by over-eating: the Gaucher disease associated gene GBA1, identified in a screen for mediators of autophagic cell death, is necessary for developmental cell death in *Drosophila* midgut, Cell Cycle 16 (21) (2017) 2003–2010.

[83] J.-Y. Byun, C.-H. Yoon, S. An, I.-C. Park, C.-M. Kang, M.-J. Kim, S.-J. Lee, The Rac1/MKK7/JNK pathway signals upregulation of Atg5 and subsequent autophagic cell death in response to oncogenic Ras, Carcinogenesis 30 (11) (2009) 1880–1888.

[84] A. Fernández-Medarde, E. Santos, Ras in cancer and developmental diseases, Genes Cancer 2 (3) (2011) 344–358.

[85] M. Elgendy, C. Sheridan, G. Brumatti, S.J. Martin, Oncogenic Ras-induced expression of Noxa and Beclin-1 promotes autophagic cell death and limits clonogenic survival, Mol. Cell 42 (1) (2011) 23–35.

[86] C. Ploner, R. Kofler, A. Villunger, Noxa: at the tip of the balance between life and death, Oncogene 27 (1) (2008) S84–S92.

[87] R. Kang, H.J. Zeh, M.T. Lotze, D. Tang, The Beclin 1 network regulates autophagy and apoptosis, Cell Death Differ. 18 (4) (2011) 571–580.

[88] J.Y. Guo, H.-Y. Chen, R. Mathew, J. Fan, A.M. Strohecker, G. Karsli Uzunbas, J.J. Kamphorst, G. Chen, J.M.S. Lemons, V. Karantza, H.A. Coller, R.S. DiPaola, C. Gelinas, J.D. Rabinowitz, E. White, Activated Ras requires autophagy to maintain oxidative metabolism and tumorigenesis, Genes Dev. 25 (5) (2011) 460–470.

[89] G. Boulay, G.J. Sandoval, N. Riggi, S. Iyer, R. Buisson, B. Naigles, M.E. Awad, S. Rengarajan, A. Volorio, M.J. McBride, L.C. Broye, L. Zou, I. Stamenkovic, C. Kadoch, M.N. Rivera, Cancer-specific retargeting of BAF complexes by a prion-like domain, Cell 171 (1) (2017) 163–178. e19.

[90] X. Su, J.A. Ditlev, E. Hui, W. Xing, S. Banjade, J. Okrut, D.S. King, J. Taunton, M.K. Rosen, R.D. Vale, Phase separation of signaling molecules promotes T cell receptor signal transduction, Science 352 (6285) (2016) 595–599.
[91] A. Boija, I.A. Klein, R.A. Young, Biomolecular condensates and cancer, Cancer Cell 39 (2) (2021) 174–192.
[92] M.R. Mansour, B.J. Abraham, L. Anders, A. Berezovskaya, A. Gutierrez, A.D. Durbin, J. Etchin, L. Lawton, S.E. Sallan, L.B. Silverman, M.L. Loh, S.P. Hunger, T. Sanda, R.A. Young, A.T. Look, An oncogenic super-enhancer formed through somatic mutation of a noncoding intergenic element, Science 346 (6215) (2014) 1373–1377.

CHAPTER 7

Therapeutics targeting BCL2 family proteins

Haiming Dai[1,2], X. Wei Meng[3,4], Kaiqin Ye[1,2], Jia Jia[1,2] and Scott H. Kaufmann[3,4]

[1]Department of Pathology and Laboratory Medicine, Hefei Cancer Hospital, Chinese Academy of Sciences, Hefei, Anhui, China; [2]Anhui Province Key Laboratory of Medical Physics and Technology, Institute of Health & Medical Technology, Hefei Institutes of Physical Science, Chinese Academy of Sciences, Hefei, Anhui, China; [3]Department of Molecular Pharmacology and Experimental Therapeutics, Mayo Clinic, Rochester, MN, United States; [4]Division of Oncology Research, Mayo Clinic, Rochester, MN, United States

List of abbreviations

Apaf1	apoptotic protease-activating factor 1;
BAD	BCL2-associated death promoter protein;
BAK	BCL2 homologous antagonist/killer protein;
BAX	BCL2-associated X protein;
BCL2	B-cell lymphoma two protein;
BCLX$_L$	B-cell lymphoma-extra-large protein;
BH	BCL2 homology domain;
BID	BH3 interacting-domain death agonist;
BIM	BCL2 interacting mediator of cell death;
BOK	BCL2-related ovarian killer;
cGAS	cyclic GMP-AMP synthase;
CLL	chronic lymphocytic leukemia;
DAMP	damage-associated molecular patterns;
IRF3	interferon regulatory factor 3;
MAVS	mitochondrial antiviral signaling protein;
MCL1	myeloid cell leukemia sequence 1 protein;
MEF	mouse embryonic fibroblast;
MOMP	mitochondrial outer membrane permeabilization;
PUMA	p53 up-regulated modulator of apoptosis protein;
SAHB	stabilized α-helices of BCL2 domains;
SCLC	small cell lung carcinoma

1. Apoptotic pathways

1.1 Apoptosis definition and function

Apoptosis is a type of programmed cell death that is defined by its specific morphological and biochemical characteristics [1–4]. Morphologically, it is

Mechanisms of Cell Death and Opportunities for Therapeutic Development
ISBN 978-0-12-814208-0
https://doi.org/10.1016/B978-0-12-814208-0.00007-5

characterized by cell shrinkage, blebbing, and chromatin condensation, followed by the formation of membrane-enclosed fragments called apoptotic bodies. Biochemically, it is characterized by the extensive degradation of chromosomal DNA into oligomers of roughly 180 base pairs, which reflect activation of endogenous endonucleases, such as the caspase-activated DNase [5]. Apoptotic cells can then be recognized and cleared by phagocytes, principally macrophages, through the "eat me" signal on the cell surface [2,3]. Most of the apoptosis hallmarks, for example, chromatin condensation, protein degradation, and DNA fragmentation, are generated as a consequence of the proteolytic activity of apoptotic caspases, particularly caspase-3 [6]. Thus, apoptotic caspases are essential for the processes associated with dismantling of the cell and the formation of apoptotic bodies. Most of the time, integrity of the plasma membrane is maintained during the process of apoptosis until the apoptotic cell is phagocytosed. As a result, the release of degrading cellular contents is prevented, resulting in minimal inflammatory response by and disruption of the surrounding tissue. Recent studies have indicated that caspase-mediated degradation of several DAMP-sensing receptors, such as cGAS, IRF3, and MAVS, also plays an important role in inhibiting the inflammatory response [7].

Apoptosis occurs in multicellular organisms and plays an important role in normal physiological processes, such as development, immune response, and tissue homeostasis [8,9]. Moreover, during the development of cancer, apoptosis is often dysregulated. For example, the antiapoptotic proteins BCL2, $BCLX_L$, and MCL1 are often overexpressed in a variety of neoplasms, such as lymphoma, leukemia, melanoma, lung, breast, and prostate cancers [10–12]. Mutations or deletions are also found in genes encoding pro-apoptotic proteins in some types of cancer [13,14]. Furthermore, apoptosis was also found to play important roles in degenerative diseases, especially neurodegenerative diseases [9].

1.2 Two different apoptotic pathways

Apoptosis is generally triggered through one of two different pathways (Fig. 7.1), the extrinsic apoptotic pathway and the intrinsic apoptotic pathway [15,16]. The extrinsic apoptotic pathway, also called the death receptor pathway, is initiated when certain members of the tumor necrosis factor (TNF) receptor family bind to their ligands; for example, binding of FAS (CD95) to the FAS ligand (CD95L) initiates extrinsic apoptosis.

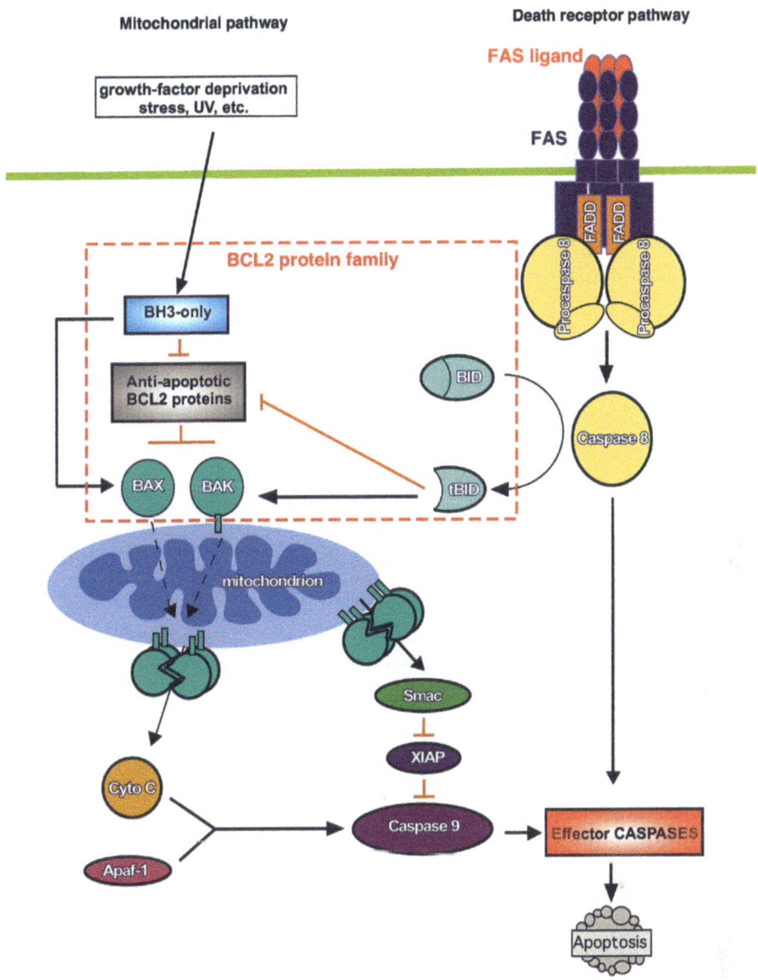

Figure 7.1 ***Two distinct apoptosis pathways: Death receptor (extrinsic) and mitochondrial (intrinsic) pathways.*** The death receptor apoptotic pathway is initiated by proteins of tumor necrosis factor superfamily, which bind to their receptors and cause the activation of caspase 8. The mitochondrial apoptotic pathway is controlled by BCL2 family proteins. The signaling activated by the death receptor pathway can be amplified as a consequence of activation of the mitochondrial pathway through BID cleavage. As part of mitochondrial pathway activation, the mitochondrial outer membrane (MOM) will be permeabilized by BAK and/or BAX, which will release the mitochondrial intermembrane proteins, including cytochrome *c* and Smac. Cytochrome *c* and Smac will activate the downstream caspases to generate the apoptotic features of a dying cell.

The TNF receptor superfamily includes the TNF receptor, FAS, and the TRAIL receptors DR4 and DR5. The ligation of these death receptors triggers the formation of intracellular death-inducing signaling complex (DISC) consisting of death receptors, death receptor adaptor proteins and caspase precursors, followed by the induction of pro-caspase cleavage, resulting in the activation of a caspase cascade and subsequent cell death [17–19].

The intrinsic pathway, also called the mitochondrial apoptosis pathway, is controlled by the BCL2 family of proteins. As shown in Fig. 7.2A, the BCL2 protein family is divided into three different subfamilies: the pro-apoptotic multidomain proteins BAK and BAX, which form oligomers after activation and directly permeabilize the MOM; the antiapoptotic multidomain BCL2 family members, including $BCLX_L$, BCLw, BCLB, MCL1, BCL2A1 (A1 in mouse), and BCL2 itself, which inhibit apoptosis through protein–protein interactions; and the pro-apoptotic BH3-only proteins, including BIM, PUMA, BID, NOXA, BAD, BIK, BMF, and HRK, which share similarity with the other BCL2 family members in only about 15 amino acids. Recent studies also indicate that BOK, which has a structure similar to that of BAK and BAX, can directly permeabilize the MOM [20]. Thus, BOK also belongs to the pro-apoptotic multidomain BCL2 subfamily.

The BH3-only protein BID, which is cleaved by the activated caspase 8 after the activation of death receptor pathway to generate a truncated BID (tBID) molecule with an exposed BH3 domain, links the signal transduction of the death receptor (extrinsic) pathway with the mitochondrial (intrinsic) pathway (Fig. 7.1) [21,22]. According to the dependence on BID cleavage during death receptor-induced apoptosis, cells are divided into two types, type I cells (such as lymphocytes) in which BID cleavage is not required for apoptosis and type II cells (such as hepatocytes) in which BID cleavage is essential [23]. The reason that BID cleavage is essential for death receptor-induced apoptosis in type II cells might be due to the low expression levels of death receptors in these cells, leading to low levels of active caspase 8 when death receptors on these cells are ligated [24].

1.3 Downstream of the mitochondrial apoptotic pathway

In a living cell, the pro- and antiapoptotic BCL2 protein family proteins are well balanced. Under some conditions, for example, growth-factor deprivation, stresses, some anticancer drug treatments, or UV irradiation, the

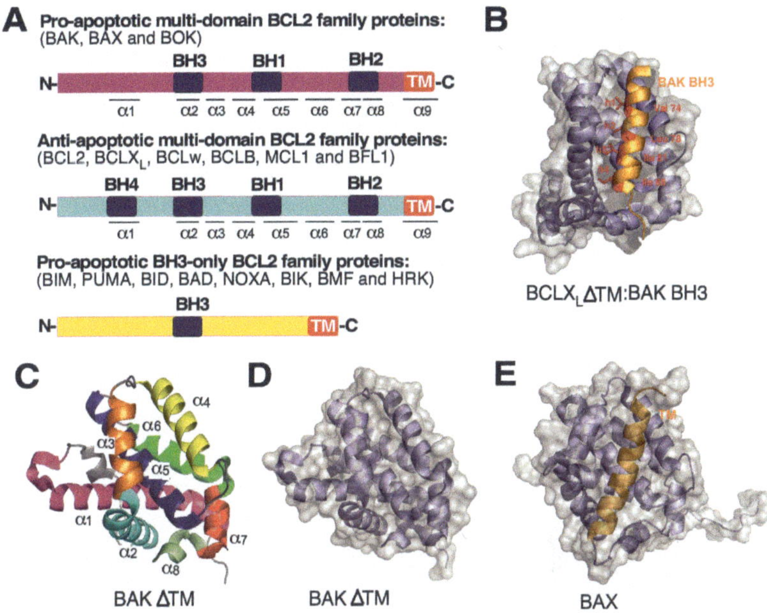

Figure 7.2 ***Structure of BCL2 family proteins.*** (A) BCL2 family proteins are divided into three subfamilies: The pro-apoptotic multi-domain BCL2 subfamily, the anti-apoptotic multi-domain BCL2 subfamily, and the pro-apoptotic BH3-only proteins. The pro-apoptotic multi-domain BCL2 family proteins contain three BCL2 homology (BH) domains; the anti-apoptotic multi-domain BCL2 family proteins contain four BH domains; and the pro-apoptotic BH3-only domain family proteins only have the BH3 domain in common with the other two subgroups. Most of the secondary structures of BCL2 family proteins are helixes, which are indicated under the BH domains. All three subfamilies of proteins also have a transmembrane (TM) domain to help target them to the MOM under certain conditions. BFL1 is also known as BCL2A1. (B) the structure and key interactions of anti-apoptotic protein BCLX$_L$ (white) in complex with the BH3 domain (orange) of BAK (PDB: 5FMK, Lee et al., 2016). Four hydrophobic residues (h1-4) along the BAK BH3 domain lying in the hydrophobic pockets of BCLX$_L$ were highlighted: Val 74, Leu 78, Ile 81 and Ile 85. (C, D) Ribbon (C) and Ribbon with Surface (D) diagrams show that BAK without transmembrane domain (BAKΔTM) has eight helices (PDB: 2IMS). (E) the structure of BAX with TM (orange) has nine helixes (PDB: 1F16). Direct comparison of the structure of monomer BAK (D) and BAX (E) shows that α9 (TM) of BAX sits in the groove of itself, while the TM domain of BAK usually extends out.

balance among the BCL2 family proteins is disrupted through molecular signaling pathways. Downstream of these pathways, BAK and/or BAX are activated, the mitochondrial outer membrane is permeabilized, and scores or hundreds of proteins that reside between the mitochondrial outer membrane and the inner membrane are released into the cytoplasm (Fig. 7.1). At least two of these proteins play important roles in mediating apoptosis. One is cytochrome *c*, which binds to the cytoplasmic scaffolding protein APAF1 (apoptotic protease activating factor 1) to induce oligomerization of APAF1, resulting in the recruitment of procaspase-9 to form a complex called the apoptosome and the subsequent activation of caspase-9 in an ATP- or dATP-dependent manner [25,26]. Another is SMAC (second mediator of activated caspases, also known as DIABLO), which binds to X-linked inhibitor of apoptosis (XIAP) protein to prevent XIAP from inhibiting caspase activation [27–29]. After activation through the concerted action of cytochrome *c* and SMAC, caspase-9 cleaves the downstream procaspases, particularly procaspase-3, and subsequently induces the morphological and biochemical characteristics of apoptosis. A recent study indicated that the RNase PNPT1 is also released from mitochondria after MOMP to mediate RNA decay, which also plays an important role in promoting apoptosis [30]. Readers are also referred to Chapters 1, 2, 3 and 5 of this volume for topics related to apoptosis.

2. Structure and function of BCL2 family proteins

2.1 The discovery of BCL2 family proteins

The *BCL2* (B cell lymphoma 2) gene was first identified because of the t(14; 18) chromosome translocation that is present in greater than 90% of follicular B cell lymphomas [11,31,32]. This translocation leads to the juxtaposition of the immunoglobulin heavy chain promoter and enhancer on chromosomes 14 with the protein coding sequence of *BCL2*, which is originally located on chromosome 18, resulting in excessive BCL2 expression and suggesting a possible role for BCL2 in the occurrence of these lymphomas. After a seminal study found that BCL2 could provide a survival signal and cooperate with c-myc to promote proliferation of B-cell precursors [33], subsequent investigations found that BCL2 overexpression can inhibit programmed cell death in a wide variety of cells, including B and T cells [34,35] as well as neurons and fibroblasts [36,37]. More importantly, the BCL2-mediated inhibition of cell death was shown to depend on inhibition of MOMP [38,39], which is mediated by BAX or BAK as indicated above, but not on inhibition of changes in mitochondrial respiration [40].

After the discovery of BCL2, additional antiapoptotic BCL2 family members, including $BCLX_L$, MCL1, BCLw, BCL2A1, and BCLb

(BCL2L10 in mice), have been identified. Like BCL2, all five of these proteins can inhibit apoptosis [16,41,42]. Moreover, all of these anti-apoptotic BCL2 family proteins have similar structural features (Fig. 7.2A), which include four conserved BH (BCL2 homology) domains named BH1-4 and a transmembrane (TM) domain at the C terminus that can target them to the MOM.

The proapoptotic multidomain BCL2 family proteins BAX and BAK were identified based on their abilities to bind BCL2 or $BCLX_L$ [43,44]. This subgroup of proteins contains BH domains one to three at the N-terminus and a TM domain at the C-terminus (Fig. 7.2A). In contrast to the antiapoptotic BCL2 family members, BAX and BAK induce cell death upon overexpression rather than inhibit it [43,45,46]. Later studies suggest that BOK also shares a similar domain structure and has pro-apoptotic function [47].

The pro-apoptotic BH3-only proteins, including BAD, BIK, and BIM, were also found because of their ability to form heterodimers with anti-apoptotic BCL2 family members [48–50]. This heterodimer formation displaces pro-apoptotic BAK and/or BAX from antiapoptotic BCL2 proteins, contributing to BAX/BAK activation. Structurally different from the other two subgroups, BH3-only proteins share similarity with other BCL2 family members only in the ~15 amino acid segment corresponding to the BH3 domain of BCL2 [49]. More importantly, the BH3 domains of the BH3-only proteins are critical for their binding to multidomain BCL2 proteins and their pro-apoptotic function [48,49].

2.2 Structures of BCL2 family proteins

2.2.1 Structure of antiapoptotic BCL2 family members

The antiapoptotic BCL2 proteins are typically comprised of 9 or 10 α helices that adopt a globular fold termed the BCL2 fold (Fig. 7.2B). The last α helix is typically separate from the globular core and inserts into the MOM, providing a TM tail-anchor. The BCL2 fold of BCL2, $BCLX_L$, MCL1, BCL2A1, and BCLb has 8 α helices, while in BCLw the BCL2 fold also includes an additional short C-terminal helix $\alpha9$ [42]. In each of these proteins, $\alpha5$ sits in the middle of the globular protein and is surrounded by the other helices. Each of the BH domains spans one or two helices surrounding $\alpha5$. As shown in Fig. 7.2A, for the multidomain BCL2 family proteins, the BH4 domain is part of the $\alpha1$ helix, and the BH3 domain is the $\alpha2$ helix. The BH1 domain bridges the $\alpha4$ and $\alpha5$ helices, and the BH2 domain spans the $\alpha7$ and $\alpha8$ helices. All of these antiapoptotic BCL2 proteins also include a canonical hydrophobic groove, formed by helices α3, α4, and α5, which constitute the binding site for pro-apoptotic BH3 domains. As shown in Fig. 7.2B, for example, the BH3 domain of the pro-apoptotic protein BAK binds to the canonical hydrophobic groove of the antiapoptotic $BCLX_L$ [51]. Along the hydrophobic groove, there are four pockets that accommodate four hydrophobic residues termed h1, h2, h3, and h4 along the BH3 domain that binds the groove [52]. Because the detailed structures of the hydrophobic grooves differ among the anti-apoptotic BCL2 family proteins, the structural features of these pockets are very important for the development of drugs selectively targeting the antiapoptotic BCL2 family proteins (see below).

2.2.2 Structure of pro-apoptotic BAK/BAX

The BAK or BAX protein in its monomeric state has a structure similar to that of the antiapoptotic BCL2 family proteins (Figs. 7.2C–E) and exists in solution as a globular structure consisting of nine helices. In the BAK monomer, the TM domain (α9) extends from its main globular structure and targets BAK to the MOM. As shown in Fig. 7.2C, the central hydrophobic core helix (α5) of BAK is also surrounded by seven alpha helices [53]. Among these helices, four of them (α1, α3, α4, and α6) are long helices that form a circle around the central helix, while the others (α2, α7, and α8) are shorter helices that link either the longer helices or the main structure to the TM domain. The structure of the BAX monomer is different from that of BAK [53,54]. As shown in Fig. 7.2D and E, the hydrophobic groove, which consists of helices α3, α4, and α5, opens wider in BAX than that in BAK and, therefore, the TM domain of BAX (α9) sits inside the hydrophobic groove. Because the TM domain of BAX is sequestered, BAX is predominantly distributed in the cytoplasm; BAK, in contrast, is mostly bound to the MOM [54,55]. When activated, BAX undergoes a conformational change that exposes the TM and allows localization of BAX to the MOM [56,57]. Moreover, through an interaction with $BCLX_L$, BAX could shuttle on and off the MOM to control mitochondrial priming [58].

2.2.3 Structure of BH3-only proteins

The BH3-only proteins are less similar to the other BCL2 family members and to each other. Most BH3-only proteins have coiled-coil structures and are tail-anchored to the MOM, which is independent of the binding capacity to antiapoptotic BCL2 proteins [59,60]. While the BH3-only domains have high similarities, the hydrophobic tails are different among the BH3-only proteins. BIM, BIK, and HRK possess a hydrophobic tail sequence that allows them to interact with the MOM [50]; NOXA contains a mitochondrial targeting sequence at its C-terminus that is not found in other BH3-only proteins [61,62]. In addition, some of the BH3-only proteins are normally targeted to cytoskeleton proteins before translocating to mitochondria. For example, BIM is sequestered by the microtubule-associated dynein motor complex through its interaction with dynein light chain proteins, which also form large complexes with BIM on mitochondria [63,64]. BMF is sequestered to myosin V motors through its interaction with dynein light chain 2 [65]. In response to certain apoptotic stimuli, such as certain microtubule- or actin-targeting agents, these BH3-only proteins are released from their interaction partners [65] and traffic to mitochondria, where they interact with anti-apoptotic BCL2 family members as well as BAK and BAX to initiate apoptosis, as will be described below.

The structure of BID is different from that of the other BH3-only proteins. BID possesses a complete BCL2 fold and has eight helices in its native state [66,67]. BID is cleaved by caspase-8 to generate its activated form termed truncated BID (tBID) upon the activation of the death receptor pathway [23,68]. tBID has six helices and contains a flexible loop between the helices $\alpha 4$ and $\alpha 5$. When interacting with MOM, tBID will expose all six helices to activate the mitochondrial apoptotic pathway [69].

2.3 Interactions between BCL2 family proteins

BCL2 family members control the intrinsic apoptotic pathway through their interactions. In a living cell, the effector pro-apoptotic BCL2 family proteins BAK and BAX are not activated until the balance between the antiapoptotic proteins and the pro-apoptotic proteins is disrupted. Before activation, BAK can form complexes with VDAC2 or the antiapoptotic proteins $BCLX_L$ and MCL1 [70–72]. Although BAX is mostly likely in an unbound state before activation, BAX in an activated conformation is also inhibited by the antiapoptotic BCL2 family proteins before MOMP [43,73]. Some of the BH3 only proteins, such as BIM, BAD, and NOXA, are also bound and inhibited by the antiapoptotic BCL2 proteins in normal living cells [39,74]. Therefore, antiapoptotic proteins act as gatekeepers to block apoptosis by inhibiting either activated BAK/BAX or the BH3-only proteins [75].

Not all of the BCL2 family member interactions are equal. Instead, different antiapoptotic BCL2 family proteins have different abilities to restrain activated BAK and BAX. For example, $BCLX_L$ and MCL1 are most effective in restraining BAK, while BCLw, BCL2, and BCL2A1 have a much lower affinity for BAK [76]. Some naturally existing BCL2 variants also show moderate effects in restraining BAK [77]. The abilities of anti-apoptotic BCL2 proteins in restraining BAX are also different. BCL2 and $BCLX_L$ have a higher affinity for BAX, whereas MCL1 showed a much lower affinity [75]. Unlike BAK and BAX, which can be inhibited by antiapoptotic BCL2 family members, BOK does not interact with any of the antiapoptotic BCL2 family proteins except for a marginal interaction with MCL1 [78]. Therefore, BOK-mediated MOMP is more likely regulated by ER-associated degradation than by the antiapoptotic BCL2 proteins [20].

The pro-apoptotic BH3-only proteins also selectively bind to anti-apoptotic BCL2 proteins. In general, the BH3-only proteins BIM and PUMA bind to and are inhibited by all the antiapoptotic BCL2 family members at low nanomolar affinities [74,79]. BAD binds to BCL2, $BCLX_L$, and BCLw selectively, while NOXA specifically binds to MCL1 and BCL2A1. BMF, BIK, and HRK exhibit moderate affinities for all of the antiapoptotic BCL2 family members, among which BMF more selectively interacts with BCL2, $BCLX_L$, and BCLw, while BIK and HRK display a higher affinity for $BCLX_L$, BCLw, and BCL2A1 [79]. tBID has low nanomolar affinities for $BCLX_L$, BCLw, and BCL2A1 but a much weaker affinity for BCL2 and MCL1 [79].

2.4 BAK/BAX activation and MOMP

2.4.1 BH3-only proteins and BAK/BAX activation

The most important function of BCL2 family proteins is to control MOMP, which is the fatal and usually irreversible step of the mitochondrial apoptosis pathway. The importance of BAK and BAX in this process is highlighted by the observation that cells lacking BAK and BAX are resistant to most apoptotic stimuli [80,81]. More recent studies have demonstrated that BOK can also directly mediate MOMP; however, its expression is limited to only a few tissues, such as ovary, uterus, and testis [20,47,82].

The activation of BAK and BAX requires BH3-only proteins in most cases [83]. Although different models have been proposed for BAK and BAX activation [73,83,84], a unified model is now widely accepted [75]. In this unified model, the BH3-only proteins play two different roles; one is to directly activate BAK or BAX, and the other is to displace the activated BAK and BAX from antiapoptotic BCL2 family members. According to their ability to activate BAK or BAX directly, BH3-only proteins are divided into two groups: activators, which can directly bind and activate BAK or BAX, and sensitizers, which neutralize antiapoptotic proteins to release BAK, BAX, and the sequestered BH3-only activators [80,84,85]. With the exception of BAD, all BH3-only proteins have been reported to be activators [24,56,86–89]. BIM, PUMA, and tBID were initially found to be direct BAK/BAX activators that can directly bind to BAK and/or BAX to induce their oligomerization and activation [56,86]. When murine genes encoding the three activators Bid, Bim, and Puma are deleted, the triple knockout mice exhibited developmental defects similar to that of *Bak/Bax* double knockout mice, suggesting essential roles for Bid, Bim, and Puma in the apoptosis that occurs during development [90]. Moreover, cells from the *Bid*, *Bcl2l11* (Bim), and *Bbc3* (Puma) triple knockout mice are resistant to most apoptotic stimuli, emphasizing the important roles of these BH3-only proteins in Bak and Bax activation. Whether NOXA is a direct activator is controversial. While some studies show NOXA as a direct activator [87,91], other studies report that the BH3 domain of NOXA cannot directly activate BAK or BAX [92–94]. The BH3-only proteins BIK, BMF, and HRK were also shown to have the abilities to directly activate BAK or BAX [89]. Several sites on both BAK and BAX have been found to initiate activation. For example, the BH3-only proteins BIM and PUMA initiate BAK activation at its canonical BH3 binding groove [87,88,94], while BMF and HRK initiate BAK activation through binding to a groove formed by α4, α6, and α7 [95]. The BH3-only proteins BIM and PUMA also initiate BAX activation at its canonical BH3 binding groove [92]. However, several other studies have suggested that the BH3-only proteins bind to the N-terminus of BAX to induce a conformational change that releases the TM domain from BAX BH3 binding groove [56,96].

2.4.2 Concentration-dependent BAK auto-activation

In addition to the BH3-only proteins, the BH3 domains of BAK or BAX have also been found to act as activators [94]. Thus, in some cell lines with high BAK levels, BAK appears to become activated independent of the activator BH3-only proteins [72]. This might be explained by occasional BH3 domain exposure in some BAK proteins leading to BAK auto-activation. This kind of auto-activation does not occur with BAX, which usually localizes to the cytoplasm and has a much lower local concentration. More importantly, BAK auto-activation depends on the presence of the mitochondrial outer membrane and is concentration-dependent [72,97]. In cells lacking BID, BIM, PUMA, and p53, or cells deficient of all eight BH3-only proteins (OctaKO), cells still undergo apoptosis after $BCLX_L$ and MCL1 inactivation [97,98], further supporting the auto-activation mechanism. After auto-activation, BAK is inhibited by the antiapoptotic BCL2 family proteins, such as $BCLX_L$ and MCL1. Thus, in cells with auto-activated BAK, BH3-only proteins can displace the auto-activated BAK from $BCLX_L$ and MCL1 to initiate apoptosis [99]. The binding status of BAK after its auto-activation determines the sensitivity of cells to certain BCL2 family protein inhibitors: Cells with constitutive $BCLX_L$/BAK complexes are particularly sensitive to the $BCLX_L$/BCL2/BCLW inhibitor navitoclax, while cells with constitutive MCL1/BAK complexes are more sensitive to MCL1 antagonist A1210477 [72].

2.4.3 Model of activated BAK or BAX and MOMP

When activated, BAK or BAX can undergo a conformational change, oligomerize, and permeabilize the MOM [3,100,101]. Earlier studies have shown that the BH3 domains of BAK and BAX are essential for this activation [102]. Later studies revealed two interaction surfaces involved in the homo-oligomerization of BAK and BAX. One is the BH3 domain (α2 helix):BH3 binding groove (α3–α5 helices) interaction surface; and the second might involve the dimerization interface of helix α6 [103–105]. While the dimerization surface involving the BH3 domain:BH3 binding groove has been confirmed by structural studies [92,94], the detailed structure of the other surface of the BAK or BAX higher order oligomer is not available yet. It has been suggested that this second binding surface might involve helix α9 in addition to α6 [106–108]. The mutations of both BH3 domain and BH3-binding groove can inhibit BAK/BAX-mediated MOMP [87,103–105]; however, no essential secondary interface has been identified that can be disrupted by mutagenesis. Thus, how the effector proteins permeabilize MOM is still not clear. Several different models have been proposed to explain this mechanism.

One of the models is the hairpin insertion model [109,110]. Because the monomer structures of BAK and BAX have a fold similar to that of the channel-forming domains of bacterial toxins, such as diphtheria toxin and colicin E1 [53,54,111,112], BAK and BAX were initially thought to use similar hairpin insertion mechanisms to permeabilize the MOM.

Evidence supporting this model includes not only the assay showing the inability to label the residues in the α5 and α6 region with hydrophilic reagents after activation [110], but also the finding that the pore-forming peptides (α5 or α5 plus α6) of BAK or BAX can independently form large lipid pores and permeabilize liposomes in vitro [113,114]. Another model was built based on the observations that the core domain (α2–α5) and the latch domain (α6–α8) of BAK and BAX were found to separate during the process of activation [94,95], which suggested an asymmetric oligomerization model. Based on this model, together with the long-distance restraints obtained from double electron-electron resonance spectroscopy, a 3D clamp model for active BAX on the MOM has also been proposed [115]. In this model, the core domains (α2–α5) from two activated BAX monomers form a dimer on the rim of the mitochondrial pore, while two α6 helices from two BAX molecules lie on the surface of the MOM. While BAX was activated on the MOM, the structure of rings and arcs could be found under a dual-color single-molecule localization based on super-resolution microscopy [116,117], further supporting this clamp like model of BAK and BAX activation.

The third model was proposed based on the fact that not all the residues within the α5 and α6 helices in BAK and BAX were found to be embedded in the membrane during MOMP [118]. Moreover, the α1 domain and the α1–α2 loop are mobile after BAK activation, while the core domain is rigid, suggesting an in-plane model, in which activated BAK and BAX use disordered clusters to permeabilize the MOM after activation [119]. Further structural study indicated some specific lipids can promote the formation of high-order BAK oligomers [120]. In a separate study, the mitochondrial lipid cardiolipin was shown to be involved in tBID-induced BAX activation and further cytochrome *c* release [121]. Thus, after the core domain and latch domain separation from each other during activation followed by the formation of a BH3 in groove dimer of BAK/BAX, how BAK and BAX permeabilize MOM remains to be elucidated and requires further investigation.

2.5 Other cellular functions related to BCL2 family proteins

In addition to controlling MOMP, BCL2 family proteins have also been reported to regulate other important functions, such as autophagy, mitochondrial fission and fusion, the DNA damage response, ER stress, and transcription. However, whether BCL2 family proteins regulate transcription and DNA damage response or not requires further study, and the molecular mechanisms underlying the other functions also remain to be clarified.

2.5.1 Regulation of autophagy

About two decades ago, BECLIN1 was found to play a critical role in autophagy in mammalian cells [122]. BECLIN1 contains a conserved BH3 domain that can interact with BCL2 [122]. Further studies found that BECLIN1 is a component of the class III phosphatidylinositol-3 kinase (PI3K) complex involved in autophagosome initiation; moreover, BECLIN1 is critical for early embryonic survival [123]. The interactions between BECLIN1 and antiapoptotic BCL2 family proteins represent a potential mechanism of crosstalk between apoptosis and autophagy, because BCL2 has both antiapoptotic and antiautophagic roles [124,125]. Besides BCL2, other BCL2 family members, such as $BCLX_L$ and MCL1, can also interact with BECLIN1 and inhibit autophagy. A structural study not only confirmed the interaction of $BCLX_L$ and the BH3 domain of BECLIN1 but also showed binding of the BECLIN1 BH3 domain to the canonical BH3 binding groove of the antiapoptotic protein. Moreover, mutations of the BH3 domain that inhibit the binding of BECLIN1 to antiapoptotic BCL2 proteins also abolish the antiautophagic effect of the antiapoptotic BCL2 proteins [126]. Upon nutrient withdrawal, BCL2 is phosphorylated by activated JNK1 (stress-activated c-Jun N-terminal protein kinase 1), which releases BECLIN1 from BCL2 and subsequently induces autophagy [124].

This interaction between BECLIN1 and antiapoptotic BCL2 family members can be modulated by BH3-only proteins. For example, BAD can promote autophagy through its competitive binding to BCL2 or $BCLX_L$, which displaces BECLIN1 from complexes containing BECLIN1 and BCL2 or $BCLX_L$ [126,127]. Consistent with this model, knockdown of BAD in HeLa cells or mouse embryonic fibroblasts inhibits starvation-induced autophagy, which can be rescued by addition of the BH3 mimic compound ABT-737, further supporting the important roles of the BH3-only proteins in autophagy [126]. On the other hand, BAD overexpression can promote autophagy even under normal condition [126]. Some other BH3-only proteins, such as PUMA, BIK, NOXA, and BIM, have also been implicated in the regulation of autophagy [127–129].

While studies have suggested that BCL2 negatively regulates autophagy through its interaction with BECLIN1, several recent studies have led to a different model in which anti-apoptotic BCL2 family proteins regulate autophagy downstream of BAK and BAX [130,131]. Additional studies have shown that mitochondrial DNA released after MOMP can activate a cGAS-STING dependent pathway to induce autophagy, providing a possible explanation for autophagy induction downstream of BAK/BAX by the BH3-only proteins or BH3 mimetics [132–134]. On the other hand, like BECLIN1, AMBRA1 contains a novel BH3-like domain and is a component of the Class III PIK3 complex. Under basal conditions, AMBRA1 interacts with BECLIN1 and regulates autophagy positively [135,136]. A recent study showed that AMBRA1 binding to BCL2 and MCL1 also plays an important role in regulating autophagy and apoptosis [137]. Taken together, while there is clear evidence linking BCL2 family proteins to autophagy, precisely how they regulate autophagy still requires further study.

2.5.2 Regulation of mitochondrial fission and fusion

Antiapoptotic and pro-apoptotic BCL2 family members are also involved in the regulation of mitochondrial dynamics through mitochondrial fission and fusion. The balance between mitochondrial fission and fusion determines mitochondrial morphology. Mitochondrial fission is mainly mediated by the large GTPase protein Drp1 (Dynamic related protein 1). Mitochondrial fusion is controlled by Mfn1 (Mitofusion 1) and Mfn2 (Mitofusion 2), which are located on the MOM and mediate fusion, and Opa1 (optic atrophy 1), which controls mitochondrial inner membrane fusion. Several studies have indicated that BCL2 family proteins interact with mitochondrial fission and fusion factors to regulate mitochondrial morphogenesis [138–141]. For example, in healthy cells, BAK or BAX is required for the regulation of mitochondrial morphology [138]. During apoptosis, BAX colocalizes with Drp1 and Mfn2 and regulates mitochondrial morphology [142]. Two antiapoptotic BCL2 family members, $BCLX_L$ and BCLw, are also involved in the regulation of mitochondrial morphology [140,141]. Using fluorescence microscopy and computational strategies, $BCLX_L$ was found to increase the rate of mitochondrial fusion and fission and mitochondrial biomass in neuronal cells [141]. Moreover, a study using a chimeric $BCLX_L$/BAX protein suggests that the BCL2 family proteins might inhibit the activity of Mfn1 and Mfn2 and affect mitochondrial morphology through interaction with certain complexes [143]. All of these findings suggest that BCL2 family proteins are involved in mitochondrial fission and fusion. However, the detailed mechanism by which BCL2 family proteins regulate mitochondrial fission and fusion requires further elucidation.

2.5.3 Regulation of DNA damage response

The pro-apoptotic BH3-only protein BID was reported to be involved in the DNA damage response [144,145]. BID is phosphorylated by ATM and induces an S phase arrest following induction of DNA double-strand breaks [144]. However, in another study, BID deficient cells underwent DNA damage-induced cell-cycle arrest in a manner indistinguishable from control WT cells, arguing against a role for BID in the DNA damage response [146]. Thus, it is still controversial whether BCL2 family proteins play a role in the DNA damage response [147].

2.5.4 Regulation of ER stress

The BH3 only proteins BIM and PUMA have been reported to play important roles in regulating endoplasmic reticulum (ER) stress. In ER stress induced by misfolded proteins, CHOP–C/EBPα upregulates BIM, which is phosphorylated, resulting in apoptosis [148]. In high glucose concentration-mediated ER and oxidative stress in type 2 diabetes, both BIM and PUMA are upregulated [149]. Moreover, in ER stress-regulated antibody secretion by primary B cells, BIM and PUMA reportedly interact with IRE1α and inhibit its effects on XBP-1 mRNA splicing [150].

2.5.5 Regulation of transcription

Antiapoptotic BCL2 family proteins are implicated in regulating the transcription process directly [151]. Using a gain-of-function screen, three antiapoptotic proteins, MCL1, $BCLX_L$, and BCL2, have been reported to regulate GLI-mediated transcriptional activity. These antiapoptotic proteins directly inhibit the interactions between SUFU (Suppressor of Fused) and GLI, thereby promoting the activity of GLI. Subsequently, GLI induces the transcription of *MCL1*, *BCLX*$_L$, and *BCL2* genes, which forms a feed-forward loop to promote cancer cell survival and growth. Moreover, BH3 mimetics disrupt the GLI activity and inhibit the feed forward signaling driven by BCL2 family proteins.

3. BCL2 family proteins and diseases

As stated above, BCL2 family proteins not only play an important role in controlling mitochondrial apoptosis, but also participate in the regulation of autophagy, mitochondrial fission, and fusion, ER stress and possibly other important cellular functions. Therefore, when genes encoding BCL2 family proteins are mutated, silenced, altered in copy number, overexpressed at the protein level or dysregulated by miRNA, cellular functions can be disrupted. Dysregulation of BCL2 family proteins may lead to certain diseases, including cancer, neurodegenerative diseases, and autoimmune disorders.

3.1 BCL2 family proteins and tumorigenesis

Cancer is often caused by disturbance of the homeostatic balance between cell growth and cell death [152]. Under some conditions, such as certain genetic alterations, apoptosis is dysregulated and leads to enhanced cell survival, which eventually might contribute to cancer, for example, by allowing cells with damaged DNA to persist when they otherwise might not be able to survive. Three types of changes in BCL2 family proteins that dysregulate apoptosis are seen in various cancers: gene alterations inducing upregulation of antiapoptotic BCL2 proteins, including BCL2, $BCLX_L$, and MCL1; mutation or deletion of pro-apoptotic BH3-only genes; mutation or deletion of genes expressing the pro-apoptotic multidomain BCL2 proteins BAK and BAX. These alterations not only contribute to tumorigenesis but also lead to resistance of cancer cells to a variety of treatments.

3.1.1 Antiapoptotic BCL2 family proteins and tumorigenesis

3.1.1.1 BCL2

BCL2 protein overexpression has been found in many types of cancer, including small cell lung cancer [152,153], melanoma [154], breast cancer [155], prostate cancer [156], colon cancer [157], bladder cancer [158], and human lymphoid malignancies [159–161]. BCL2 overexpression not only plays an important role in cancer development but also contributes to treatment resistance [159,160]. BCL2 protein level up-regulations can be seen in quite a few different situations. In about 90% of follicular B cell lymphomas and about 30% diffuse large B cell lymphomas, the presence of the t(14,18) chromosomal translocation [162–164] places the *BCL2* gene from chromosome 18 next to the enhancer elements of the immunoglobulin heavy chain promoter on chromosome 14, thereby leading to extraordinarily high BCL2 overexpression [11,31,32]. Besides the t(14,18) translocation, other *BCL2* gene rearrangements and epigenetic changes may also upregulate BCL2 protein levels in different cancers [165,166]. In addition, BCL2 expression levels are regulated by miRNAs. For example, chromosome deletions and mutations affecting regions encoding miR-15a and miR-16, which target and repress *BCL2* mRNA, occur in about 50% of chronic lymphocytic leukemia, resulting in BCL2 upregulation [167]. BCL2 upregulation can also occur through dysregulation of ubiquitination and protein degradation. Mutations of *KEAP1*, which encodes the inhibitor of the nuclear factor E2 related factor 2 (INrf2) that interacts with the BH2 domain of BCL2 to facilitate BCL2 ubiquitination and degradation, were found in cell lines from lung cancer and other cancer types, resulting in BCL2 overexpression [168].

BCL2 upregulation also plays an important role in conferring resistance to a variety of cancer treatments. The increased expression of BCL2 in acute myeloid leukemia was associated with poor response to chemotherapy [169]. On the other hand, by down-regulating BCL2, some miRNAs, including miR-195, miR-24-2, mir-365-2, and miR-204, have been reported to increase chemotherapy sensitivity in certain cancer cells [170,171].

3.1.1.2 $BCLX_L$

As an important antiapoptotic protein, $BCLX_L$ also plays an important role in tumorigenesis, especially in the tumors driven by Myc. In a *myc*-driven lymphomagenesis mouse model, loss of even a single *Bcl2l1* allele encoding $BCLX_L$ delays tumorigenesis, while a normal level of $BCLX_L$ is sufficient to accelerate this process [172]. $BCLX_L$ upregulation has also been seen in several other tumor types, including hormone refractory prostate cancer [173] and mesenchymal breast cancers [174]. Several mechanisms, including gene amplification, increased transcription, and decreased miRNA-mediated mRNA degradation, can lead to $BCLX_L$ upregulation.

The genomic region containing *BCL2L1* is amplified in many malignancies, contributing to increased transcript levels [12]. Interestingly, the *MYC* gene is also included in these regions, suggesting an important role of $BCLX_L$ upregulation in *MYC*-driven tumors [12]. In chronic myeloid leukemia (CML), the product of the fusion gene *BCR-ABL* can phosphorylate STAT5, which subsequently binds to the promoter of *BCL2L1* gene to drive its overexpression [175]. Several miRNAs, including miR-let7c, can target the 3′-UTR region of $BCLX_L$ mRNA and decrease $BCLX_L$ expression. Thus, down-regulation of miR-let7c, which has been found in hepatocellular carcinomas, induces $BCLX_L$ upregulation [176]. Increased $BCLX_L$ confers resistance to anticancer reagents such as cisplatin [177]. Similar to BCL2, $BCLX_L$ protein levels are also regulated by the INrf2-Cul3-Rbx1 E3 ubiquitin ligase complex [168]. Consequently, *KEAP1* mutations lead to increased $BCLX_L$ protein levels.

3.1.1.3 MCL1

MCL1 upregulation also contributes to tumorigenesis in many situations. B cell lymphoma develops at a very high incidence (85%) over a period of

2 years in *Mcl1* transgenic mice [178], providing evidence that MCL1 can act as an oncogene. MCL1 upregulation confers chemotherapy resistance [179], whereas MCL1 down-regulation promotes drug-induced cell death in many cancer cells [180,181], suggesting a critical role for MCL1 in cancer cell survival. Conditional knockout of *Mcl1* in a mouse model of acute myeloid leukemia (AML) induces the death of transformed AML cells and eradication of AML in afflicted mice, further supporting an essential role for MCL1 in AML development [182]. Moreover, MCL1 levels were also found to correlate with cancer progression [183].

MCL1 upregulation has also been observed in many types of solid tumors [12,184] or at the time of relapse in leukemia patients [185]. MCL1 is regulated at multiple levels. *MCL1* gene amplification, subsequently leading to MCL1 overexpression, has been observed in lung and breast cancer cell lines [12]. MCL1 is also regulated by miRNAs. For example, downregulation of miR-29, which targets MCL1 mRNA and inhibits its translation, has been found in a variety of tumors [186,187]. Decreased expression of two other miRNAs, miR-125b, and miR-133b, which target 3′-UTR region of MCL1 mRNA and inhibit its expression, is found in liver cancer [188]. Finally, MCL1 levels are also regulated by ubiquitination and degradation. The MCL1 deubiquitinase USP9X, which removes ubiquitin from MCL1 and protects it from degradation, was found to be overexpressed in follicular lymphoma (FL), diffuse large B cell lymphoma, and multiple myeloma [188].

The other antiapoptotic BCL2 family proteins, such as BCLw, BCLB, and BFL1 (BCL2A1), also play important roles in tumorigenesis and confer chemotherapy resistance to cancer cells. Readers are referred to other references for details [4,189].

3.1.2 BH3-only proteins in tumorigenesis

Mutations or deletions in genes encoding BH3-only proteins have also been found in several kinds of tumors. Moreover, genetic studies in mice have provided experimental evidence that certain BH3-only proteins can serve as tumor suppressors [190].

BIM (BCL2L11) is found to be associated with tumorigenesis. In a genetic study using Eμ-*Myc* transgenic mice, even deletion of a single *Bim* allele could accelerate epithelial tumorigenesis [191]. Moreover, gene deletion or down-regulation caused by promoter hypermethylation of *BIM* is found in many neoplasms. For example, homozygous deletion of *BIM* is observed in mantle cell lymphoma and *BIM* promoter hypermethylation is found in Burkitt lymphoma [14,192]. Point mutations in *MYC*, which prevent MYC from activating *BIM* and thereby cause BIM down-regulation, are also found in certain Burkitt lymphomas [193]. miRNAs can also regulate BIM levels and participate in tumorigenesis. For example, miR-17-92, which represses *BIM* expression, is often overexpressed or amplified in human cancers [194].

Loss of PUMA and NOXA is also associated with tumorigenesis. Because both *PUMA* and *NOXA* are transcriptional target genes of the *TP53* gene, which is widely mutated in human tumors, more than 50% of human tumors exhibit impaired ability to induce PUMA or NOXA expression in response to genotoxic damage or oncogenic stress. Although a genetic study indicates that $Noxa^{-/-}Puma^{-/-}$ mice do not develop tumors [195], *Puma* or *Noxa* deletions can accelerate tumorigenesis under certain conditions. For example, *Puma* deletion accelerates B lymphomagenesis in Eμ-*Myc* mice [196,197]. *Noxa* deletion accelerates γ-radiation-induced thymic lymphomagenesis [198]. Moreover, *PUMA* and *NOXA* deletion or down-regulations is found in many cancer types. For example, loss of heterozygosity or silencing of *PUMA* is found in Burkitt lymphomas, head and neck squamous cell carcinoma and lung cancer [196,199]. In cutaneous melanoma, *PUMA* mRNA is decreased [200]. *NOXA* has also been found to be mutated and silenced in diffuse large B-cell lymphoma [14]. miRNAs can also regulate PUMA levels and participate in tumorigenesis. MiR-34a, which can repress PUMA expression, is overexpressed in nasopharyngeal and gastric carcinoma [201].

Other BH3-only proteins also have tumor suppressor functions. Although most *Bad*-deficient mice develop normally, a higher incidence of diffuse large B-cell lymphoma was seen with age in $Bad^{-/-}$ mice when compared to wild-type control [202]. Mice deficient of both *Bim* and *Bad* has accelerated development of γ-radiation-induced thymic lymphomas [203]. Loss of *Bmf* also accelerates lymphomagenesis in mouse models [204,205]. Moreover, about 50% of *Bid*-deficient mice develop a chronic myelomonocytic leukemia-like syndrome by 2 years of age [206]. Genes encoding these BH3-only proteins are often mutated or silenced in several cancer types. For example, loss of function mutations of *BAD* was found in colon adenocarcinoma [207]. In patients with Burkitt lymphoma, BMF was also found to be downregulated [205].

3.1.3 BAK/BAX and tumorigenesis

In some situations, the pro-apoptotic multiple-domain BCL2 family proteins BAX and BAK inhibit tumorigenesis. *Bax*-deficient mice did not display an increased incidence of cancer [208]; however, in another study, the development of epithelial tumors by a *Tp53* dominant-negative mutant appeared to require the loss of both *Bax* and *Bak*, suggesting that under some conditions *Bax* and *Bak* can suppress tumorigenesis [209]. Moreover, loss of *Bax* or *Bak* can accelerate the progression of tumorigenesis in different experimental models. For example, loss of *Bax* accelerates the development of SV40-T antigen-driven mouse brain tumors [210]; even the loss of just one *Bax* allele can accelerate tumor progression [211]. Loss of *Bax* also promotes B cell lymphoma progression in Eμ-*Myc* transgenic mice [212]. Interestingly, *Bax* suppressed tumorigenesis of certain cell types more than in others. For example, when *Bax* was deleted in *Arf*-deficient mice, the spectrum of tumors that developed shifted from lymphoma only toward multiple primary tumors, including lymphoma, sarcoma, and carcinoma [213].

BAX or *BAK* mutations or deletions were also observed in multiple cancers. For example, loss-of-function *BAX* [13,214] and *BAK* mutations [215] are found in human colorectal carcinomas, gastric carcinomas, or hematopoietic cancers. In more than a half of the gastrointestinal tumors of the microsatellite mutator phenotype, at least one copy of *BAX* is mutationally inactivated [216].

3.2 BCL2 family proteins and neurodegenerative diseases

Caspase-mediated apoptosis also contributes to both acute and chronic neurodegenerative diseases [9,217]. Many studies have shown that the activation of caspases during neuronal cell death is regulated by BCL2 family proteins. For example, overexpression of BCL2 significantly increases the number of neurons in mice [218]. Moreover, overexpression of BCL2 homologue in *C. elegans* can prevent apoptosis of sympathetic neurons that is generally induced by nerve growth factor deprivation [219]. On the other hand, although Bcl2-deficient mice have normal development of the prenatal nervous system, loss of motor, sensory, and sympathetic neurons is found in these mice after birth, suggesting that BCL2 is important for maintaining specific types of neurons at specific times [220]. While BCL2 family proteins are known to be important in neuronal development and neural cell death, how BCL2 family proteins contribute to neurodegenerative diseases is still not completely understood.

Increased neuronal apoptosis is detected in Alzheimer's disease [221−223], a devastating neurodegenerative disease affecting millions of people, especially people over the age of 60 years. In many cases, Alzheimer's disease appears to be an apoptosis-related phenomenon due to caspase activation. The pathological characteristics of Alzheimer's disease include extracellular senile plaques, consisting predominantly of amyloid β (Aβ) peptide fibrils, and intracellular neurofibrillary tangles (NFTs). The NFTs, which are accumulations of the microtubule-associated protein tau, play an important role in the development of Alzheimer's disease. A recent study found that BCL2-regulated cellular resistance to apoptosis inhibits the degeneration triggered by the accumulation of the amyloid fibrils [224]. Moreover, CDK5-mediated phosphorylation of MCL1 on Thr92 causes mitochondrial dysfunction, promoting neurotoxicity in Alzheimer's disease [225]. On the other hand, nicergoline, a drug used to protect neurons againstβ-amyloid toxicity in the Alzheimer's disease, can upregulate the BCL2 protein expression and reduce BAX levels [226].

In addition to directly controlling MOMP, the regulation of ER-stress by BCL2 family proteins also contributes to neurodegenerative diseases. Aberrantly aggregated proteins induce elevated ER stress, which has been observed in dying neurons in several animal disease models, including ischemia, Parkinson's disease, Huntington's disease, and Alzheimer's disease. The BH3-only proteins BIM and PUMA play an essential role in C/EBP homologous protein (CHOP)-mediated ER stress and induce neuronal cell apoptosis, suggesting a possible pathway contributing to neurodegenerative diseases [227].

Autophagy, which is also regulated by BCL2 family proteins as described above, might also contribute to neurodegenerative diseases. Mutations in genes encoding mitophagy-regulating proteins, especially in the Parkin/PINK1 signaling pathway, contribute to the development of Parkinson's disease [228]. Studies have found that mice lacking phosphoglycerate mutase family member 5 (PGAM5), which inhibits mitophagy, display a Parkinson's-like movement disorder [229]. Through a direct interaction with PGAM5, a mitochondrion-localized phosphatase, $BCLX_L$ blocks FUNDC1 dephosphorylation to promote mitophagy [230].

3.3 BCL2 family proteins and auto-immune diseases

The regulation of mitochondrial apoptosis by BCL2 family proteins is also essential in preventing auto-immune diseases [231]. For example, *Bim*$^{-/-}$ mice exhibit prominent lymphoid hyperplasia, suggesting that BIM regulates lymphoid homeostasis [232]. Moreover, lymphoid and myeloid cells accumulate in *Bim*$^{-/-}$ mice, and T-cell development is also perturbed, which can lead to autoimmune kidney disease [232]. Conversely, BIM is required for the deletion of autoreactive thymocytes during thymocyte-negative selection [233]. Collectively, these observations implicate BIM as a barrier to autoimmunity. Although inhibition of CD95 death receptor-induced apoptosis is involved in the development of autoimmune lymphoproliferative syndrome, NRAS mutation-induced BIM downregulation has also been reported in a lymphoproliferative syndrome patient whose CD95-signaling pathway was intact [234]. Furthermore, BIM is essential in the depletion of autoreactive B cells [235] and required for the termination of acute and chronic immune responses, again indicating that BIM is important for the prevention of autoimmunity [236,237].

BIM also participates in the death of other immune cells. In NK cells, IL-15 upregulates MCL1, which interacts with BIM (and NOXA) to prevent cell death [238]. The fate of dendritic cells, the most important antigen-presenting cells in the immune system, is also tightly regulated by apoptosis. Accordingly, $Bim^{-/-}$ mice possess elevated numbers of DCs, which contribute to auto-immune diseases [239].

Antiapoptotic BCL2 proteins also play an important role in autoimmune diseases. For example, antiapoptotic BCL2 family proteins are critical for the survival of lymphocytes [218]. $BCLX_L$ and BCL2 have important roles in promoting cell survival at different stages of lymphoid development. Overexpression of Bcl2 in mice can trigger a systemic lupus erythematosus (SLE)-like disease [240]. Furthermore, cytokines released in SLE-patients such as IL-17 or BAFF (TNFRSF13C) can upregulate the expressions of pro-survival BCL2 family proteins, including BCL2 and BCL2A1, supporting the idea that antiapoptotic BCL2 family proteins contribute to autoimmunity [241]. Consistent with this hypothesis, genes encoding BCL2, MCL1, and BCL2A1 are reportedly overexpressed in lymphoid cells from SLE patients [242–244].

In some situations, effector BCL2 family proteins also contribute to the control of auto-immune diseases. For example, as a result of dysregulated apoptosis, type I diabetes is often caused by aberrant T-cell activation-induced cell death and defective peripheral tolerance. Codelivery of cDNA encoding BAX and a DNA vaccine encoding a secreted form of the pancreatic beta cell antigen glutamic acid decarboxylase can help recruit dendritic cells and enhance prevention of autoimmune diabetes in mice [245].

4. Therapeutics targeting the BCL2 family

Because BCL2 family proteins control MOMP and are associated with the development of several diseases, especially cancers, agents modulating MOMP by targeting members of the BCL2 family offers an appealing approach for treating these diseases. As a result, there have been significant efforts to develop inhibitors of the antiapoptotic BCL2 family proteins and to identify direct activators of BAK and BAX. The inhibitors of anti-apoptotic BCL2 proteins include antisense oligonucleotides, α/β foldamers, stapled peptides, and small molecules. These agents have been tested in preclinical studies and clinical trials in several diseases, most notably cancer.

4.1 Antisense oligonucleotides oblimersen

Antisense oligonucleotides were the first agents developed to target the mitochondrial apoptotic pathway. Oblimersen sodium (also named G3139 and Genasense), developed by Genta Incorporated, was an antisense oligonucleotide with 18 nucleotides complementary to the first six codons of BCL2 mRNA [246]. In cancer cell lines, oblimersen decreased BCL2 expression [247] and induced BAX-dependent apoptosis. Moreover, oblimersen showed preclinical activity in many cancer models, including melanoma [248], prostate cancer [249], gastric cancer [250], ovarian cancer [251], NSCLC [252], breast cancer [253,254], and bladder cancer [255].

Most of the phase I/II studies of oblimersen were done in combination with other anticancer agents. For example, oblimersen was tested in combination with docetaxel in castration refractory prostate cancer and breast cancer [256–258], with cytarabine and daunorubicin in acute myeloid leukemia (AML) [259,260], with anti-CD20 antibody rituximab in non-Hodgkin's lymphoma (NHL) [261] and with paclitaxel and carboplatin/etoposide in SCLC [262–264]. In some of the phase II studies, combination therapy with oblimersen did not show any improvement over chemotherapy alone, for example, when combined with carboplatin/etoposide in SCLC patients, or with docetaxel in castration–resistant prostate cancer [263,265]. However, oblimersen showed promising results in other studies. In a multicenter phase II study, treatment with oblimersen in combination with rituximab achieved 42% overall response rate in patients with NHL [261]. In another phase II study in patients with multiple myeloma, oblimersen combined with dexamethasone and thalidomide showed encouraging results [266].

Phase III studies of oblimersen in combination with other anticancer agents were completed in various cancer types. For example, a phase III study of fludarabine and cyclophosphamide (FC) with or without oblimersen in relapsed or refractory chronic lymphocytic leukemia (CLL) patients showed no significant difference in 5-year survival between the two arms in an intent-to-treat analysis [267]. Likewise, a phase III study comparing oblimersen plus dexamethasone versus dexamethasone alone in patients with relapsed or refractory myeloma showed no differences in time to progression (TTP) or objective response rate between the two groups [268]. While most of the phase III studies failed, in a phase III study of oblimersen plus dacarbazine (DTIC) versus DTIC alone in advanced melanoma showed significant improvement in multiple clinical outcomes. Among the 771 randomized patients, there was a trend toward improved survival at 24-month minimum follow-up (median, 9.0 vs. 7.8 months; $P = .077$), a significant increase in progression-free survival (median, 2.6 vs. 1.6 months, $P < .001$), and a significant benefit in overall survival in patients without elevated baseline serum LDH (lactate dehydrogenase) (median overall survival, 11.4 vs. 9.7 months; $P = .02$) on the oblimersen arm [269].

Although some studies have shown oblimersen had anticancer activity, it did not meet the criteria to be approved by the United States Food and Drug Administration (FDA); and further development of this agent was halted. Among these studies, only a small number of patients treated with oblimersen showed decreased levels of BCL2 protein; thus, oblimersen's failure could perhaps because of the failure to successfully modulate the intended target rather than the choice of the the wrong drug target [266,270,271].

4.2 α/β foldamers

As discussed above, the BH3 domain is essential for the interaction of BH3-only proteins, BAX and BAK with antiapoptotic BCL2 family members. Synthetic BH3 peptides consisting of only 15 to 25 amino acids retain the ability to bind the antiapoptotic BCL2 family members [79,84,87], raising the possibility that these peptides could be used to saturate the binding capacity of antiapoptotic BCL2 family members and facilitate the action of BH3-only proteins, BAX and BAK. However, peptides have pharmacological liabilities, such as low stability and poor cell membrane permeability, that prevent them from being good drug candidates. Therefore, great effort has gone into strategies aiming to increase the binding affinity and proteolytic stability of these sequences [272]. One strategy is to incorporate α/β-amino acid to enhance the helicity and proteolytic stability of the peptides [272,273]. The α/β-peptide of the PUMA BH3 domain mimicked the structure of the PUMA BH3 domain and showed similar affinities for BCL2, BCLw, and $BCLX_L$ when compared to PUMA BH3; however, its affinity to MCL1 was decreased about 50-fold [274]. Consistent with its decreased affinity for MCL1, this peptide induced apoptosis in *Mcl1*$^{-/-}$ mouse embryonic fibroblast (MEF) cells, but not in wild-type MEFs. Further efforts generated a series of PUMA BH3-based foldamers with increased affinity for MCL1, which mimicked the selectivity of PUMA BH3 [275]. The α/β-peptides of the BIM BH3 domain had similar affinities for BCL2, $BCLX_L$, and MCL1 compared to BIM BH3 peptide but also showed increased helicity and nearly 100-fold more resistance to proteolysis than the BIM BH3 peptide [276,277]. Importantly, these α/β-peptides were able to inhibit antiapoptotic BCL2 family proteins and induce cancer cell death [277]. Interestingly, when β3 was changed to β2 substitutions, the affinities of the α/β-peptides to antiapoptotic BCL2 proteins were also changed due to a change in helicity [278].

4.3 Stapled peptides

To increase the pharmacological properties of the BH3 peptides, stapled peptides have been explored [272]. Specifically, a chemical strategy termed hydrocarbon stapling was used to generate stabilized alpha-helix of BCL2 domain (SAHB) constructs [279]. The stabilized BH3 peptides showed improved pharmacological properties, such as enhanced helicity, protease resistance, and increased cell permeability, compared to nonstapled peptides. SAHBs of the BID BH3 domain can interact with the BH3 binding pocket of antiapoptotic BCL2 family proteins [279] and directly activate BAX [280]. Moreover, SAHBs of the BID BH3 domain induced mitochondrial apoptosis and suppressed tumor growth in human leukemia xenografts, providing a useful strategy for experimental and therapeutic modulation by BH3 peptides [279]. Subsequently, stapled peptides of other BH3 peptides, such as BIM SAHB, have also been tested, and the results suggested similar mechanisms [281]. These stapled peptides also increased the sensitivity of cancer cells to other anticancer drugs [281]. A further study indicated that stapling the BH3 peptides drives cellular uptake through extending the hydrophobic surface in addition to optimizing peptide hydrophobicity and α-helicity [282]. Interestingly, the SAHB of BH3 domain of the antiapoptotic protein MCL1 can be a MCL1 inhibitor and overcome MCL1-mediated drug resistance [283]. Although the stapled peptides of BH3 domains showed promising results in preclinical studies, other studies reported that the staple itself could affect the affinity and might not increase the cell permeability [284]. Thus, while SAHB peptides are incredibly useful biologic probes, whether they will be good drug candidates still requires further investigation.

4.4 Small molecules

As previously discussed, overexpression of BCL2 family proteins contributes to the pathogenesis of various cancers, including lymphoid neoplasms, particularly FL and CLL [11,31,32], and possibly also solid tumors [285]; therefore, the possibility of developing small molecules targeting BCL2 family proteins has received substantial attention. Nuclear magnetic resonance (NMR)-based screening and computational screening along with further structure-based drug derivation have been used to obtain small molecule inhibitors [286–288]. As shown in Fig. 7.3, quite a few different types of small molecules have been identified [272], including gossypol (AT-101, Ascenta Therapeutics, Malvern, PA, USA) and its derivatives, obatoclax (GX-15-070, Gemin X Biotechnologies, Montreal, QC, Canada), and acyl sulfonamide series (ABT-737, ABT-263 and ABT-199).

Because BAK and BAX are the executioner proteins that directly permeabilize the MOM, for a small molecule to be truly a BH3 mimetic, it must meet the following conditions: (1) exhibit high affinity for at least one of the multidomain BCL2 family members; (2) induce cell death in a manner depending on BAK and/or BAX; and (3) induce cytotoxicity that correlates with its binding to BCL2 family protein(s). Some compounds originally thought to be BH3 mimetics, for example, BH3I class compounds, HA14-1, antimycin A, and purpurogallin, are no longer believed to be bona fide BH3 mimetics because they only display low affinity for the BCL2 family proteins and/or kill cells independent of BAK and BAX [189,289–292].

Among the small molecules that satisfy the conditions for being a true BH3 mimetic, some display low micromolar or even submicromolar affinities for multiple antiapoptotic proteins. These molecules, which include gossypol and its derivatives as well as obatoclax, are often termed

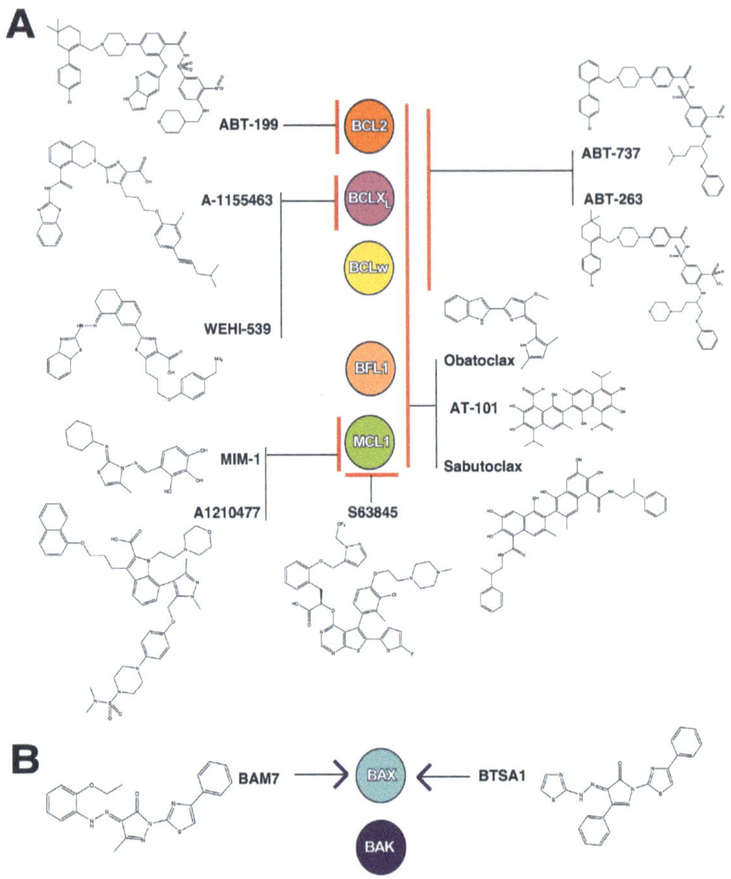

Figure 7.3 ***Small molecules targeting BCL2 family proteins.*** (A) molecular structures of the small molecules that inhibit anti-apoptotic BCL2 family proteins, and their targeting proteins are shown. Among these molecules, ABT-263 and ABT-737 inhibit BCL2, BCLX$_L$ and BCLw; obatoclax, AT-101 and sabutoclax interact with all the anti-apoptotic BCL2 proteins with moderate affinities; ABT-199 (venetoclax) selectively interacts with BCL2; A-1155463 and WEHI-539 selectively inhibit BCLX$_L$; and MIM-1, A-1210477, and S63845 selectively inhibit MCL1. (B) molecular structures of the small molecules that activate BAX are shown. Both BTSA1 and BAM7 are reportedly to activate BAX selectively.

pan-BCL2 inhibitors [293]. Other small molecules exhibit specificity for a single BCL-2 family member, such as ABT-199, which specifically inhibits BCL2, and S63845, which specifically inhibits MCL1.

4.4.1 Gossypol and its derivatives

One family of pan-BCL2 inhibitors includes the compound gossypol and its derivatives, including AT-101 (apogossypol, R(−)-gossypol), TW-37,

ApoG2 (apogossypolone), and subutoclax (BI−97C1). Through a natural product screen, gossypol was found to interact directly with $BCLX_L$ with a K_i of about 0.5 μM to displace BH3 peptides from $BCLX_L$ [287]. While two reactive aldehydic groups were present in gossypol, apogossypol was derived as an enantiomer, although its K_i to displace BH3 peptides from $BCLX_L$ increased to about 2.3 μM [294]. Apogossypol and its derivatives were found to interact with not only $BCLX_L$, but also other anti-apoptotic BCL2 family members, including BCL2, MCL1 and BFL1 [295,296]. These apogossypol derivatives showed efficacy in vivo, and induced cancer cell apoptosis in a BAK/BAX-dependent manner [294−296]. TW-37, which is also derived from gossypol, exhibits K_is for $BCLX_L$, BCL2, and MCL1 of 1.11, 0.29, and 0.26 μM, respectively [297]. ApoG2, which was derived from gossypol through the removal of hydroxyl from 1 and 1′ positions and the removal of the two reactive aldehyde groups from 8 and 8′ positions, also showed improved efficacy and reduced toxicity [298−300]. Sabutoclax, which was derived from apogossypol through structure-based derivation, can bind to $BCLX_L$, BCL2, MCL1, and BFL1 with IC_{50} values of 0.31, 0.32, 0.20, and 0.62 μM, respectively [301]. In an animal study, sabutoclax could overcome tyrosine kinase inhibitor (TKI) resistance in a CML mouse model by inhibiting the growth of leukemia stem cells [302]. Gossypol and its derivatives not only showed activity in leukemia models, but also demonstrated activity in cell lines and mouse models of head and neck cancer, breast cancer, prostate cancer, colon cancer, and NSCLC [303−306].

Although most studies showed that gossypol and its derivatives induce cancer cell death by inducing BAK- and/or BAX-dependent apoptosis [295,296,307], some studies have suggested different mechanisms. For example, gossypol treatment of primary lymphocytes taken from CLL patients induces a caspase-independent cell death, raising the possibility that gossypol might have other targets beyond BCL2 family members [308]. In addition, gossypol can induce BAK- and BAX-independent cell death in MEFs [292]. Moreover, several other studies have reported that gossypol or its derivatives suppress tumor growth via antiangiogenic effects, which are independent of its function as an antagonist of antiapoptotic BCL2 family members [309]. Finally, although AT-101 and sabutoclax decrease platelets and trigger phosphatidylserine exposure on platelets, which is a phenomenon seen with $BCLX_L$ inhibition, the platelet death caused by AT-101 and sabutoclax is independent of caspases [310], again suggesting the possibility that cell death is occurring by a process other than apoptosis.

Among gossypol and its derivatives, only AT-101 has undergone clinical trials in several types of cancer. In a phase II study of AT-101 in combination with rituximab for relapsed/refractory CLL, an ORR of 44% with no CRs was reported [311]. While AT-101 exhibits efficacy in CLL, it did not show enough activity in other cancers. In a phase I/II trial in patients with small cell lung carcinoma (SCLC), AT-101 did not prove to be efficacious either as monotherapy or in combination with topotecan or docetaxel [312–314]. In a more recent phase II study of AT-101 as a single agent in metastatic, recurrent or primarily unresectable adrenal cortical carcinoma, none of the first 21 patients attained a PR as assessed by RECIST (Response Evaluation Criteria In Solid Tumors) [315]. AT-101 also was not efficacious when combined with other anticancer agents. For example, in a phase II study in patients with newly diagnosed castration-sensitive metastatic prostate cancer, AT-101 in combination with androgen deprivation therapy (ADT) did not meet the prespecified criteria [316]. Likewise, in a phase II trial in patients with head and neck cancer, AT-101 did not demonstrate much efficacy in combination with docetaxel [317]. In contrast, however, a more recent pilot clinical trial that enrolled 13 patients with esophageal cancer who received AT-101 concurrently with chemoradiation revealed dramatic CRs and encouraging overall survival [318]. Thus, whether the gossypol derivatives have some effects in patients still merits further study.

4.4.2 Obatoclax

Obatoclax (GX15-070) is a small molecular indole bipyrrole compound developed by GeminX Pharmaceuticals (Malvern, PA, USA) to target antiapoptotic BCL2 family proteins [319]. It binds to all antiapoptotic BCL2 family members, including $BCLX_L$, BCL2, MCL1, BCLB, BCLw and BFL-1, with low micromolar affinities [293], meeting the definition of a pan-BCL2 inhibitor. In preclinical studies, obatoclax showed ability to disrupt MCL1/BAK complexes and overcome BCL2- and MCL1-mediated drug resistance [319,320]. Moreover, obatoclax displayed anticancer activity in several types of cancers, including AML, lymphoma, melanoma, SCLC, cholangiocarcinoma, and pancreatic cancer, in preclinical studies [321–323]. Several studies found that obatoclax could synergize with other ant-cancer agents such as rituximab, 5-fluorouracil, and BEZ235 (a dual PI3K and mTOR kinase inhibitor) to induce cancer cell death [324–328]. While most studies have shown the cytotoxicity of obatoclax is the result of inhibiting the antiapoptotic BCL2 proteins, other studies have shown obatoclax might induce cell death through other mechanisms. It is interesting that obatoclax-induced cell death not only depends on anti-apoptotic BCL2 proteins such as MCL1, but also depends on key autophagy components, suggesting that obatoclax can induce autophagic cell death [329]. Consistent with this possibility, it has been reported that enhanced killing in some of the combination studies involves the autophagy-inducing ability of obatoclax [326,328]. Moreover, a recent study found that obatoclax triggers RIP1- and RIP3-dependent cell death, suggesting that obatoclax may induce necroptosis in rhabdomyosarcoma cells [330]. In a separate study, obatoclax-induced killing of thyroid cancer cells was linked to its lysosomal accumulation, suggesting a necrosis-inducing effect [331].

Obatoclax has demonstrated modest clinical activity in some types of cancer. In a phase I study of obatoclax monotherapy in patients with leukemia and myelodysplastic syndrome (MDS), one patient with AML achieved a complete remission (CR), while 3 out of 14 patients with MDS showed hematologic improvements [332]. In another phase I study of obatoclax in 26 patients with advanced chronic lymphocytic leukemia (CLL), one partial response (PR) was observed and 18 out of 26 patients showed improvements in blood counts [267]. More importantly, in patient samples obtained after treatment, BAX and BAK were found to be activated, suggesting that obatoclax at least partially acts on the mitochondrial apoptotic pathway in vivo. Although obatoclax treatment showed some efficacy in phase I studies, the results of several phase II studies in hematological malignancies were all disappointing. For example, in a study of obatoclax in older patients with AML, no objective responses were observed [333]. In another phase II study of obatoclax treatment in 22 patients with myelofibrosis, no PRs or CRs were observed [334]. In combination with bortezomib in patients with relapsed or refractory mantle cell lymphoma, obatoclax did not show obvious synergistic effect [335].

The clinical studies of obatoclax in solid tumors were also disappointing. In a phase I study of obatoclax in combination with topotecan in 14 patients with solid tumors, two patients with small cell lung cancer (SCLC) achieved PRs [336]; however, in a subsequent phase II study in nine patients with recurrent SCLC, no PRs or CRs were seen [337]. In a separate phase II study, 155 SCLC patients were randomized to treatment with carboplatin and etoposide plus obatoclax (CbEOb) versus carboplatin and etoposide (CbE) [338]. While the overall response rate (ORR) (62% vs. 53%, $P = .143$), median PFS (progression-free survival) (5.8 months vs.

5.2 months), and median OS (overall survival) (10.5 months vs. 9.8 months, $P = .121$) all showed a trend of improvement in the CbEOb group, the changes were not statistically significant. More recently, in another phase I/II study of obatoclax in combination with docetaxel in relapsed non-small cell lung cancer (NSCLC), PRs were observed in only 3 out 32 patients [339].

4.4.3 ABT-737 and ABT-263

ABT-737 and its derivatives also target BCL2 family proteins. ABT-737 was developed by high-throughput NMR-based screening of a small molecule fragment library, and two small fragments were identified that bind different sites along the BH3 binding groove of $BCLX_L$, followed by optimization with a linker to improve its affinity [286]. ABT-737 was found to bind to $BCLX_L$, BCL2, and BCLw with very high affinity ($K_i \leq 1$ nM), while it binds BCLb, MCL1, and BCL2A1 with much lower affinity. This compound inhibited tumor cell growth in vitro as well as tumor growth in xenograft models [286]. Further structural analysis showed that the two large hydrophobic chemical groups of ABT-737 are embedded in two of the four key hydrophobic pockets (P2 and P4) on the surface of $BCLX_L$ [340]. Several other preclinical studies also demonstrated that ABT-737 induced apoptosis in several cancer cell lines or animal models, including AML, CLL, multiple myeloma, SCLC, pancreatic, and ovarian cancer [341–346]. Whereas ABT-737 showed promising results as an anticancer agent in preclinical studies, its further clinical use was hampered by its poor physiochemical and pharmaceutical properties, for example, poor oral bioavailability and low aqueous solubility, which led to the development of ABT-263 [347].

ABT-263 (Navitoclax) is an orally bioavailable analog of ABT-737 that shares many properties with ABT-737, including high affinities for BCL2, $BCLX_L$, and BCLw but not MCL1 and BCL2A1 [347]. ABT-263 also has exhibited anticancer activity in a series of cancer cell lines and animal models, including SCLC, AML, ALL, and lymphoma [347,348]. Moreover, synergistic effects of ABT-263 with classical cytotoxic agents and molecularly targeted drugs have been observed. For example, ABT-263 can overcome paclitaxel or gemcitabine resistance that is mediated by BCL2 or $BCLX_L$ in ovarian cancer and non-small cell lung cancer models [349–351]. In another study, ABT-263 enhances sensitivities of rituximab, bortezomib, and etoposide in several lymphoma animal models [352]. Moreover, ABT-263 can overcome chemoresistance induced by epithelial-mesenchymal transition in mesenchymal lung cancer cells [353]. Many studies have suggested the cytotoxic effects of ABT-737 and ABT-263 depend on BAK and/or BAX, as one would expect if death were triggered by inhibition of antiapoptotic BCL2 family members [292,341,347].

ABT-263 has been tested clinically for the treatment of hematological malignancies, SCLC, and other advanced solid tumors [354–358]. Because of its low nanomolar potency in inhibiting BCL2 and $BCLX_L$, ABT-263 displays good efficacy in patients with CLL and SCLC [355,356]; however, thrombocytopenia (low platelet count) as a consequence of therapy has limited the widespread use of ABT-263. For example, patients with relapsed or refractory CLL receiving ABT-263 at more than 10 mg/d on the 14- or 21-day schedule all experienced acute but transient reductions in platelet counts [355]. Because this reflects an on-target effect that limits platelet survival, transfused platelets suffer the same fate, making the reversal of thrombocytopenia potentially difficult. In a phase II study of ABT-263 monotherapy in SCLC patients, grade III-IV thrombocytopenia was found in about 41% of patients, which limited the dosage of ABT-263 in these patients [356]. In a further study of ABT-263 in combination with rituximab in previously untreated CLL, the combination yielded a higher response rate than rituximab alone and resulted in prolonged progression-free survival when the treatment continued beyond 12 weeks [358]. Because $BCLX_L$ restrains BAK to maintain platelet survival [359,360], the reduction of platelets in these patients also implies the on-target effect of ABT-263 from another point of view. Further clinical trials of ABT-263 are still ongoing.

4.4.4 ABT-199 (Venetoclax)

The thrombocytopenia described above has limited the dose escalation of ABT-263 and its widespread use in the clinic. Because $BCLX_L$ plays an important role in the maintenance of platelet survival [348,360], the strategy of selectively targeting BCL2 while diminishing binding to $BCLX_L$ was devised to avoid severe thrombocytopenia. ABT-199 (also named venetoclax) has these desired characteristics. Structural comparison shows that the amino acid in the P4 groove in $BCLX_L$ is GLU96, which has a bigger side chain than that of the corresponding amino acid ASP103 in BCL2. Therefore, when the bigger piperidyl pyrimidine ring was introduced and the smaller thiophenyl unit was removed from ABT-263 to generate ABT-199, the binding affinity to $BCLX_L$ decreased more than 100-fold. Surprisingly, compared to ABT-263, ABT-199 also displayed increased affinity for BCL2 and decreased affinity for BCLw [288].

Preclinical studies showed that ABT-199 could not only inhibit the growth of CLL and non-Hodgkin's lymphoma lines as well as the corresponding animal models with high levels of BCL2, but also spared human platelets [288,361]. Studies of ABT-199 also demonstrated preclinical efficacy in other tumors, including SCLC [362], breast cancer [363], myelodysplastic syndromes [364], lymphomas [365], and other hematologic malignancies [366]. Importantly, the efficacy of ABT-199 is related to high-BCL2 expression [362]. Moreover, ABT-199 in combination with other agents, such as cisplatin, paclitaxel, or MLN4924 (a potent and selective small molecule inhibitor of NEDD8-activating enzyme), showed synergistic effects in breast cancer cells [367], ovarian cancer cells [351], and acute myelogenous leukemia [368]. Importantly, the ability of ABT-199 to induce cell death was shown to depend on BAK/BAX activation [369], suggesting that ABT-199 works through the mitochondrial apoptosis pathway.

Clinical studies of ABT-199 have focused most extensively on BCL2-dependent hematological malignancies. For example, the phase II study of ABT-199 monotherapy in relapsed or refractory (R/R) CLL showed an overall response rate of 79.4% [370]. Moreover, in a study of ABT-199 monotherapy in CLL after treatment with the BTK inhibitor ibrutinib or the PI3K-δ inhibitor idelalisib, an ORR of 70% was found in patients who progressed on ibrutinib and 48% in patients who progressed on idelalisib [371]. These studies led to the accelerated approval of ABT-199 for the treatment of R/R CLL by the FDA in April 2016. While ABT-199 has been approved to treat R/R CLL, clinical studies in multiple myeloma and Non-Hodgkin lymphoma are still ongoing. For example, in a recent phase I study of ABT-199 in patients with relapsed/refractory multiple myeloma, an ORR of 40% (12/ of 30 patients) was found in t(11; 14) multiple myeloma, with eight patients achieving a very good PR or better [372]. Moreover, the ORR was about 80% in cases with a high ratio of *BCL2* expression compared to *BCL2L1*, suggesting a possible strategy for selecting patients for treatment with ABT-199. In another phase I study of ABT-199 in patients with relapsed/refractory Non-Hodgkin lymphoma [373], responses were seen in all histologies, including mantle cell lymphoma (ORR, 75%; CR, 21%), FL (ORR, 38%; CR, 14%), diffuse large B-cell lymphoma (DLBCL, ORR 18%; CR, 12%), DLBCL-Richter transformation (ORR, 43%; no CRs), marginal zone lymphoma (ORR, 67%; no CRs), and Waldenstrom macroglobulinemia (ORR, 100%; no CRs). ABT-199 can also synergize with other agents in lymphoma patients. In a recent phase II study of ABT-199 plus ibrutinib to treat patients with mantle-cell lymphoma (mostly R/R), the CRR at week 16 was 42%, which was much higher than 9% with ibrutinib treatment only [374]. Thus, it is expected that ABT-199 may be approved for more indications either as monotherapy or in combination with other drugs.

4.4.5 $BCLX_L$-selective BH3 mimetics

While many hematologic malignancies are associated with BCL2 overexpression, a large number of solid tumors rely on $BCLX_L$ for survival [12]. Thus, a compound that selectively inhibits this pro-survival protein could be useful for treating these tumors. However, due to the thrombocytopenia associated with $BCLX_L$ inhibition as discussed above, clinical use of $BCLX_L$ inhibitors might be problematic [359,360]. Through high-throughput screening, WEHI-539 was developed to selectively inhibit $BCLX_L$ with a dissociation constant below 1 nM, while having a much lower affinity for BCL2 and MCL1 [375]. This $BCLX_L$-selective inhibitor induced BAK/BAX-dependent apoptosis, suggesting the molecule works on the right target [375,376]. In a parallel study using molecular docking-based screening, two molecules, BXI-61 and BXI-72, were discovered to selectively inhibit $BCLX_L$ and they exhibited toxicity against lung cancer cells [377]. More recently, a potent and selective $BCLX_L$ inhibitor, A-1155463, was derived from NMR-based fragment screening and structure-based design [378]. Even though A-1155463 induced thrombocytopenia in mice, this effect was reversible. More importantly, A-1155463 inhibited growth of H146 small cell lung cancer xenografts, which express high levels of $BCLX_L$ [379]. Moreover, these $BCLX_L$ inhibitors sensitized cancer cells to other anticancer agents; for example, WEHI-539 sensitized osteosarcoma cells to doxorubicin [380] and A-1155463 sensitized several cancer cell lines to a mitochondrial chaperone inhibitor in murine models [381]. Whether $BCLX_L$ inhibitors can be successfully developed for use in the clinic still requires further investigation.

An alternative approach for targeting $BCLX_L$ in tumor cells involves tethering a $BCLX_L$-specific binder to an E3 ligase binder, thereby creating a protease targeting chimera (PROTAC) that induces protein degradation in cells expressing the pertinent E3 ligase. This approach has been applied using the $BCLX_L$/BCL2 binding moiety of ABT-263 and a moiety that binds the VHL E3 ligase, which is minimally expressed in platelets [382].

The resulting chimera, DT2216, causes selective $BCLX_L$ degradation in a VHL- and proteasome-dependent manner. Moreover, DT2216 exhibits some single-agent activity and markedly sensitizes a number of preclinical models to a variety of chemotherapeutic agents while sparing platelets [382], suggesting that this approach also warrants further study. Indeed, DT2216 is now being tested in a phase I clinical trial.

4.4.6 MCL1 selective BH3 mimetics

Because MCL1 is very important for the development of several types of tumors [178,182] and for anticancer drug resistance [179], there have been intense efforts to develop selective MCL1 inhibitors. Through a high-throughput screen under cell-free conditions, MIM-1 was discovered as a selective MCL1 binder with a modest affinity (IC_{50} 4.8 μM) [383]. Moreover, MIM-1 induced death of leukemic cells either as a single agent or in a combination with ABT-737. Further, MIM-1 induced apoptosis in several cancer cell lines in a BAK-dependent manner, suggesting that MIM-1 works through the mitochondrial pathway; however, its potency might need to be improved [384]. In contrast, studies from another group have questioned whether MIM-1 acts by directly binding MCL-1 in cells. Instead, it has been suggested that MIM-1 triggers the unfolded protein response, resulting in upregulation of the BH3-only protein NOXA and MCL1 neutralization [385]. Thus, the mechanisms of action of this agent are an open question.

Another group used NMR-based fragment screening to identify two lead compounds as MCL1 inhibitors. One compound optimized from indole-2-carboxylic acid displays a K_i of 0.055 μM for MCL1 [386], and the other generated from tricyclic indole 2-carboxylic acid shows a K_i of 0.003 μM for MCL1, while having much lower affinities for BCL2 and $BCLX_L$ [387]. In a different study, also using an indole-2-carboxylic acid core, A-1210477 was synthesized and shown to have a K_i of 0.43 nM for MCL1 [388]. Moreover, A-1210477 induced BAX/BAK-dependent cytotoxicity in several cancer cell lines as a single agent or in combination with ABT-263 [389]. More recently, based on NMR fragment screening and structure-guided drug derivation, S63845 was obtained with a K_D of 0.19 nM for MCL1 [390]. Like other MCL1 antagonists, S63845 showed BAK/BAX-dependent antitumor activity in several cancer cell lines and patient samples. In a preclinical study, S63845 showed synergistic activity with docetaxel in triple-negative breast cancer and with trastuzumab or lapatinib in HER2-amplified breast cancer xenograft models [391].

Several clinical trials using the MCL1 inhibitor MIK665 (also named S64315, a derivative of S63845) are currently ongoing (http://clinicaltrials.gov/) [392].

4.4.7 BFL1 (BCL2A1) inhibitors

The mouse BFL1 mRNA is expressed at particularly high levels in hematopoietic cells, while in humans a more widespread tissue expressions is found [393]. Several studies have tried to target this protein [394,395]. In contrast to the other antiapoptotic BCL2 family proteins, the BH3 binding groove of BFL1 contains a unique cysteine residue (C55) that is involved in covalently binding to Noxa [396,397]. Using a sulfhydryl tethering screen in which sulfhydryl-reactive molecules were assessed for their ability to inhibit BFL1 function in vitro, Harvey et al. identified 4E14, a sulfhydryl-containing N-acetyltryptophan analog, as a lead molecule that abrogates the ability of BFL1, but not $BCLX_L$, to bind and neutralize the active fragment of BID [395]. This is a potentially important step for selectively targeting BFL1, but it will be important to confirm this selectivity in intact cells and assess the pharmacokinetic features of 4E14 and its derivatives in future studies.

4.4.8 Small molecules targeting BAX and BAK

Changes in BCL2 family members that have been implicated in tumorigenesis include not only the overexpression of antiapoptotic BCL2 family proteins, but also mutations or deletions of some of the activator BH3-only proteins [190,191]. Needless to say, cancer cells that lack activator BH3-only proteins are resistant to anticancer agents [398]. Thus, the development of drugs that could directly activate BAK or BAX represents a different approach to activate apoptosis for treating cancer. However, the development of agents directly activating BAK and BAX has lagged significantly behind the development of drugs that inhibit their pro-survival relatives. This might be because of the complexity of structural changes during the processes of activating BAK and BAX. However, the recent structural studies in this area have provided new insights into targeting these proteins [92–94,115,118]. Although the binding modes between BH3 domains and the canonical grooves of antiapoptotic BCL2 proteins or the effector BCL2 proteins BAK/BAX are similar in a structural sense, the consequences of these interactions are much different. In the case of the antiapoptotic proteins, a stable complex will form after the binding, while in the case of BAX or BAK, interactions at this interface are transient and will initiate a conformational change that leads to homo-oligomerization and MOMP [92,94]. Therefore, the therapeutic agents directly targeting the canonical groove of BAX and BAK might have completely different cell death activities. One possibility is that BH3 groove binding agents might initiate BAX or BAK conformational change and trigger the MOMP process after transient binding. Alternatively, agents that interact with the groove might form a stable complex and inhibit other BH3-domain interactions, for example, formation of BAX or BAK symmetric in-groove dimers [103], which could instead inhibit apoptosis. The BH3 peptide of BIM has been reportedly converted from an activator to an inhibitor through structure-based design, supporting the two possibilities [399]. While the pro-apoptotic agents could be used for the treatment of cancer, the apoptosis inhibiting agents might hold promise for the treatment of conditions where excessive apoptosis leads to pathology, such as neurodegenerative disorders or cardiovascular disease. Accordingly, efforts to develop agents that directly target the BH3-binding groove of BAK or BAX are ongoing. In order to achieve selectivity of these agents for cancer cells, as opposed to normal cells expressing BAK and/or BAX, it will likely be necessary to deliver the molecules tethered to peptides or antibodies that specifically bind cancer cells. Luckily, there are ample precedents for successful development of antibody-drug conjugates bearing active moieties that interact with other drug targets [400].

On the other hand, BAX can also be activated through a second site on its N-terminal domain [56]. There are efforts to develop direct activators to activate BAX by targeting this noncanonical BH3 binding site. Through an in silico screening of 750,000 small molecules, BAM7 was identified and shown to directly bind to BAX at the rear site [401]. Moreover, biochemical and cell-based studies confirmed that BAM7 can activate BAX directly and induce BAX-dependent apoptosis. More recently, BTSA1 was found to also activate BAX through the N-terminus and suppress cancer growth in acute myeloid leukemia xenograft models while showing no toxicity to normal cells [402].

Surprisingly, a third site was also found to be involved in inhibiting BAX activation. This site, which sits on the back of the N-terminus and involves the ends of BAX helices $\alpha3$, $\alpha4$, $\alpha5$, and $\alpha6$, was reported to interact with the BH4 domain of BCL2 [403] and with a protein from human cytomegalovirus vMIA (viral mitochondria-localized inhibitor of apoptosis) to inhibit BAX activation [404]. Using NMR-based fragment screening, compound BIF-44 (BAX interacting fragment-44) was found to displace vMIA peptide from this binding site and induce a conformational change in BAX, facilitating the process of MOMP [405].

Conversely, inhibitors of BAK or BAX might have implications for the protection from side effects during cancer treatments. For example, a BAX inhibitor alleviates doxorubicin-induced cardiomyopathy in mouse models [406]. Further studies are still needed to explore clinical applications of these compounds.

5. Perspective

While the FDA approval of venetoclax marks a milestone in the therapeutic targeting BCL2 family proteins, there is still much work to be done. Results of clinical trials of venetoclax in other types of cancers, either as a single agent or in combination with other anticancer agents, are of particular interest. For example, trials of venetoclax monotherapy in *TP53* wild-type CLL, other lymphoid malignancies, SCLC, and breast cancers are still awaited. Moreover, while some combinations such as venetoclax with ibrutinib in mantle-cell lymphoma, or venetoclax with rituximab in R/R CLL have shown promising results [374,407], investigations using combinations of venetoclax with other drugs in BCL2 overexpressing cancers are still ongoing.

Although the prospect of agents that selectively target $BCLX_L$ is also appealing, this approach faces obstacles. Due to the important role of $BCLX_L$ in platelet survival, $BCLX_L$ inhibition induces thrombocytopenia, which hampers the widespread use of navitoclax in the clinic [359,360]. The other $BCLX_L$ inhibitors, WEHI-539 and A-1155463, also induce thrombocytopenia [378], indicating that this is a class-specific problem rather than a drug-specific problem. Thus, a clinically viable strategy is urgently required to avoid or overcome this on-target side effect while still maintaining anticancer effects. Whether the $BCLX_L$-selective PROTAC described above will prove effective remains to be seen. As for MCL1 inhibitors, while some MCL1 inhibitors including A-1210477 and S63845, showed promising results in preclinical studies [389,390], the results of clinical trials of the MCL1 inhibitors in lymphoma, multiple myeloma, and AML have not been reported and are awaited anxiously. Additional antiapoptotic BCL2 members, such as BCLw and BFL1, although not as widely overexpressed in cancers as BCL2, $BCLX_L$, and MCL1, have been less studied but are still worthy of selectively targeting in some specific cancers.

Resistance of cancer cells to antiapoptotic BCL2 protein inhibitors might reflect a lack of autoactivated BAK or BAX [72]. On the other hand, the absence of the complexes of activator BH3-only proteins with anti-apoptotic BCL2 proteins also results in drug resistance [398]. In both situations, compounds that directly activate BAX or BAK may be beneficial. Although compounds that directly activate BAX have been developed [401,402], they have not yet been tested in clinical trials. Moreover, there are still no drugs directly targeting BAK. Whether such compounds could directly target BAK or BAX in a way that allows selective killing of cancer cells or whether they will be protective also remains to be determined.

Acknowledgments

This work was supported by grants from the Natural Science Foundation of China (NFSC: 21772201; NFSC: 81572948) and the Hundred-Talent Program of Chinese Academy of Sciences to H.D.

References

[1] M.O. Hengartner, The biochemistry of apoptosis, Nature 407 (6805) (2000) 770–776.
[2] X.W. Meng, S.H. Lee, S.H. Kaufmann, Apoptosis in the treatment of cancer: a promise kept? Curr. Opin. Cell Biol. 18 (6) (2006) 668–676.
[3] R.C. Taylor, S.P. Cullen, S.J. Martin, Apoptosis: controlled demolition at the cellular level, Nat. Rev. Mol. Cell Biol. 9 (3) (2008) 231–241.
[4] H. Dai, W.X. Meng, S.H. Kaufmann, BCL2 family, mitochondrial apoptosis, and beyond, Cancer Trans. Med. 2 (1) (2016) 7–20.
[5] M. Enari, H. Sakahira, H. Yokoyama, K. Okawa, A. Iwamatsu, S. Nagata, A caspase-activated DNase that degrades DNA during apoptosis, and its inhibitor ICAD, Nature 391 (6662) (1998) 43–50.
[6] A.G. Porter, R.U. Jänicke, Emerging roles of caspase-3 in apoptosis, Cell Death Differ. 6 (2) (1999) 99–104.
[7] F.J. Bock, S.W.G. Tait, Mitochondria as multifaceted regulators of cell death, Nat. Rev. Mol. Cell Biol. 21 (2) (2020) 85–100.
[8] J.C. Martinou, R.J. Youle, Mitochondria in apoptosis: bcl-2 family members and mitochondrial dynamics, Dev. Cell 21 (1) (2011) 92–101.
[9] B.T. Hyman, J. Yuan, Apoptotic and non-apoptotic roles of caspases in neuronal physiology and pathophysiology, Nat. Rev. Neurosci. 13 (6) (2012) 395–406.
[10] C.B. Thompson, Apoptosis in the pathogenesis and treatment of disease, Science 267 (5203) (1995) 1456–1462.
[11] Y. Tsujimoto, J. Cossman, E. Jaffe, C.M. Croce, Involvement of the bcl-2 gene in human follicular lymphoma, Science 228 (4706) (1985) 1440–1443.

[12] R. Beroukhim, C.H. Mermel, D. Porter, G. Wei, S. Raychaudhuri, J. Donovan, et al., The landscape of somatic copy-number alteration across human cancers, Nature 463 (7283) (2010) 899–905.
[13] N. Rampino, H. Yamamoto, Y. Ionov, Y. Li, H. Sawai, J.C. Reed, et al., Somatic frameshift mutations in the BAX gene in colon cancers of the microsatellite mutator phenotyp, Science 275 (5302) (1997) 967–969.
[14] C. Mestre-Escorihuela, F. Rubio-Moscardo, J.A. Richter, R. Siebert, J. Climent, V. Fresquet, et al., Homozygous deletions localize novel tumor suppressor genes in B-cell lymphomas, Blood 109 (1) (2007) 271–280.
[15] C. Wang, R.J. Youle, The role of mitochondria in apoptosis, Annu. Rev. Genet. 43 (1) (2009) 11–22.
[16] P.E. Czabotar, G. Lessene, A. Strasser, J.M. Adams, Control of apoptosis by the BCL-2 protein family: implications for physiology and therapy, Nat. Rev. Mol. Cell Biol. 15 (1) (2014) 49–63.
[17] A. Ashkenazi, Targeting death and decoy receptors of the tumour-necrosis factor superfamily, Nat. Rev. Cancer 2 (6) (2002) 420–430.
[18] S.H. Kaufmann, D.P. Steensma, On the TRAIL of a new therapy for leukemia, Leukemia 19 (12) (2005) 2195–2202.
[19] N. Yan, Y. Shi, Mechanisms of apoptosis through structural biology, Cell Dev. Biol. 21 (21) (2005) 35–56.
[20] F. Llambi, Y.M. Wang, B. Victor, M. Yang, D.M. Schneider, S. Gingras, et al., BOK is a non-canonical BCL-2 family effector of apoptosis regulated by ER-associated degradation, Cell 165 (2) (2016) 421–433.
[21] X. Luo, I. Budihargjo, H. Zou, C. Slaughter, X. Wan, Bid, a Bcl2 interacting protein, mediates cytochrome c release from mitochondria in response to activation of cell surface death receptors, Cell 94 (4) (1998) 481–490.
[22] H. Li, H. Zhu, C.J. Xu, J. Yuan, Cleavage of BID by caspase 8 mediates the mitochondrial damage in the Fas pathway of apoptosis, Cell 94 (4) (1998) 491–501.
[23] C. Scaffidi, S. Fulda, A. Srinivasan, C. Friesen, F. Li, K.J. Tomaselli, et al., Two CD95 (APO-1/Fas) signaling pathways, EMBO J. 17 (6) (1998) 1675–1687.
[24] X.W. Meng, K.L. Peterson, H. Dai, P. Schneider, S.H. Lee, J.S. Zhang, et al., High cell surface death receptor expression determines type I versus type II signaling, J. Biol. Chem. 286 (41) (2011) 35823–35833.
[25] Y. Pang, X.C. Bai, C. Yan, Q. Hao, Z. Chen, J.W. Wang, et al., Structure of the apoptosome: mechanistic insights into activation of an initiator caspase from *Drosophila*, Genes Dev. 29 (3) (2015) 277–287.
[26] P. Li, D. Nijhawan, I. Budihardjo, S.M. Srinivasula, M. Ahmad, E.S. Alnemri, Cytochrome c and dATP-dependent formation of Apaf-1/caspase-9 complex initiates an apoptotic protease cascade, Cell 91 (4) (1997) 479–489.
[27] Q.L. Deveraux, R. Takahashi, G.S. Salvesen, J.C. Reed, X-linked IAP is a direct inhibitor of cell-death proteases, Nature 388 (6639) (1997) 300–304.
[28] C. Du, M. Fang, Y. Li, L. Li, X. Wang, Smac, a mitochondrial protein that promotes cytochrome c-dependent caspase activation by eliminating IAP inhibition, Cell 102 (1) (2000) 33–42.
[29] A.M. Verhagen, P.G. Ekert, M. Pakusch, J. Silke, L.M. Connolly, G.E. Reid, Identification of DIABLO, a mammalian protein that promotes apoptosis by binding to and antagonizing IAP proteins, Cell 102 (1) (2000) 43–53.
[30] X. Liu, R. Fu, Y. Pan, K.F. Meza-Sosa, Z. Zhang, J. Lieberman, PNPT1 release from mitochondria during apoptosis triggers decay of poly(A) RNAs, Cell 174 (1) (2018) 187–201.

[31] A. Bakhshi, J.P. Jensen, P. Goldman, J.J. Wright, O.W. McBride, A.L. Epstein, et al., Cloning the chromosomal breakpoint of t(14;18) human lymphomas: clustering around JH on chromosome 14 and near a transcriptional unit on 18, Cell 41 (3) (1985) 899–906.
[32] M.L. Cleary, S.D. Smith, J. Sklar, Cloning and structural analysis of cDNAs for bcl-2 and a hybrid bcl-2/immunoglobulin transcript resulting from the t (14;18) translocation, Cell 47 (1) (1986) 19–28.
[33] D.L. Vaux, S. Cory, J.M. Adams, Bcl-2 gene promotes haemopoietic cell survival and cooperates with c-myc to immortalize pre-B cells, Nature 335 (6189) (1988) 440–442.
[34] C.L. Sentman, J.R. Shutter, D. Hockenbery, O. Kanagawa, S.J. Korsmeyer, bcl-2 inhibits multiple forms of apoptosis but not negative selection in thymocytes, Cell 67 (5) (1991) 879–888.
[35] S. Henderson, M. Rowe, C. Gregory, D. Croom-Carter, F. Wang, R. Longnecker, et al., Induction of bcl-2 expression by epstein-barr virus latent membrane protein 1 protects infected B cells from programmed cell death, Cell 65 (7) (1991) 1107–1115.
[36] I. Garcia, I. Martinou, Y. Tsujimoto, J.C. Martinou, Prevention of programmed cell death of sympathetic neurons by the bcl-2 proto-oncogene, Science 258 (5080) (1992) 302–304.
[37] T. Hennet, G. Bertoni, C. Richter, E. Peterhans, Expression of BCL-2 protein enhances the survival of mouse fibrosarcoid cells in tumor necrosis factor-mediated cytotoxicity, Cancer Res. 53 (6) (1993) 1456–1460.
[38] A. Gross, J. Jockel, M.C. Wei, S.J. Korsmeyer, Enforced dimerization of BAX results in its translocation, mitochondrial dysfunction and apoptosis, EMBO J. 17 (14) (1998) 3878–3885.
[39] E.H. Cheng, M.C. Wei, S. Weiler, R.A. Flavell, T.W. Mak, T. Lindsten, et al., BCL-2, BCL-XL sequester BH3 domain-only molecules preventing BAX- and BAK-mediated mitochondrial apoptosis, Mol. Cell. 8 (3) (2001) 705–711.
[40] M.D. Jacobson, J.F. Burne, M.P. King, T. Miyashita, J.C. Reed, M.C. Raff, Bcl-2 blocks apoptosis in cells lacking mitochondrial DNA, Nature 361 (6410) (1993) 365–369.
[41] R.J. Youle, A. Strasser, The BCL-2 protein family: opposing activities that mediate cell death, Nat. Rev. Mol. Cell Biol. 9 (1) (2008) 47–59.
[42] T. Moldoveanu, A.V. Follis, R.W. Kriwacki, D.R. Green, Many players in BCL-2 family affairs, Trends Biochem. Sci. 39 (3) (2014) 101–111.
[43] Z.N. Oltvai, C.L. Milliman, S.J. Korsmeyer, bcl-2 heterodimerizes in vivo with a conserved homolog, bax, that accelerates programed cell death, Cell 74 (4) (1993) 609–619.
[44] S.N. Farrow, J.H. White, I. Martinou, T. Raven, K.T. Pun, C.J. Grinham, et al., Cloning of a bcl-2 homologue by interaction with adenovirus E1B 19K, Nature 374 (6524) (1995) 731–733.
[45] M.C. Kiefer, M.J. Brauer, V.C. Powers, J.J. Wu, S.R. Umansky, L.D. Tomei, et al., Modulation of apoptosis by the widely distributed Bcl-2 homologue Bak, Nature 374 (6524) (1995) 736–739.
[46] T. Chittenden, E.A. Harrington, R. O'Connor, C. Flemington, R.J. Lutz, G.I. Evan, et al., Induction of apoptosis by the bcl-2 homologue bak, Nature 374 (6524) (1995) 733–736.
[47] S.Y. Hsu, A. Kaipia, E. McGee, M. Lomeli, A.J. Hsueh, Bok is a pro-apoptotic Bcl-2 protein with restricted expression in reproductive tissues and heterodimerizes with selective anti-apoptotic Bcl-2 family members, Proc. Natl. Acad. Sci. U.S.A. 94 (23) (1997) 12401–12406.

[48] E. Yang, J. Zha, J. Jockel, L.H. Boise, C.B. Thompson, S.J. Korsmeyer, Bad, a heterodimeric partner for Bcl-XL and Bcl-2, displaces Bax and promotes cell death, Cell 80 (2) (1995) 285–291.
[49] J.M. Boyd, G.J. Gallo, B. Elangovan, A.B. Houghton, S. Malstrom, B.J. Avery, et al., Bik, a novel death-inducing protein shares a distinct sequence motif with Bcl-2 family proteins and interacts with viral and cellular survival-promoting proteins, Oncogene 11 (9) (1995) 1921–1928.
[50] L. O'Connor, A. Strasser, L.A. O'Reilly, G. Hausmann, J.M. Adams, S. Cory, et al., Bim: a novel member of the Bcl-2 family that promotes apoptosis, EMBO J. 17 (2) (1998) 384–395.
[51] E.F. Lee, S. Grabow, S. Chappaz, G. Dewson, C. Hockings, R.M. Kluck, et al., Physiological restraint of Bak by Bcl-xL is essential for cell survival, Genes Dev. 30 (10) (2016) 1240–1250.
[52] P.E. Czabotar, E.F. Lee, G.V. Thompson, A.Z. Wardak, W.D. Fairlie, P.M. Colman, Mutation to bax beyond the BH3 domain disrupts interactions with pro-survival proteins and promotes apoptosis, J. Biol. Chem. 286 (9) (2011) 7123–7131.
[53] T. Moldoveanu, Q. Liu, A. Tocilj, M. Wattson, G. Shore, K. Gehring, The X-ray structure of a BAK homodimer reveals an inhibitory zinc binding site, Mol. Cell 24 (5) (2006) 677–688.
[54] M. Suzuki, R.J. Youle, N. Tjandra, Structure of Bax: coregulation of dimer formation and intracellular localization, J. Palest. Stud. 41 (2) (2012) 128–129.
[55] A. Nechushtan, C.L. Smith, Y.T. Hsu, R.J. Youle, Conformation of the Bax C-terminus regulates subcellular location and cell death, EMBO J. 18 (9) (1999) 2330–2341.
[56] E. Gavathiotis, M. Suzuki, M.L. Davis, K. Pitter, G.H. Bird, S.G. Katz, et al., BAX activation is initiated at a novel interaction site, Nature 455 (7216) (2008) 1076–1081.
[57] E. Gavathiotis, D.E. Reyna, M.L. Davis, G.H. Bird, L.D. Walensky, BH3-triggered structural reorganization drives the activation of proapoptotic BAX, Mol. Cell 40 (3) (2010) 481–492.
[58] F. Edlich, S. Banerjee, M. Suzuki, M.M. Cleland, D. Arnoult, C. Wang, et al., Bcl-x(L) retrotranslocates Bax from the mitochondria into the cytosol, Cell 145 (1) (2011) 104–116.
[59] M. Doerflinger, J.A. Glab, H. Puthalakath, BH3-only proteins: a 20-year stock-take, FEBS J. 282 (6) (2015) 1006–1016.
[60] F. Wilfling, A. Weber, S. Potthoff, F.N. Vögtle, C. Meisinger, S.A. Paschen, et al., BH3-only proteins are tail-anchored in the outer mitochondrial membrane and can initiate the activation of Bax, Cell Death Differ. 19 (8) (2012) 1328–1336.
[61] C. Ploner, R. Kofler, A. Villunger, Noxa: at the tip of the balance between life and death, Oncogene 27 (Suppl. 1) (2008) S84–S92.
[62] Y.W. Seo, J.N. Shin, K.H. Ko, J.H. Cha, J.Y. Park, B.R. Lee, et al., The molecular mechanism of noxa-induced mitochondrial dysfunction in p53-mediated cell death, J. Biol. Chem. 278 (48) (2003) 48292–48299.
[63] H. Puthalakath, D.C. Huang, L.A. O'Reilly, S.M. King, A. Strasser, The proapoptotic activity of the Bcl-2 family member Bim is regulated by interaction with the dynein motor complex, Mol. Cell 3 (3) (1999) 287–296.
[64] P.K. Singh, A. Roukounakis, D.O. Frank, S. Kirschnek, K.K. Das, S. Neumann, et al., Dynein light chain 1 induces assembly of large Bim complexes on mitochondria that stabilize Mcl-1 and regulate apoptosis, Genes Dev. 31 (17) (2017) 1754–1769.

[65] H. Puthalakath, A. Villunger, L.A. O'Reilly, J.G. Beaumont, L. Coultas, R.E. Cheney, et al., Bmf: a proapoptotic BH3-only protein regulated by interaction with the myosin V actin motor complex, activated by anoikis, Science 293 (5536) (2001) 1829–1832.
[66] J.J. Chou, H. Li, G.S. Salvesen, J. Yuan, G. Wagner, Solution structure of BID, an intracellular amplifier of apoptotic signaling, Cell 96 (5) (1999) 615–624.
[67] J.M. McDonnell, D. Fushman, C.L. Milliman, S.J. Korsmeyer, D. Cowburn, Solution structure of the proapoptotic molecule BID: a structural basis for apoptotic agonists and antagonists, Cell 96 (5) (1999) 625–634.
[68] A. Gross, X.M. Yin, K. Wang, M.C. Wei, J. Jockel, C. Milliman, et al., Caspase cleaved BID targets mitochondria and is required for cytochrome c release, while BCL-XL prevents this release but not tumor necrosis factor-R1/Fas death, J. Biol. Chem. 274 (2) (1999) 1156–1163.
[69] A. Shamas-Din, S. Bindner, W. Zhu, Y. Zaltsman, C. Campbell, A. Gross, et al., tBid undergoes multiple conformational changes at the membrane required for Bax activation, J. Biol. Chem. 288 (30) (2013) 22111–22127.
[70] E.H. Cheng, T.V. Sheiko, J.K. Fisher, W.J. Craigen, S.J. Korsmeyer, VDAC2 inhibits BAK activation and mitochondrial apoptosis, Science 301 (5632) (2003) 513–517.
[71] M. Lazarou, D. Stojanovski, A.E. Frazier, A. Kotevski, G. Dewson, W.J. Craigen, et al., Inhibition of Bak activation by VDAC2 is dependent on the Bak transmembrane anchor, J. Biol. Chem. 285 (47) (2010) 36876–36883.
[72] H. Dai, H. Ding, X.W. Meng, K.L. Peterson, P.A. Schneider, J.E. Karp, et al., Constitutive BAK activation as a determinant of drug sensitivity in malignant lymphohematopoietic cells, Genes Dev. 29 (20) (2015) 2140–2152.
[73] S.N. Willis, J.I. Fletcher, T. Kaufmann, M.F. van Delft, L. Chen, P.E. Czabotar, et al., Apoptosis initiated when BH3 ligands engage multiple Bcl-2 homologs, not Bax or Bak, Science 315 (5813) (2007) 856–859.
[74] M. Certo, V. Del Gaizo Moore, M. Nishino, G. Wei, S. Korsmeyer, S.A. Armstrong, et al., Mitochondria primed by death signals determine cellular addiction to antiapoptotic BCL-2 family members, Cancer Cell 9 (5) (2006) 351–365.
[75] F. Llambi, T. Moldoveanu, S.W. Tait, L. Bouchier-Hayes, J. Temirov, L.L. McCormick, A unified model of mammalian BCL-2 protein family interactions at the mitochondria, Mol. Cell 44 (4) (2011) 517–531.
[76] S.N. Willis, L. Chen, G. Dewson, A. Wei, E. Naik, J.I. Fletcher, et al., Proapoptotic Bak is sequestered by Mcl-1 and Bcl-xL, but not Bcl-2, until displaced by BH3-only proteins, Genes Dev. 19 (11) (2005) 1294–1305.
[77] H. Dai, X.W. Meng, S.H. Lee, P.A. Schneider, S.H. Kaufmann, Context-dependent bcl-2/bak interactions regulate lymphoid cell apoptosis, J. Biol. Chem. 284 (27) (2009) 18311–18322.
[78] N. Echeverry, D. Bachmann, F. Ke, A. Strasser, H.U. Simon, T. Kaufmann, Intracellular localization of the BCL-2 family member BOK and functional implications, Cell Death Differ. 20 (6) (2013) 785–799.
[79] L. Chen, S.N. Willis, A. Wei, B.J. Smith, J.I. Fletcher, M.G. Hinds, et al., Differential targeting of prosurvival Bcl-2 proteins by their BH3-only ligands allows complementary apoptotic function, Mol. Cell 17 (3) (2005) 393–403.
[80] M.C. Wei, T. Lindsten, V.K. Mootha, S. Weiler, A. Gross, M. Ashiya, et al., tBID, a membrane-targeted death ligand, oligomerizes BAK to release cytochrome c, Genes Dev. 14 (16) (2000) 2060–2071.
[81] W.X. Zong, T. Lindsten, A.J. Ross, G.R. MacGregor, C.B. Thompson, BH3-only proteins that bind pro-survival Bcl-2 family members fail to induce apoptosis in the absence of Bax and Bak, Genes Dev. 15 (12) (2001) 1481–1486.

[82] S. Einsele-Scholz, S. Malmsheimer, K. Bertram, D. Stehle, J. Johänning, M. Manz, et al., Bok is a genuine multi-BH-domain protein that triggers apoptosis in the absence of Bax and Bak, J. Cell Sci. 129 (11) (2016) 2213–2223.
[83] S.N. Willis, J.M. Adams, Life in the balance: how BH3-only proteins induce apoptosis, Curr. Opin. Cell Biol. 17 (6) (2005) 617–625.
[84] A. Letai, M.C. Bassik, L.D. Walensky, M.D. Sorcinelli, S. Weiler, S.J. Korsmeyer, Distinct BH3 domains either sensitize or activate mitochondrial apoptosis serving as prototype cancer therapeutics, Cancer Cell 2 (3) (2002) 183–192.
[85] T. Kuwana, L. Bouchier-Hayes, J.E. Chipuk, C. Bonzon, B.A. Sullivan, D.R. Green, et al., BH3 domains of BH3-only proteins differentially regulate bax-mediated mitochondrial membrane permeabilization both directly and indirectly, Mol. Cell 17 (4) (2005) 525–535.
[86] H. Kim, M. Rafiuddin-Shah, H.C. Tu, J.R. Jeffers, G.P. Zambetti, J.J. Hsieh, et al., Hierarchical regulation of mitochondrion-dependent apoptosis by BCL-2 subfamilies, Nat. Cell Biol. 8 (12) (2006) 1348–1358.
[87] H. Dai, A. Smith, X.W. Meng, P.A. Schneider, Y.P. Pang, S.H. Kaufmann, Transient binding of an activator BH3 domain to the Bak BH3-binding groove initiates Bak oligomerization, J. Cell Biol. 194 (2011) 39–48.
[88] H. Dai, Y.P. Pang, M. Ramirez-Alvarado, S.H. Kaufmann, Evaluation of the BH3-only protein Puma as a direct bak activator, J. Biol. Chem. 289 (1) (2014) 89–99.
[89] C. Hockings, K. Anwari, R.L. Ninnis, J. Brouwer, M. O'Hely, M. Evangelista, et al., Bid chimeras indicate that most BH3-only proteins can directly activate Bak and Bax, and show no preference for Bak versus Bax, Cell Death Dis. 6 (2015) e1735.
[90] D. Ren, H.C. Tu, H. Kim, G.X. Wang, G.R. Bean, O. Takeuchi, et al., BID, BIM, and PUMA are essential for activation of the BAX- and BAK-dependent cell death program, Science 330 (2010) 1390–1393.
[91] H. Du, J. Wolf, B. Schafer, T. Moldoveanu, J.E. Chipuk, T. Kuwana, BH3 domains other than Bim and Bid can directly activate Bax/Bak, J. Biol. Chem. 286 (2011) 491–501.
[92] P.E. Czabotar, D. Westphal, G. Dewson, S. Ma, C. Hockings, W.D. Fairlie, et al., Bax crystal structures reveal how BH3 domains activate Bax and nucleate its oligomerization to induce apoptosis, Cell 152 (3) (2013) 519–531.
[93] T. Moldoveanu, C.R. Grace, F. Llambi, A. Nourse, P. Fitzgerald, K. Gehring, et al., BID-induced structural changes in BAK promote apoptosis, Nat. Struct. Mol. Biol. 20 (5) (2013) 589–597.
[94] J.M. Brouwer, D. Westphal, G. Dewson, A.Y. Robin, R.T. Uren, R. Bartolo, et al., Bak core and latch domains separate during activation, and freed core domains form symmetric homodimers, Mol. Cell 55 (6) (2014) 938–946.
[95] K. Ye, W.X. Meng, H. Sun, B. Wu, M. Chen, Y.P. Pang, et al., Characterization of an alternative BAK-binding site for BH3 peptides, Nat. Commun. 11 (1) (2020) 3301.
[96] H. Kim, H.C. Tu, D. Ren, O. Takeuchi, J.R. Jeffers, G.P. Zambetti, et al., Stepwise activation of BAX and BAK by tBID, BIM, and PUMA initiates mitochondrial apoptosis, Mol. Cell 36 (3) (2009) 487–499.
[97] K.L. O'Neill, K. Huang, J. Zhang, Y. Chen, X. Luo, Inactivation of prosurvival Bcl-2 proteins activates Bax/Bak through the outer mitochondrial membrane, Genes Dev 30 (8) (2016) 973–988.
[98] J. Zhang, K. Huang, K.L. O'Neill, X. Pang, X. Luo, Bax/Bak activation in the absence of bid, Bim, Puma, and p53, Cell Death Dis. 7 (2016) e2266.
[99] K. Huang, K.L. O'Neill, J. Li, W. Zhou, N. Han, X. Pang, et al., BH3-only proteins target BCL-xL/MCL1, not BAX/BAK, to initiate apoptosis, Cell Res. 29 (11) (2019) 942–952.

[100] X. Jiang, X. Wang, Cytochrome C-mediated apoptosis, Annu. Rev. Biochem. 73 (73) (2004) 87–106.
[101] P.G. Ekert, D.L. Vaux, The mitochondrial death squad: hardened killers or innocent bystanders? Curr. Opin. Cell Biol. 17 (6) (2005) 626–630.
[102] K. Wang, A. Gross, G. Waksman, S.J. Korsmeyer, Mutagenesis of the BH3 domain of BAX identifies residues critical for dimerization and killing, Mol. Cell. Biol. 18 (10) (1999) 6083–6089.
[103] G. Dewson, T. Kratina, H.W. Sim, H. Puthalakath, J.M. Adams, P.M. Colman, et al., To trigger apoptosis, bak exposes its BH3 domain and homodimerizes via BH3:groove interactions, Mol. Cell 30 (3) (2008) 369–380.
[104] G. Dewson, T. Kratina, P. Czabotar, C.L. Day, J.M. Adams, R.M. Kluck, Bak activation for apoptosis involves oligomerization of dimers via their alpha 6 helices, Mol. Cell 36 (4) (2009) 696–703.
[105] Z. Zhang, W. Zhu, S.M. Lapolla, Y. Miao, Y. Shao, M. Falcone, et al., Bax forms an oligomer via separate, yet interdependent, surfaces, J. Biol. Chem. 285 (23) (2010) 17614–17627.
[106] S. Lyer, F. Bell, D. Westphal, K. Anwari, J. Gulbis, B.J. Smith, et al., Bak apoptotic pores involve a flexible C-terminal region and juxtaposition of the C-terminal transmembrane domains, Cell Death Differ. 22 (10) (2015) 1665–1675.
[107] R.F. Gahl, Y. He, S. Yu, N. Tjandra, Conformational rearrangements in the pro-apoptotic protein, Bax, as it inserts into mitochondria: a cellular death switch, J. Biol. Chem. 289 (47) (2014) 32871–32882.
[108] Z. Zhang, S. Subramaniam, J. Kale, C. Liao, B. Huang, H. Brahmbhatt, et al., BH3-in-groove dimerization initiates and helix 9 dimerization expands Bax pore assembly in membranes, EMBO J. 35 (2) (2016) 208–236.
[109] M.W. Parker, S.C. Feil, Pore-forming protein toxins: from structure to function, Prog. Biophys. Mol. Biol. 88 (1) (2005) 91–142.
[110] M.G. Annis, E.L. Soucie, P.J. Dlugosz, J.A. Cruz-Aguado, L.Z. Penn, B. Leber, et al., Bax forms multispanning monomers that oligomerize to permeabilize membranes during apoptosis, EMBO J. 24 (12) (2005) 2096–2103.
[111] S. Choe, M.J. Bennett, G. Fujii, P.M. Curmi, K.A. Kantardjieff, R.J. Collier, et al., The crystal structure of diphtheria toxin, Nature 357 (6375) (1992) 216–222.
[112] P. Elkins, A. Bunker, W.A. Cramer, C.V. Stauffacher, A mechanism for toxin insertion into membranes is suggested by the crystal structure of the channel-forming domain of colicin E1, Structure 5 (3) (1997) 443–458.
[113] S. Qian, W. Wang, L. Yang, H.W. Huang, Structure of transmembrane pore induced by Bax-derived peptide: evidence for lipidic pores, Proc. Natl. Acad. Sci. U.S.A. 105 (2008) 17379–17383.
[114] A.J. Garcia-Saez, M. Coraiola, M.D. Serra, I. Mingarro, P. Muller, J. Salgado, et al., Peptides corresponding to helices 5 and 6 of Bax can independently form large lipid pores, FEBS J. 273 (5) (2006) 971–981.
[115] S. Bleicken, G. Jeschke, C. Stegmueller, R. Salvador-Gallego, A.J. García-Sáez, E. Bordignon, et al., Structural model of active bax at the membrane, Mol. Cell 56 (4) (2014) 496–505.
[116] R. Salvador-Gallego, M. Mund, K. Cosentino, J. Schneider, J. Unsay, U. Schraermeyer, et al., Bax assembly into rings and arcs in apoptotic mitochondria is linked to membrane pores, EMBO J. 35 (4) (2016) 389–401.
[117] L. Große, C.A. Wurm, C. Brüser, D. Neumann, D.C. Jans, S. Jakobs, Bax assembles into large ring-like structures remodeling the mitochondrial outer membrane in apoptosis, EMBO J. 35 (4) (2016) 402–413.

[118] D. u, G. Dewson, M. Menard, P. Frederick, S. Lyer, R. Bartolo, et al., Apoptotic pore formation is associated with in-plane insertion of Bak or Bax central helices into the mitochondrial outer membrane, Proc. Natl Acad. Sci. U.S.A. 111 (39) (2014) E4076–E4085.

[119] R.T. Uren, M. O'Hely, S. Lyer, R. Bartolo, M.X. Shi, J.M. Brouwer, et al., Disordered clusters of Bak dimers rupture mitochondria during apoptosis, Elife 6 (2017) e19944.

[120] A.D. Cowan, N.A. Smith, J.J. Sandow, E.A. Kapp, Y.H. Rustam, J.M. Murphy, et al., BAK core dimers bind lipids and can be bridged by them, Nat. Struct. Mol. Biol. 27 (11) (2020) 1024–1031.

[121] E. Raemy, S. Montessuit, S. Pierredon, A.H. van Kampen, F.M. Vaz, J.-C. Martinou, Cardiolipin or MTCH2 can serve as tBID receptors during apoptosis, Cell Death Differ. 23 (7) (2016) 1165–1174.

[122] X.H. Liang, L.K. Kleeman, H.H. Jiang, G. Gordon, J.E. Goldman, G. Berry, et al., Protection against fatal sindbis virus encephalitis by Beclin, a novel bcl-2-interacting protein, J. Virol. 72 (11) (1998) 8586–8596.

[123] Z. Yue, S. Jin, C. Yang, A.J. Levine, N. Heintz, Beclin 1, an autophagy gene essential for early embryonic development, is a haploinsufficient tumor suppressor, Proc. Natl. Acad. Sci. U.S.A. 100 (25) (2003) 15077–15082.

[124] Y. Wei, S.C. Sinha, B. Levine, Dual Role of JNK1-mediated phosphorylation of Bcl-2 in autophagy and apoptosis regulation, Autophagy 4 (7) (2008) 949–951.

[125] H.L. Xiao, M.D. Schneider, R. Garuti, N. Mizushima, M. Packer, S. Pattingre, et al., Bcl-2 antiapoptotic proteins inhibit Beclin 1-dependent autophagy, Cell 122 (6) (2005) 927–939.

[126] M.C. Maiuri, G. Le Toumelin, A. Criollo, J.C. Rain, F. Gautier, P. Juin, et al., Functional and physical interaction between Bcl-X (L) and BH3-like domain in Beclin-1, EMBO J. 26 (10) (2007) 2527–2539.

[127] R. Rashmi, S. Pillai, S. Vijayalingam, J. Ryerse, G. Chinnadurai, BH3-only protein BIK induces caspase-independent cell death with autophagic features in Bcl-2 null cells, Oncogene 27 (10) (2008) 1366–1375.

[128] K.S. Yee, S. Wilkinson, J. James, K.M. Ryan, K.H. Vousden, PUMA- and Bax-induced autophagy contributes to apoptosis, Cell Death Differ. 16 (8) (2009) 1135–1145.

[129] H. Zhang, M. Bosch-Marce, L. Shimoda, Y. Tan, J. Baek, J. Wesley, et al., Mitochondrial autophagy is an HIF-1-dependent adaptive metabolic response to hypoxia, J. Biol. Chem. 283 (16) (2008) 10892–10903.

[130] L.M. Lindqvist, M. Heinlein, D.C. Huang, D.L. Vaux, Prosurvival Bcl-2 family members affect autophagy only indirectly, by inhibiting Bax and Bak, Proc. Natl. Acad. Sci. U.S.A. 111 (23) (2014) 8512–8517.

[131] L.M. Lindqvist, D.L. Vaux, BCL2 and related prosurvival proteins require BAK1 and BAX to affect autophagy, Autophagy 10 (8) (2014) 1474–1475.

[132] X. Gui, H. Yang, T. Li, X. Tan, P. Shi, M. Li, et al., Autophagy induction via STING trafficking is a primordial function of the cGAS pathway, Nature 567 (7747) (2019) 262–266.

[133] D. Liu, H. Wu, C. Wang, Y. Li, H. Tian, S. Siraj, et al., STING directly activates autophagy to tune the innate immune response, Cell Death Differ. 26 (9) (2019) 1735–1749.

[134] L.M. Lindqvist, D. Frank, K. Mcarthur, T.A. Dite, M. Lazarou, J.S. Oakhill, et al., Autophagy induced during apoptosis degrades mitochondria and inhibits type I interferon secretion, Cell Death Differ. 25 (4) (2018) 782–794.

[135] S. Di Bartolomeo, M. Corazzari, F. Nazio, S. Oliverio, G. Lisi, M. Antonioli, et al., The dynamic interaction of AMBRA1 with the dynein motor complex regulates mammalian autophagy, J. Cell Biol. 191 (2010) 155–168.
[136] G.M. Fimia, A. Stoykova, A. Romagnoli, L. Giunta, S. Di Bartolomeo, R. Nardacci, et al., Ambra1 regulates autophagy and development of the nervous system, Nature 447 (7148) (2007) 1121–1125.
[137] F. Strappazzon, A. Di Rita, V. Cianfanelli, M.D. Orazio, F. Nazio, G.M. Fimia, et al., Prosurvival AMBRA1 turns into a proapoptotic BH3-like protein during mitochondrial apoptosis, Autophagy (2016) 963–975.
[138] M. Karbowski, K.L. Norris, M.M. Cleland, S.Y. Jeong, R.J. Youle, Role of bax and bak in mitochondrial morphogenesis, Nature 443 (7112) (2006) 658–662.
[139] P. Delivani, C. Adrain, R.C. Taylor, P.J. Duriez, S.J. Martin, Role for CED-9 and Egl-1 as regulators of mitochondrial fission and fusion dynamics, Mol. Cell 21 (6) (2006) 761–773.
[140] Q.A. Liu, H. Shio, H. Mitochondrial Morphogenesis, Dendrite development, and synapse formation in cerebellum require both bcl-w and the glutamate receptor delta2, PLoS Genet. 4 (6) (2008) e1000097.
[141] S.B. Berman, Y.B. Chen, B. Qi, J.M. McCaffery, E.B. Rucker, S. Goebbels, et al., Bcl-xL increases mitochondrial fission, fusion, and biomass in neurons, J. Cell Biol. 184 (5) (2009) 707–719.
[142] M. Karbowski, Y.J. Lee, B. Gaume, S.Y. Jeong, S. Frank, A. Nechushtan, et al., Spatial and temporal association of Bax with mitochondrial fission sites, Drp1, and Mfn2 during apoptosis, J. Cell Biol. 159 (6) (2002) 931–938.
[143] M.M. Cleland, K.L. Norris, M. Karbowski, C. Wang, D.F. Suen, S. Jiao, et al., Bcl-2 family interaction with the mitochondrial morphogenesis machinery, Cell Death Differ. 18 (2) (2011) 235–247.
[144] I. Kamer, R. Sarig, Y. Zaltsman, H. Niv, G. Oberkovitz, L. Regev, et al., Proapoptotic BID is an ATM effector in the DNA-damage response, Cell 122 (4) (2005) 593–603.
[145] Y. Liu, C.C. Bertram, Q. Shi, S.S. Zinkel, Proapoptotic Bid mediates the Atr-directed DNA damage response to replicative stress, Cell Death Differ. 18 (5) (2011) 841–852.
[146] T. Kaufmann, L. Tai, P.G. Ekert, D.C. Huang, F. Norris, R.K. Lindemann, et al., The BH3-only protein bid is dispensable for DNA damage- and replicative stress-induced apoptosis or cell-cycle arrest, Cell 129 (2) (2007) 423–433.
[147] A. Gross, Y. Zaltsman, M. Maryanovich, The ATM-BID pathway plays a critical role in the DNA damage response by regulating mitochondria metabolism, Cell Death Differ. 23 (1) (2016), 182.
[148] H. Puthalakath, L.A. O'Reilly, P. Gunn, L. Lee, P.N. Kelly, N.D. Huntington, et al., ER stress triggers apoptosis by activating BH3-only protein Bim, Cell 129 (7) (2007) 1337–1349.
[149] J.A. Wali, D. Rondas, M.D. McKenzie, Y. Zhao, L. Elkerbout, S. Fynch, et al., The proapoptotic BH3-only proteins Bim and Puma are downstream of endoplasmic reticulum and mitochondrial oxidative stress in pancreatic islets in response to glucotoxicity, Cell Death Dis. 5 (3) (2014) e1124.
[150] D.A. Rodriguez, S. Zamorano, F. Lisbona, D. Rojas-Rivera, H. Urra, J.R. Cubillos-Ruiz, et al., BH3-only proteins are part of a regulatory network that control the sustained signalling of the unfolded protein response sensor IRE1α, EMBO J. 31 (10) (2012) 2322–2335.
[151] X. Wu, L.S. Zhang, J. Toombs, Y.C. Kuo, J.T. Piazza, R. Tuladhar, et al., Extra-mitochondrial prosurvival BCL-2 proteins regulate gene transcription by inhibiting the SUFU tumour suppressor, Nat. Cell Biol. 19 (10) (2017) 1226–1236.

[152] D. Hanahan, R.A. Weinberg, Hallmarks of cancer: the next generation, Cell 144 (2011) 646–674.
[153] S.X. Jiang, Y. Sato, S. Kuwao, T. Kameya, Expression of bcl-2 oncogene protein is prevalent in small cell lung carcinomas, J. Pathol. 177 (2) (1995) 135–138.
[154] R. Grover, G.D. Wilson, Bcl-2 expression in malignant melanoma and its prognostic significance, Eur. J. Surg. Oncol. 22 (4) (1996) 347–349.
[155] H. Joensuu, L. Pylkkänen, S. Toikkanen, Bcl-2 protein expression and long-term survival in breast cancer, Am. J. Pathol. 145 (5) (1994) 1191–1198.
[156] T.J. McDonnell, P. Troncoso, S.M. Brisbay, C. Logothetis, L.W. Chung, J.T. Hsieh, et al., Expression of the protooncogene bcl-2 in the prostate and its association with emergence of androgen-independent prostate cancer, Cancer Res. 52 (24) (1992) 6940–6944.
[157] F.A. Sinicrope, J. Hart, F. Michelassi, J.J. Lee, Prognostic value of bcl-2 oncoprotein expression in stage II colon carcinoma, Clin. Cancer Res. 1 (10) (1995) 1103–1110.
[158] P. Gazzaniga, A. Gradilone, R. Vercillo, O. Gandini, I. Silvestri, M. Napolitano, et al., Bcl-2/bax mRNA expression ratio as prognostic factor in low-grade urinary bladder cancer, Int. J. Cancer 69 (2) (1996) 100–104.
[159] K.W. Yip, J.C. Reed, Bcl-2 family proteins and cancer, Oncogene 27 (50) (2008) 6398–6406.
[160] J.C. Reed, Bcl-2-family proteins and hematologic malignancies: history and future prospects, Blood 111 (2008) 3322–3330.
[161] L. Scarfò, P. Ghia, Reprogramming cell death: BCL2 family inhibition in hematological malignancies, Immunol. Lett. 155 (1–2) (2013) 36–39.
[162] L.M. Weiss, R.A. Warnke, J. Sklar, M.L. Cleary, Molecular analysis of the t(14;18) chromosomal translocation in malignant lymphomas, N. Engl. J. Med. 317 (19) (1987) 1185–1189.
[163] B.Y. Ngan, Z. Chen-Levy, L.M. Weiss, R.A. Warnke, M.L. Cleary, Expression in non-Hodgkin's lymphoma of the bcl-2 protein associated with the t(14;18) chromosomal translocation, N. Engl. J. Med. 318 (25) (1988) 1638–1644.
[164] D.P. Lebrun, B.Y. Ngan, L.M. Weiss, P. Huie, R.A. Warnke, M.L. Cleary, The bcl-2 oncogene in Hodgkin's disease arising in the setting of follicular non-Hodgkin's lymphoma, Blood 83 (1) (1994) 223–230.
[165] Y. Chen, D. Wang, H. Peng, X. Chen, X. Han, J. Yu, et al., Epigenetically upregulated oncoprotein PLCE1 drives esophageal carcinoma angiogenesis and proliferation via activating the PI-PLCε-NF-κB signaling pathway and VEGF-C/Bcl-2 expression, Mol. Cancer 18 (1) (2019) 1.
[166] P.H. Rao, J. Houldsworth, K. Dyomina, N.Z. Parsa, J.C. Cigudosa, D.C. Louie, et al., Chromosomal and gene amplification in diffuse large B-cell lymphoma, Blood 92 (1) (1998) 234–240.
[167] A. Cimmino, G.A. Calin, M. Fabbri, M.V. Iorio, M. Ferracin, M. Shimizu, et al., miR-15 and miR-16 induce apoptosis by targeting BCL2, Proc. Natl. Acad. Sci. U.S.A. 102 (39) (2005) 13944–13949.
[168] S.K. Niture, A.K. Jaiswal, Inhibitor of Nrf2 (INrf2 or Keap1) protein degrades bcl-xL via phosphoglycerate mutase 5 and controls cellular apoptosis, J. Biol. Chem. 286 (2011) 44542–44556.
[169] C. Bincoletto, S.T. Saad, E.S. da Silva, M.L. Queiroz, Haematopoietic response and bcl-2 expression in patients with acute myeloid leukaemia, Eur. J. Haematol. 62 (1) (1999) 38–42.
[170] R. Singh, N. Saini, Downregulation of BCL2 by miRNAs augments drug-induced apoptosis - a combined computational and experimental approach, J. Cell Sci. 125 (2012) 1568–1578.

[171] Y. Kuwano, K. Nishida, K. Kajita, Y. Satake, Y. Akaike, K. Fujita, et al., Transformer 2 beta and miR-204 regulate apoptosis through competitive binding to 3 ' UTR of BCL2 Mrna, Cell Death Differ. 22 (5) (2015) 815–825.
[172] P.N. Kelly, S. Grabow, A.R. Delbridge, A. Strasser, J.M. Adams, Endogenous Bcl-xL is essential for Myc-driven lymphomagenesis in mice, Blood 118 (24) (2011) 6380–6386.
[173] C. Castilla, B. Congregado, D. Chinchon, F.J. Torrubia, M.A. Japon, C. Saez, Bcl-xL is overexpressed in hormone-resistant prostate cancer and promotes survival of LNCaP cells via interaction with proapoptotic Bak, Endocrinology 10 (2006) 4960–4967.
[174] U. Keitel, A. Scheel, J. Thomale, R. Halpape, S. Kaulfuss, C. Scheel, et al., Bcl-xL mediates therapeutic resistance of a mesenchymal breast cancer cell subpopulation, Oncotarget 5 (23) (2014) 11778–11791.
[175] S. Gutierrez-Castellanos, M. Cruz, L. Rabelo, R. Godinez, E. Reyes-Maldonado, C. Riebeling-Navarro, Differences in BCL-X-L expression and STAT5 phosphorylation in chronic myeloid leukaemia patients, Eur. J. Haematol. 72 (4) (2004) 231–238.
[176] S. Shimizu, T. Takehara, H. Hikita, T. Kodama, T. Miyagi, A. Hosui, et al., The let-7 family of microRNAs inhibits Bcl-xL expression and potentiates sorafenib-induced apoptosis in human hepatocellular carcinoma, J. Hepatol. 52 (5) (2010) 698–704.
[177] M. Zhan, Q. Qu, G. Wang, H. Zhou, Let-7c sensitizes acquired cisplatin-resistant A549 cells by targeting ABCC2 and Bcl-XL, Pharmazie 68 (12) (2013) 955–961.
[178] P. Zhou, N.B. Levy, H. Xie, L. Qian, C.Y. Lee, R.D. Gascoyne, et al., MCL1 transgenic mice exhibit a high incidence of B-cell lymphoma manifested as a spectrum of histologic subtypes, Blood 97 (12) (2001) 3902–3909.
[179] X.W. Meng, S.H. Lee, H. Dai, D. Loegering, C. Yu, K. Flatten, et al., Mcl-1 as a buffer for proapoptotic Bcl-2 family members during TRAIL-induced apoptosis: a mechanistic basis for sorafenib (Bay 43-9006)-induced TRAIL sensitization, J. Biol. Chem. 282 (41) (2007) 29831–29846.
[180] S.R. Hussain, C.M. Cheney, A.J. Johnson, T.S. Lin, M.R. Grever, M.A. Caligiuri, et al., Mcl-1 is a relevant therapeutic target in acute and chronic lymphoid malignancies: down-regulation enhances rituximab-mediated apoptosis and complement-dependent cytotoxicity, Clin. Cancer Res. 13 (7) (2007) 2144–2150.
[181] S. Kitada, J. Andersen, S. Akar, J.M. Zapata, S. Takayama, S. Krajewski, et al., Expression of apoptosis-regulating proteins in chronic lymphocytic leukemia: correlations with in vitro and in vivo chemoresponses, Blood 91 (9) (1998) 3379–3389.
[182] S.P. Glaser, E.F. Lee, E. Trounson, P. Bouillet, A. Wei, W.D. Fairlie, et al., Anti-apoptotic Mcl-1 is essential for the development and sustained growth of acute myeloid leukemia, Genes Dev. 26 (2) (2012) 120–125.
[183] B.A. Quinn, R. Dash, B. Azab, S. Sarkar, S.K. Das, S. Kumar, et al., Targeting Mcl-1 for the therapy of cancer, Expet Opin. Invest. Drugs 20 (10) (2011) 1397–1411.
[184] Q. Ding, X. He, W. Xia, J.M. Hsu, C.T. Chen, L.Y. Li, et al., Myeloid cell leukemia-1 inversely correlates with glycogen synthase kinase-3β activity and associates with poor prognosis in human breast cancer, Cancer Res. 67 (10) (2007) 4564–4571.
[185] S.H. Kaufmann, J.E. Karp, P.A. Svingen, S. Krajewski, P.J. Burke, S.D. Gore, et al., Elevated expression of the apoptotic regulator Mcl-1 at the time of leukemic relapse, Blood 91 (3) (1998) 991–1000.
[186] Y. Xiong, J.H. Fang, J.P. Yun, J. Yang, Y. Zhang, W.H. Jia, et al., Effects of MicroRNA-29 on apoptosis, tumorigenicity and prognosis of hepatocellular carcinoma, Hepatology 51 (2010) 836–845.

[187] R. Steele, J.L. Mott, R.B. Ray, MBP-1 upregulates miR-29b that represses Mcl-1, collagens, and matrix-metalloproteinase-2 in prostate cancer cells, Genes Cancer 1 (4) (2010) 381–387.
[188] M. Schwickart, X. Huang, J.R. Lill, J. Liu, R. Ferrando, D.M. French, et al., Deubiquitinase USP9X stabilizes MCL1 and promotes tumour cell survival, Nature 463 (7277) (2010) 103–107.
[189] C. Correia, S.H. Lee, X.W. Meng, N.D. Vincelette, K.L.B. Knorr, H. Ding, et al., Emerging understanding of Bcl-2 biology: implications for neoplastic progression and treatment, Biochim. Biophys. Acta 1853 (7) (2015) 1658–1671.
[190] L. Happo, A. Strasser, S. Cory, BH3-only proteins in apoptosis at a glance, J. Cell Sci. 125 (5) (2012) 1081–1087.
[191] T.T. Tan, K. Degenhardt, D.A. Nelson, B. Beaudoin, W. Nieves-Neira, P. Bouillet, et al., Key roles of BIM-driven apoptosis in epithelial tumors and rational chemotherapy, Cancer Cell 7 (3) (2005) 227–238.
[192] H. Tagawa, S. Karnan, R. Suzuki, K. Matsuo, X. Zhang, A. Ota, et al., Genome-wide array-based CGH for mantle cell lymphoma: identification of homozygous deletions of the proapoptotic gene BIM, Oncogene 24 (2005) 1348–1358.
[193] M.T. Hemann, A. Bric, J. Teruya-Feldstein, A. Herbst, J.A. Nilsson, C. Cordon-Cardo, et al., Evasion of the p53 tumour surveillance network by tumour-derived MYC mutants, Nature 436 (7052) (2005) 807–811.
[194] C. Xiao, L. Srinivasan, D.P. Calado, H.C. Patterson, B. Zhang, J. Wang, et al., Lymphoproliferative disease and autoimmunity in mice with increased miR-17-92 expression in lymphocytes, Nat. Immunol. 9 (4) (2008) 404–414.
[195] E.M. Michalak, A. Villunger, J.M. Adams, A. Strasser, In several cell types tumour suppressor p53 induces apoptosis largely via Puma but Noxa can contribute, Cell Death Differ. 15 (6) (2008) 1019–1029.
[196] S.P. Garrison, J.R. Jeffers, C. Yang, J.A. Nilsson, M.A. Hall, J.E. Rehg, et al., Selection against PUMA gene expression in Myc-driven B-cell lymphomagenesis, Mol. Cell Biol. 28 (17) (2008) 5391–5402.
[197] M.T. Hemann, J.T. Zilfou, Z. Zhao, D.J. Burgess, G.J. Hannon, S.W. Lowe, Suppression of tumorigenesis by the p53 target PUMA, Proc. Natl. Acad. Sci. U.S.A. 101 (25) (2004) 9333–9338.
[198] E.M. Michalak, C.J. Vandenberg, A.R.D. Delbridge, L. Wu, C.L. Scott, J.M. Adams, et al., Apoptosis-promoted tumorigenesis: -irradiation-induced thymic lymphomagenesis requires Puma-driven leukocyte death, Genes Dev. 24 (15) (2010) 1608–1613.
[199] M.O. Hoque, S. Begum, M. Sommer, T. Lee, B. Trink, E. Ratovitski, et al., PUMA in head and neck cancer, Cancer Lett. 199 (1) (2003) 75–81.
[200] A.M. Karst, D.L. Dai, M. Martinka, G. Li, PUMA expression is significantly reduced in human cutaneous melanomas, Oncogene 24 (2005) 1111–1116.
[201] M. Yamakuchi, M. Ferlito, C.J. Lowenstein, miR-34a repression of SIRT1 regulates apoptosis, Proc. Natl. Acad. Sci. U.S.A. 105 (36) (2008) 13421–13426.
[202] A.M. Ranger, J. Zha, H. Harada, S.R. Datta, N.N. Danial, A.P. Gilmore, et al., Bad-deficient mice develop diffuse large B cell lymphoma, Proc. Natl. Acad. Sci. U.S. A. 100 (2003) 9324–9329.
[203] P.N. Kelly, M.J. White, M.W. Goschnick, K.A. Fairfax, D.M. Tarlinton, S.A. Kinkel, et al., Individual and overlapping roles of BH3-only proteins Bim and Bad in apoptosis of lymphocytes and platelets and in suppression of thymic lymphoma development, Cell Death Differ. 17 (10) (2010) 1655–1664.
[204] V. Labi, M. Erlacher, S. Kiessling, C. Manzl, A. Frenzel, L. O'Reilly, et al., Loss of the BH3-only protein Bmf impairs B cell homeostasis and accelerates γ irradiation–induced thymic lymphoma development, J. Exp. Med. 205 (3) (2008) 641–655.

[205] A. Frenzel, V. Labi, W. Chmelewskij, C. Ploner, S. Geley, H. Fiegl, et al., Suppression of B-cell lymphomagenesis by the BH3-only proteins Bmf and Bad, Blood 115 (2010) 995–1005.
[206] S.S. Zinkel, C.C. Ong, D.O. Ferguson, H. Iwasaki, K. Akashi, R.T. Bronson, et al., Proapoptotic BID is required for myeloid homeostasis and tumor suppression, Genes Dev. 17 (2) (2003) 229–239.
[207] J.W. Lee, Y.H. Soung, S.Y. Kim, S.W. Nam, C.J. Kim, Y.G. Cho, et al., Inactivating mutations of proapoptotic Bad gene in human colon cancers, Carcinogenesis 25 (8) (2004) 1371–1376.
[208] C.M. Knudson, G.M. Johnson, Y. Lin, S.J. Korsmeyer, Bax accelerates tumorigenesis in p53-deficient mice, Cancer Res. 61 (2) (2001) 659–665.
[209] K. Degenhardt, G. Chen, T. Lindsten, E. White, BAX and BAK mediate p53-independent suppression of tumorigenesis, Cancer Cell 2 (2002) 193–203.
[210] C. Yin, C.M. Knudson, S.J. Korsmeyer, T. Van Dyke, Bax suppresses tumorigenesis and stimulates apoptosis in vivo, Nature 385 (6617) (1997) 637–640.
[211] M.A. Shibata, M.L. Liu, M.C. Knudson, E. Shibata, K. Yoshidome, T. Bandey, et al., Haploid loss of bax leads to accelerated mammary tumor development in C3(1)/SV40-TAg transgenic mice : reduction in protective apoptotic response at the preneoplastic stage, EMBO J. 18 (1999) 2692–2701.
[212] C.M. Eischen, M.F. Roussel, S.J. Korsmeyer, J.L. Cleveland, Bax loss impairs myc-induced apoptosis and circumvents the selection of p53 mutations during myc-mediated lymphomagenesis, Mol. Cell Biol. 21 (22) (2001) 7653–7662.
[213] C.M. Eischen, J.E. Rehg, S.J. Korsmeyer, J.L. Cleveland, Loss of Bax alters tumor spectrum and tumor numbers in ARF-deficient mice, Cancer Res. 62 (7) (2002) 2184–2191.
[214] J.P. Meijerink, E.J. Mensink, K. Wang, T.W. Sedlak, A.W. Slöetjes, T. de Witte, et al., Hematopoietic malignancies demonstrate loss-of-function mutations of BAX, Blood 91 (8) (1998) 2991–2997.
[215] S. Kondo, Y. Shinomura, Y. Miyazaki, T. Kiyohara, S. Tsutsui, S. Kitamura, et al., Mutations of the bak gene in human gastric and colorectal cancers, Cancer Res. 60 (16) (2000) 4328–4330.
[216] Y. Ionov, H. Yamamoto, S. Krajewski, J.C. Reed, M. Perucho, Mutational inactivation of the proapoptotic gene BAX confers selective advantage during tumor clonal evolution, Proc. Natl. Acad. Sci. U.S.A. 97 (20) (2000) 10872–10877.
[217] E. Radi, P. Formichi, C. Battisti, A. Federico, Apoptosis and oxidative stress in neurodegenerative diseases, J. Alzheim. Dis. 42 (Suppl. 3) (2014) S125–S152.
[218] J.T. Opferman, A. Kothari, Anti-apoptotic BCL-2 family members in development, Cell Death Differ. 25 (1) (2018) 37–45.
[219] V. Gagliardini, P.A. Fernandez, R.K. Lee, H.C. Drexler, R.J. Rotello, M.C. Fishman, et al., Prevention of vertebrate neuronal death by the crmA gene, Science 263 (5148) (1994) 826–828.
[220] D.J. Veis, C.M. Sorenson, J.R. Shutter, S.J. Korsmeyer, Bcl-2-deficient mice demonstrate fulminant lymphoid apoptosis, polycystic kidneys, and hypopigmented hair, Cell 75 (1993) 229–240.
[221] Z.S. Nagy, M.M. Esiri, Apoptosis-related protein expression in the hippocampus in Alzheimer's disease, Neurobiol. Aging 18 (6) (1997) 565–571.
[222] M. Yaar, S. Zhai, P.F. Pilch, S.M. Doyle, P.B. Eisenhauer, R.E. Fine, et al., Binding of beta-amyloid to the p75 neurotrophin receptor induces apoptosis. A possible mechanism for Alzheimer's disease, J. Clin. Invest. 100 (9) (1997) 2333–2340.
[223] E. Masliah, M. Mallory, M. Alford, S. Tanaka, L.A. Hansen, Caspase dependent DNA fragmentation might Be associated with excitotoxicity in Alzheimer disease, J. Neuropathol. Exp. Neurol. 57 (11) (1998) 1041–1052.

[224] C.C. Chang, Y.T. Chang, C.W. Huang, S.J. Tsai, S.W. Hsu, S.H. Huang, et al., Associations of Bcl-2 rs956572 genotype groups in the structural covariance network in early-stage Alzheimer's disease, Alzheimer Res. Ther. 10 (1) (2018) 17.
[225] K. Nikhil, K. Shah, The Cdk5-Mcl-1 axis promotes mitochondrial dysfunction and neurodegeneration in a model of Alzheimer's disease, J. Cell Sci. 130 (18) (2017) 3023–3029.
[226] F. Caraci, M. Chisari, G. Frasca, P.L. Canonico, A. Battaglia, M. Calafiore, et al., Nicergoline, a drug used for age-dependent cognitive impairment, protects cultured neurons against β-amyloid toxicity, Brain Res. 1047 (1) (2005) 30–37.
[227] A.P. Ghosh, B.J. Klocke, M.E. Ballestas, K.A. Roth, CHOP potentially Co-operates with FOXO3a in neuronal cells to regulate PUMA and BIM expression in response to ER stress, PLoS One 7 (6) (2012) e39586.
[228] A.M. Pickrell, R.J. Youle, The roles of PINK1, parkin, and mitochondrial fidelity in Parkinson's disease, Neuron 85 (2) (2015) 257–273.
[229] W. Lu, S.S. Karuppagounder, D.A. Springer, M.D. Allen, L. Zheng, B. Chao, et al., Genetic deficiency of the mitochondrial protein PGAM5 causes a Parkinson's-like movement disorder, Nat. Commun. 5 (2014) e4930.
[230] H. Wu, D. Xue, G. Chen, Z. Han, L. Huang, C. Zhu, et al., The BCL2L1 and PGAM5 axis defines hypoxia-induced receptor-mediated mitophagy, Autophagy 10 (10) (2014) 1712–1725.
[231] D. Tischner, C. Woess, E. Ottina, A. Villunger, Bcl-2-regulated cell death signaling in the prevention of autoimmunity, Cell Death Dis. 1 (2010) e48.
[232] P. Bouillet, D. Metcalf, D.C.S. Huang, D.M. Tarlinton, T.W.H. Kay, F. Köntgen, et al., Proapoptotic bcl-2 relative Bim required for certain apoptotic responses, leukocyte homeostasis, and to preclude autoimmunity, Science 286 (5445) (1999) 1735–1738.
[233] P. Bouillet, J.F. Purton, D.I. Godfrey, L.C. Zhang, L. Coultas, H. Puthalakath, et al., BH3-only Bcl-2 family member Bim is required for apoptosis of autoreactive thymocytes, Nature 415 (2002) 922–926.
[234] J.B. Oliveira, N. Bidere, J.E. Niemela, L. Zheng, K. Sakai, C.P. Nix, et al., NRAS mutation causes a human autoimmune lymphoproliferative syndrome, Proc. Natl. Acad. Sci. U.S.A. 104 (2007) 8953–8958.
[235] A. Enders, P. Bouillet, H. Puthalakath, Y. Xu, D.M. Tarlinton, A. Strasser, Loss of the pro-apoptotic BH3-only Bcl-2 family member Bim inhibits BCR stimulation-induced apoptosis and deletion of autoreactive B cells, J. Exp. Med. 198 (7) (2003) 1119–1126.
[236] M. Pellegrini, G. Belz, P. Bouilet, A. Strasser, Shutdown of an acute T cell immune response to viral infection is mediated by the proapoptotic Bcl-2 homology 3-only protein Bim, Proc. Natl. Acad. Sci. U.S.A. 100 (24) (2003) 14175–14180.
[237] P.D. Hughes, G.T. Belz, K. Fortner, R.C. Budd, A. Strasser, P. Bouillet, Apoptosis regulators Fas and Bim cooperate in shutdown of chronic immune responses and prevention of autoimmunity, Immunity 28 (2) (2008) 197–205.
[238] N.D. Huntington, H. Puthalakath, P. Gunn, E. Naik, E.M. Michalak, M. Smyth, et al., Interleukin 15-mediated survival of natural killer cells is determined by interactions among Bim, Noxa and Mcl-1, Nat. Immunol. 8 (8) (2007) 856.
[239] M. Chen, L. Huang, J. Wang, Deficiency of Bim in dendritic cells contributes to overactivation of lymphocytes and autoimmunity, Blood 109 (10) (2007) 4360–4367.
[240] A. Strasser, S. Whittingham, D.L. Vaux, M.L. Bath, J.M. Adams, S. Cory, et al., Enforced BCL2 expression in B-lymphoid cells prolongs antibody responses and elicits autoimmune disease, Proc. Natl. Acad. Sci. U.S.A. 88 (19) (1991) 8661–8665.

[241] M. Batten, J. Groom, T.G. Cachero, F. Qian, P. Schneider, J. Tschopp, et al., Baff mediates survival of peripheral immature B lymphocytes, J. Exp. Med. 192 (10) (2000) 1453–1466.
[242] R. Mehrian, F.P. Quismorio Jr., G. Strassmann, M.M. Stimmler, D.A. Horwitz, R.C. Kitridou, et al., Synergistic effect between IL and bclgenotypes in determining susceptibility to systemic lupus erythematosus, Arthritis Rheum. 41 (4) (1998) 596–602.
[243] J. Hutcheson, J.C. Scatizzi, A.M. Siddiqui, G.K.H. Iii, T. Wu, Q.Z. Li, et al., Combined deficiency of proapoptotic regulators Bim and Fas results in the early onset of systemic autoimmunity, Immunity 28 (2) (2008) 206–217.
[244] J. Andre, R. Cimaz, B. Ranchin, C. Galambrun, Y. Bertrand, R. Bouvier, et al., Overexpression of the antiapoptotic gene Bfl-1 in B cells from patients with familial systemic lupus erythematosus, Lupus 16 (2) (2007) 95–100.
[245] A.F. Li, J. Hough, D. Henderson, A. Escher, Co-delivery of pro-apoptotic BAX with a DNA vaccine recruits dendritic cells and promotes efficacy of autoimmune diabetes prevention in mice, Vaccine 22 (13–14) (2004) 1751–1763.
[246] N. Dias, C.A. Stein, Potential roles of antisense oligonucleotides in cancer therapy, the example of Bcl-2 antisense oligonucleotides, Eur. J. Pharm. Biopharm. 54 (3) (2002) 263–269.
[247] J. Ramanarayanan, F.J. Hernandez-Ilizaliturri, A. Chanan-Khan, M.S. Czuczman, Pro-apoptotic therapy with the oligonucleotide GenasenseTM (oblimersen sodium) targeting Bcl-2 protein expression enhances the biological anti-tumour activity of rituximab, Br. J. Haematol. 127 (5) (2004) 519–530.
[248] M.D. Cesare, P. Perego, S.C. Righetti, G. Pratesi, N. Carenini, L. Rivoltini, et al., Enhanced antitumour efficacy of gimatecan in combination with Bcl-2 antisense oligonucleotide in human melanoma xenografts, Eur. J. Cancer 41 (8) (2005) 1213–1222.
[249] C. Leonetti, A. Biroccio, C. D'Angelo, S.C. Semple, M. Scarsella, G. Zupi, Therapeutic integration of c-myc and bcl-2 antisense molecules with docetaxel in a preclinical model of hormone-refractory prostate cancer, Prostate 67 (13) (2007) 1475–1485.
[250] V. Wacheck, E. Heere-Ress, J. Halaschek-Wiener, T. Lucas, H. Meyer, H.G. Eichler, et al., Bcl-2 antisense oligonucleotides chemosensitize human gastric cancer in a SCID mouse xenotransplantation model, J. Mol. Med. 79 (10) (2001) 587–593.
[251] C. Bartholomeusz, H. Itamochi, L.X. Yuan, F.J. Esteva, C.G. Wood, N. Terakawa, et al., Bcl-2 antisense oligonucleotide overcomes resistance to E1A gene therapy in a low HER2-expressing ovarian cancer xenograft model, Cancer Res. 65 (18) (2005) 8406–8413.
[252] Y. Hu, G. Bebb, S. Tan, R. Ng, H. Yan, J.R. Sartor, et al., Antitumor efficacy of oblimersen Bcl-2 antisense oligonucleotide alone and in combination with vinorelbine in xenograft models of human non-small cell lung cancer, Clin. Cancer Res. 10 (22) (2004) 7662–7670.
[253] M. Emi, R. Kim, K. Tanabe, Y. Uchida, T. Toge, Targeted therapy against Bcl-2 related proteins in breast cancer cells, Breast Cancer Res. 7 (2005) 940–952.
[254] L. de Menezes, N. Hudon, N. McIntosh, L.D. Mayer, Molecular and pharmacokinetic properties associated with the therapeutics of bcl-2 antisense oligonucleotide G3139 combined with free and liposomal doxorubicin, Clin. Cancer Res. 6 (7) (2000) 2891–2902.

[255] B.J. Duggan, P. Maxwell, J.D. Kelly, P. Canning, N.H. Anderson, P.F. Keane, et al., The effect of antisense Bcl-2 oligonucleotides on Bcl-2 protein expression and apoptosis in human bladder transitional cell carcinoma, J. Urol. 166 (3) (2001) 1098–1105.
[256] A.W. Tolcher, J. Kuhn, G. Schwartz, A. Patnaik, L.A. Hammond, I. Thompson, et al., A Phase I pharmacokinetic and biological correlative study of oblimersen sodium (genasense, g3139), an antisense oligonucleotide to the bcl-2 mRNA, and of docetaxel in patients with hormone-refractory prostate cancer, Clin. Cancer Res. 10 (15) (2004) 5048–5057.
[257] A.W. Tolcher, K. Chi, J. Kuhn, M. Gleave, A. Patnaik, C. Takimoto, et al., A phase II, pharmacokinetic, and biological correlative study of oblimersen sodium and docetaxel in patients with hormone-refractory prostate cancer, Clin. Cancer Res. 11 (10) (2005) 3854–3861.
[258] J. Marshall, H. Chen, D. Yang, M. Figueira, K.B. Bouker, Y. Ling, A phase I trial of a Bcl-2 antisense (G3139) and weekly docetaxel in patients with advanced breast cancer and other solid tumors, Ann. Oncol. 15 (2004) 1274–1283.
[259] G. Marcucci, J.C. Byrd, G. Dai, M.I. Klisovic, P.J. Kourlas, D.C. Young, et al., Phase 1 and pharmacodynamic studies of G3139, a Bcl-2 antisense oligonucleotide, in combination with chemotherapy in refractory or relapsed acute leukemia, Blood 101 (2) (2003) 425–432.
[260] G. Marcucci, W. Stock, G. Dai, R.B. Klisovic, S. Liu, M.I. Klisovic, et al., Phase I study of oblimersen sodium, an antisense to Bcl-2, in untreated older patients with acute myeloid leukemia: pharmacokinetics, pharmacodynamics, and clinical activity, J. Clin. Oncol. 23 (15) (2005) 3404–3411.
[261] B. Pro, B. Leber, M. Smith, L. Fayad, J. Romaguera, F. Hagemeister, et al., Phase II multicenter study of oblimersen sodium, a Bcl-2 antisense oligonucleotide, in combination with rituximab in patients with recurrent B-cell non-Hodgkin lymphoma, Br. J. Haematol. 143 (3) (2008) 355–360.
[262] C.M. Rudin, M. Kozloff, P.C. Hoffman, M.J. Edelman, R. Karnauskas, R. Tomek, et al., Phase I study of G3139, a bcl-2 antisense oligonucleotide, combined with carboplatin and etoposide in patients with small-cell lung cancer, J. Clin. Oncol. 22 (6) (2004) 1110–1117.
[263] C.M. Rudin, R. salgia, X. Wang, L.D. Hodgson, G.A. Masters, M. Green, et al., Randomized phase II study of carboplatin and etoposide with or without the bcl-2 antisense oligonucleotide oblimersen for extensive-stage small-cell lung cancer: CALGB 30103, J. Clin. Oncol. 26 (6) (2008) 870–876.
[264] C.M. Rudin, G.A. Otterson, A.M. Mauer, M.A. Villalona-Calero, R. Tomek, B. Prange, et al., A pilot trial of G3139, a bcl-2 antisense oligonucleotide, and paclitaxel in patients with chemorefractory small-cell lung cancer, Ann. Oncol. 13 (4) (2002) 539–545.
[265] C.N. Sternberg, H. Dumez, H. Van Poppel, I. Skoneczna, A. Sella, G. Daugaard, et al., Docetaxel plus oblimersen sodium (Bcl-2 antisense oligonucleotide): an EORTC multicenter, randomized phase II study in patients with castration-resistant prostate cancer, Ann. Oncol. 20 (7) (2009) 1264–1269.
[266] A.Z. Badros, O. Goloubeva, A.P. Rapoport, B. Ratterree, N. Gahres, B. Meisenberg, et al., Phase II study of G3139, a Bcl-2 antisense oligonucleotide, in combination with dexamethasone and thalidomide in relapsed multiple myeloma patients, J. Clin. Oncol. 23 (18) (2005) 4089–4099.

[267] S. O'Brien, J.O. Moore, T.E. Boyd, L.M. Larratt, A.B. Skotnicki, B. Koziner, et al., 5-year survival in patients with relapsed or refractory chronic lymphocytic leukemia in a randomized, phase III trial of fludarabine plus cyclophosphamide with or without oblimersen, J. Clin. Oncol. 27 (31) (2009) 5208–5212.
[268] A.A. Chanan-Khan, R. Niesvizky, R.J. Hohl, T.M. Zimmerman, N.P. Christiansen, G.J. Schiller, et al., Phase III randomised study of dexamethasone with or without oblimersen sodium for patients with advanced multiple myeloma, Leuk. Lymphoma 50 (4) (2009) 559–565.
[269] A.Y. Bedikian, M. Millward, H. Pehamberger, R. Conry, M. Gore, U. Trefzer, et al., Bcl-2 antisense (oblimersen sodium) plus dacarbazine in patients with advanced melanoma: the Oblimersen Melanoma Study Group, J. Clin. Oncol. 24 (29) (2006) 4738–4745.
[270] G. Dai, K.K. Chan, S. Liu, D. Hoyt, S. Whitman, M. Klisovic, et al., Cellular uptake and intracellular levels of the bcl-2 antisense G3139 in cultured cells and treated patients with acute myeloid leukemia, Clin. Cancer Res. 11 (8) (2005) 2998–3008.
[271] M.S. Davids, A. Letai, Targeting the B-cell lymphoma/leukemia 2 family in cancer, J. Clin. Oncol. 30 (25) (2012) 3127–3135.
[272] M.J. Roy, A. Vom, P.E. Czabotar, G. Lessene, Cell death and the mitochondria: therapeutic targeting of the BCL-2 family-driven pathway, Br. J. Pharmacol. 171 (8) (2014) 1973–1987.
[273] L.M. Johnson, D.E. Mortenson, H.G. Yun, W.S. Horne, T.J. Ketas, M. Lu, et al., Enhancement of α-helix mimicry by an α/β-peptide foldamer via incorporation of a dense ionic side-chain array, J. Am. Chem. Soc. 134 (17) (2012) 7317–7320.
[274] E.F. Lee, B.J. Smith, W.S. Horne, K.N. Mayer, M. Evangelista, P.M. Colman, et al., Structural basis of bcl-xL recognition by a BH3-mimetic α/β-Peptide generated by sequence-based design, Chembiochem 12 (2011) 2025–2032.
[275] B.J. Smith, E.F. Lee, J.W. Checco, M. Evangelista, S.H. Gellman, W.D. Fairlie, Structure-guided rational design of α/β-Peptide foldamers with high affinity for BCL-2 family prosurvival proteins, Chembiochem 14 (2013) 1564–1572.
[276] M.D. Boersma, H.S. Haase, K.J. Peterson-Kaufman, E.F. Lee, O.B. Clarke, P.M. Colman, et al., Evaluation of diverse α/β-backbone patterns for functional α-helix mimicry: analogues of the Bim BH3 domain, J. Am. Chem. Soc. 134 (1) (2012) 315.
[277] J.W. Checco, E.F. Lee, M. Evangelista, N.J. Sleebs, K. Rogers, A. Pettikiriarachchi, et al., α/β-peptide foldamers targeting intracellular protein-protein interactions with activity in living cells, J. Am. Chem. Soc. 137 (35) (2015) 11365–11375.
[278] G.A. Eddinger, S.H. Gellman, Differential effects of β3- versus β2- amino acid residues on the helicity and recognition properties of Bim BH3-derived α/β-peptides, Angew Chem. Int. Ed. Engl. 57 (42) (2018) 13829–13832.
[279] L.D. Walensky, A.L. Kung, I. Escher, T.J. Malia, S. Barbuto, R.D. Wright, et al., Activation of apoptosis in vivo by a hydrocarbon-stapled BH3 helix, Science 305 (5689) (2004) 1466–1470.
[280] L.D. Walensky, K. Pitter, J. Morash, K.J. Oh, S. Barbuto, J. Fisher, et al., A stapled BID BH3 helix directly binds and activates BAX, Mol. Cell 24 (2) (2006) 199–210.
[281] J.L. Labelle, S.G. Katz, G.H. Bird, E. Gavathiotis, M.L. Stewart, C. Lawrence, et al., A stapled BIM peptide overcomes apoptotic resistance in hematologic cancers, J. Clin. Investiga. 122 (6) (2012) 2018–2031.
[282] G.H. Bird, E. Mazzola, K. Opoku-Nsiah, M.A. Lammert, M. Godes, D.S. Neuberg, et al., Biophysical determinants for cellular uptake of hydrocarbon-stapled peptide helices, Nat. Chem. Biol. 12 (10) (2016) 842–852.

[283] M.L. Stewart, E. Fire, A.E. Keating, L.D. Walensky, The MCL-1 BH3 helix is an exclusive MCL-1 inhibitor and apoptosis sensitizer, Nat. Chem. Biol. 6 (8) (2010) 595–601.
[284] T. Okamoto, K. Zobel, A. Fedorova, C. Quan, H. Yang, W.J. Fairbrother, et al., Stabilizing the pro-apoptotic BimBH3 helix (BimSAHB) does not necessarily enhance affinity or biological activity, ACS Chem. Biol. 8 (2) (2013) 297–302.
[285] J.C. Reed, Dysregulation of apoptosis in cancer, J. Clin. Oncol. 17 (9) (1999) 2941–2953.
[286] T. Oltersdorf, S.W. Elmore, A.R. Shoemaker, R.C. Armstrong, D.J. Augeri, B.A. Belli, et al., An inhibitor of Bcl-2 family proteins induces regression of solid tumours, Nature 435 (7042) (2005) 677–681.
[287] S. Kitada, M. Leone, S. Sareth, D. Zhan, J.C. Reed, M. Pellecchia, Discovery, characterization, and structure-activity relationships studies of proapoptotic polyphenols targeting B-cell lymphocyte/leukemia-2 proteins, J. Med. Chem. 46 (20) (2003) 4259–4264.
[288] A.J. Souers, J.D. Leverson, E.R. Boghaert, S.L. Ackler, N.D. Catron, J. Chen, et al., ABT-199, a potent and selective BCL-2 inhibitor, achieves antitumor activity while sparing platelets, Nat. Med. 19 (2) (2013) 202–208.
[289] G. Lessene, P.E. Czabotar, P.M. Colman, BCL-2 family antagonists for cancer therapy, Nat. Rev. Drug Discov. 7 (2008) 989–1000.
[290] J.B. Baell, D.C.S. Huang, Prospects for targeting the Bcl-2 family of proteins to develop novel cytotoxic drugs, Biochem. Pharmacol. 64 (5–6) (2002) 851–863.
[291] Y.H. Han, W.H. Park, Growth inhibition in antimycin A treated-lung cancer Calu-6 cells via inducing a G1 phase arrest and apoptosis, Lung Cancer 65 (2) (2009) 150–160.
[292] M.F. van Delft, A.H. Wei, K.D. Mason, C.J. Vandenberg, L. Chen, P.E. Czabotar, et al., The BH3 mimetic ABT-737 targets selective Bcl-2 proteins and efficiently induces apoptosis via Bak/Bax if Mcl-1 is neutralized, Cancer Cell 10 (5) (2006) 389–399.
[293] D. Zhai, C. Jin, A.C. Satterthwait, J.C. Reed, Comparison of chemical inhibitors of antiapoptotic Bcl-2-family proteins, Cell Death Differ. 13 (8) (2006) 1419–1421.
[294] B. Becattini, S. Kitada, M. Leone, E. Monosov, S. Chandler, D. Zhai, et al., Rational design and real time, in-cell detection of the proapoptotic activity of a novel compound targeting Bcl-X(L), Chem. Biol. 11 (3) (2004) 389–395.
[295] J. Wei, S. Kitada, M.F. Rega, J.L. Stebbins, D. Zhai, J. Cellitti, et al., Apogossypol derivatives as pan-active inhibitors of antiapoptotic B-cell lymphoma/leukemia-2 (Bcl-2) family proteins, J. Med. Chem. 52 (14) (2009) 4511–4523.
[296] J. Wei, S. Kitada, M.F. Rega, A. Emdadi, H.B. Yuan, J. Cellitti, et al., Apogossypol derivatives as antagonists of antiapoptotic Bcl-2 family proteins, Mol. Cancer Therapeut. 8 (4) (2009) 904–913.
[297] G. Wang, Z. Nikolovska-Coleska, C.Y. Yang, R. Wang, G. Tang, J. Guo, et al., Structure-based design of potent small-molecule inhibitors of anti-apoptotic Bcl-2 proteins, J. Med. Chem. 49 (21) (2006) 6139–6142.
[298] A.A. Arnold, A. Aboukameel, J. Chen, D. Yang, S. Wang, A. Al-katib, et al., Preclinical studies of Apogossypolone: a new nonpeptidic pan small-molecule inhibitor of Bcl-2, Bcl-X L and Mcl-1 proteins in follicular small cleaved cell lymphoma model, Mol. Cancer 7 (1) (2008) 1–10.
[299] Z.Y. Hu, X.F. Zhu, Z.D. Zhong, J. Sun, J. Wang, D. Yang, et al., ApoG2, a novel inhibitor of antiapoptotic Bcl-2 family proteins, induces apoptosis and suppresses tumor growth in nasopharyngeal carcinoma xenografts, Int. J. Cancer 123 (10) (2008) 2418–2429.

[300] J. Wei, J.L. Stebbins, S. Kitada, R. Dash, D. Zhai, W.J. Placzek, et al., An optically pure apogossypolone derivative as potent pan-active inhibitor of anti-apoptotic bcl-2 family proteins, Front. Oncol. 1 (2011) 28.
[301] J. Wei, J.L. Stebbins, S. Kitada, R. Dash, W. Placzek, M.F. Rega, et al., BI-97C1, an optically pure Apogossypol derivative as pan-active inhibitor of antiapoptotic B-cell lymphoma/leukemia-2 (Bcl-2) family proteins, J. Med. Chem. 53 (10) (2010) 4166–4176.
[302] D.J. Goff, A. Court Recart, A. Sadarangani, H.J. Chun, C.L. Barrett, M. Krajewska, et al., A pan-BCL2 inhibitor renders bone-marrow-resident human leukemia stem cells sensitive to tyrosine kinase inhibition, Cell Stem Cell 12 (3) (2013) 316–328.
[303] K.G. Wolter, S.J. Wang, B.S. Henson, S. Wang, K.A. Griffith, B. Kumar, et al., (-)-gossypol inhibits growth and promotes apoptosis of human head and neck squamous cell carcinoma in vivo, Neoplasia 8 (3) (2006) 163–172.
[304] N. McGregor, L. Patel, M. Craig, S. Weidner, S. Wang, K.J. Pienta, AT-101 (R-(-)-gossypol acetic acid) enhances the effectiveness of androgen deprivation therapy in the VCaP prostate cancer model, J. Cell. Biochem. 110 (5) (2010) 1187–1194.
[305] R. Zhao, S. Zhou, B. Xia, C.Y. Zhang, P. Hai, H. Zhe, et al., AT-101 enhances gefitinib sensitivity in non-small cell lung cancer with EGFR T790M mutations, BMC Cancer 16 (1) (2016) 491.
[306] B.D. Zeitlin, E. Joo, Z. Dong, K. Warner, G. Wang, Z. Nikolovska-Coleska, et al., Antiangiogenic effect of TW37, a small-molecule inhibitor of bcl-2, Cancer Res. 66 (17) (2006) 8698–8706.
[307] K. Balakrishnan, S. Aggarwal, W. Wierda, V. Gandhi, Bax and Bak are required for apogossypolone, a BH3-mimetic, induced apoptosis in chronic lymphocytic leukemia cells, Leuk. Lymphoma 54 (5) (2013) 1097–1100.
[308] K. Balakrishnan, J.A. Burger, W.G. Wierda, V. Gandhi, AT-101 induces apoptosis in CLL B cells and overcomes stromal cell-mediated Mcl-1 induction and drug resistance, Blood 113 (1) (2009) 149–153.
[309] X. Pang, Y. Wu, Y. Wu, B. Lu, J. Chen, J. Wang, et al., (-)-Gossypol suppresses the growth of human prostate cancer xenografts via modulating VEGF signaling-mediated angiogenesis, Mol. Cancer Therapeut. 10 (5) (2011) 795–805.
[310] H. Wei, M.T. Harper, Comparison of putative BH3 mimetics AT-101, HA14-1, sabutoclax and TW-37 with ABT-737 in platelets, Platelets 32 (1) (2021) 105–112.
[311] J.E. Castro, O.J. Loria, R.A. Aguillon, A phase II, open label study of AT-101 in combination with rituximab in patients with relapsed or refractory chronic lymphocytic leukemia, Evaluation of two dose regimens, Blood 110 (11) (2007) 3119.
[312] M.Q. Baggstrom, Y. Qi, M. Koczywas, A. Argiris, E.A. Johnson, M.J. Millward, et al., A phase II study of AT-101 (Gossypol) in chemotherapy-sensitive recurrent extensive-stage small cell lung cancer, J. Thorac. Oncol. 6 (10) (2011) 1757–1760.
[313] R.S. Heist, J. Fain, B. Chinnasami, W. Khan, J.R. Molina, L.V. Sequist, et al., Phase I/II study of AT-101 with topotecan in relapsed and refractory small cell lung cancer, J. Thorac. Oncol. 5 (10) (2010) 1637–1643.
[314] N. Ready, N.A. Karaseva, S.V. Orlov, A.V. Luft, O. Popovych, J.T. Holmlund, et al., Double-blind, placebo-controlled, randomized phase 2 study of the proapoptotic agent AT-101 plus docetaxel, in second-line non-small cell lung cancer, J. Thorac. Oncol. 6 (4) (2011) 781–785.
[315] H. Xie, J. Yin, M.H. Shah, M.E. Menefee, K.C. Bible, D. Reidy-Lagunes, et al., A phase II study of the orally administered negative enantiomer of gossypol (AT-101), a BH3 mimetic, in patients with advanced adrenal cortical carcinoma, Invest. New Drugs 37 (4) (2019) 755–762.

[316] M.N. Stein, M. Hussain, W.M. Stadler, G. Liu, I.V. Tereshcheko, S. Goodin, et al., A phase II study of AT-101 to overcome bcl-2—mediated resistance to androgen deprivation therapy in patients with newly diagnosed castration-sensitive metastatic prostate cancer, Clin. Genitourin. Cancer 14 (1) (2016) 22—27.
[317] P.L. Swiecicki, E. Bellile, A.G. Sacco, A.T. Pearson, J.M. Taylor, T.L. Jackson, et al., A phase II trial of the BCL-2 homolog domain 3 mimetic AT-101 in combination with docetaxel for recurrent, locally advanced, or metastatic head and neck cancer, Invest. N. Drugs 34 (4) (2016) 481—489.
[318] S. Song, Q. Chen, Y. Li, G. Lei, A. Scott, L. Huo, et al., Targeting cancer stem cells with a pan-BCL-2 inhibitor in preclinical and clinical settings in patients with gastroesophageal carcinoma, Gut (2021), https://doi.org/10.1136/gutjnl-2020-321175.
[319] M. Nguyen, R.C. Marcellus, A. Roulston, M. Watson, L. Serfass, S.R. Murthy Madiraju, et al., Small molecule obatoclax (GX15-070) antagonizes MCL-1 and overcomes MCL-1-mediated resistance to apoptosis, Proc. Natl. Acad. Sci. U.S.A. 104 (49) (2007) 19512—19517.
[320] S. Trudel, Z.H. Li, J. Rauw, R.E. Tiedemann, X.Y. Wen, A.K. Stewart, Preclinical studies of the pan-Bcl inhibitor obatoclax (GX015-070) in multiple myeloma, Blood 109 (12) (2007) 5430—5438.
[321] S. Huang, K. Okumura, F.A. Sinicrope, BH3 mimetic obatoclax enhances TRAIL-mediated apoptosis in human pancreatic cancer cells, Clin. Cancer Res. 15 (1) (2009) 150—159.
[322] C.C. Jiang, D. Wroblewski, F. Yang, P. Hersey, X.D. Zhang, Human melanoma cells under endoplasmic reticulum stress are more susceptible to apoptosis induced by the BH3 mimetic obatoclax, Neoplasia 11 (9) (2009) 945—955.
[323] R.L. Smoot, B.R.A. Blechacz, N.W. Werneburg, S.F. Bronk, F.A. Sinicrope, A.E. Sirica, et al., A bax-mediated mechanism for obatoclax-induced apoptosis of cholangiocarcinoma cells, Cancer Res. 70 (5) (2010) 1960—1969.
[324] M.B. Gariboldi, E. Taiana, M.C. Bonzi, I. Craparotta, S. Giovannardi, M. Mancini, et al., The BH3-mimetic obatoclax reduces HIF-1 alpha levels and HIF-1 transcriptional activity and sensitizes hypoxic colon adenocarcinoma cells to 5-fluorouracil, Cancer Lett. 364 (2) (2015) 156—164.
[325] J.R. Brown, B. Tesar, L. Yu, L. Werner, N. Takebe, E. Mikler, et al., Obatoclax in combination with fludarabine and rituximab is well-tolerated and shows promising clinical activity in relapsed chronic lymphocytic leukemia, Leuk. Lymphoma 56 (12) (2015) 3336—3342.
[326] E.A. Brem, K. Thudium, S. Khubchandani, P.C. Tsai, S.H. Olejniczak, S. Bhat, et al., Distinct cellular and therapeutic effects of obatoclax in rituximab-sensitive and -resistant lymphomas, Br. J. Haematol. 153 (5) (2011) 599—611.
[327] G. Stefanzl, D. Berger, S. Cerny-Reiterer, K. Blatt, P. Valent, The pan-BCL-2-blocker obatoclax (GX15-070) and the PI3-kinase/mTOR-inhibitor BEZ235 produce cooperative growth-inhibitory effects in ALL cells, Oncotarget 8 (40) (2017) 67709—67722.
[328] Y. Wei, T. Kadia, W. Tong, M. Zhang, Y. Jia, H. Yang, et al., The combination of a histone deacetylase inhibitor with the BH3-mimetic GX15-070 has synergistic antileukemia activity by activating both apoptosis and autophagy, Autophagy 6 (7) (2010) 976—978.
[329] P. Sulkshane, T. Teni, BH3 mimetic Obatoclax (GX15-070) mediates mitochondrial stress predominantly via MCL-1 inhibition and induces autophagy-dependent necroptosis in human oral cancer cells, Oncotarget 8 (36) (2016) 60060—60079.

[330] F. Basit, S. Cristofanon, S. Fulda, Obatoclax (GX15-070) triggers necroptosis by promoting the assembly of the necrosome on autophagosomal membranes, Cell Death Differ. 20 (9) (2013) 1161–1173.
[331] D. Champa, A. Orlacchio, B. Patel, M. Ranieri, A.A. Shemetov, V.V. Verkhusha, et al., Obatoclax kills anaplastic thyroid cancer cells by inducing lysosome neutralization and necrosis, Oncotarget 7 (23) (2016) 34453–34471.
[332] A.D. Schimmer, S. O'Brien, H. Kantarjian, J. Brandwein, B.D. Cheson, M.D. Minden, et al., A phase I study of the pan bcl-2 family inhibitor obatoclax mesylate in patients with advanced hematologic malignancies, Clin. Cancer Res. 14 (24) (2008) 8295–8301.
[333] A.D. Schimmer, A. Raza, T.H. Carter, D. Claxton, H. Erba, D.J. DeAngelo, et al., A multicenter phase I/II study of obatoclax mesylate administered as a 3- or 24-hour infusion in older patients with previously untreated acute myeloid leukemia, PLoS One 9 (10) (2014).
[334] S.A. Parikh, H. Kantarjian, A. Schimmer, W. Walsh, E. Asatiani, K. El-Shami, et al., Phase II study of obatoclax mesylate (GX15-070), a small-molecule BCL-2 family antagonist, for patients with myelofibrosis, Clin. Lymphoma Myeloma Leuk. 10 (4) (2010) 285–289.
[335] A. Goy, F.J. Hernandez-Ilzaliturri, B. Kahl, P. Ford, E. Protomastro, M.A. Berger, Phase I/II study of the pan Bcl-2 inhibitor obatoclax mesylate plus bortezomib for relapsed or refractory mantle cell lymphoma, Leuk. Lymphoma 55 (12) (2014) 2761–2768.
[336] P.K. Paik, C.M. Rudin, A. Brown, N.A. Rizvi, N. Takebe, W. Travis, et al., A phase I study of obatoclax mesylate, a Bcl-2 antagonist, plus topotecan in solid tumor malignancies, Cancer Chemother. Pharmacol. 66 (6) (2010) 1079–1085.
[337] P.K. Paik, C.M. Rudin, M.C. Pietanza, A. Brown, N.A. Rizvi, N. Takebe, et al., A phase II study of obatoclax mesylate, a Bcl-2 antagonist, plus topotecan in relapsed small cell lung cancer, Lung Cancer 74 (3) (2011) 481–485.
[338] C.J. Langer, I. Albert, H.J. Ross, P. Kovacs, L.J. Blakely, G. Pajkos, et al., Randomized phase II study of carboplatin and etoposide with or without obatoclax mesylate in extensive-stage small cell lung cancer, Lung Cancer 85 (3) (2014) 420–428.
[339] A. Chiappori, C. Williams, D.W. Northfelt, J.W. Adams, S. Malik, M.J. Edelman, et al., Obatoclax mesylate, a Pan–Bcl-2 inhibitor, in combination with docetaxel in a phase 1/2 trial in relapsed non–small-cell lung cancer, J. Thorac. Oncol. 9 (1) (2014) 121–125.
[340] E.F. Lee, P.E. Czabotar, B.J. Smith, K. Deshayes, K. Zobel, P.M. Colman, et al., Crystal structure of ABT-737 complexed with Bcl-xL: implications for selectivity of antagonists of the Bcl-2 family, Cell Death Differ. 14 (9) (2007) 1711–1713.
[341] M. Konopleva, R. Contractor, T. Tsao, I. Samudia, P.P. Ruvolo, S. Kitada, et al., Mechanisms of apoptosis sensitivity and resistance to the BH3 mimetic ABT-737 in acute myeloid leukemia, Cancer Cell 10 (5) (2006) 375–388.
[342] D. Chauhan, M. Velankar, M. Brahmandam, T. Hideshima, K. Podar, P. Richardson, et al., A novel Bcl-2/Bcl-X(L)/Bcl-w inhibitor ABT-737 as therapy in multiple myeloma, Oncogene 26 (16) (2007) 2374–2380.
[343] V. Del Gaizo Moore, J.R. Brown, M. Certo, T.M. Love, C.D. Novina, A. Letai, Chronic lymphocytic leukemia requires BCL2 to sequester prodeath BIM, explaining sensitivity to BCL2 antagonist ABT-737, J. Clin. Invest. 117 (1) (2007) 112–121.
[344] J. Witham, M.R. Valenti, A.K. De-Haven-Brandon, S. Vidot, S.A. Eccles, S.B. Kaye, et al., The Bcl-2/Bcl-XL family inhibitor ABT-737 sensitizes ovarian cancer cells to carboplatin, Clin. Cancer Res. 13 (23) (2007) 7191–7198.

[345] S.K. Tahir, X. Yang, M.G. Anderson, S.E. Morgan-Lappe, A.V. Sarthy, J. Chen, et al., Influence of bcl-2 family members on the cellular response of small-cell lung cancer cell lines to ABT-737, Cancer Res. 67 (3) (2007) 1176–1183.
[346] S. Huang, F.A. Sinicrope, BH3 mimetic ABT-737 potentiates TRAIL-mediated apoptotic signaling by unsequestering Bim and bak in human pancreatic cancer cells, Cancer Res. 68 (8) (2008) 2944–2951.
[347] C. Tse, A.R. Shoemaker, J. Adickes, M.G. Anderson, J. Chen, S. Jin, et al., ABT-263: a potent and orally bioavailable Bcl-2 family inhibitor, Cancer Res. 68 (9) (2008) 3421–3428.
[348] A.R. Shoemaker, M.J. Mitten, J. Adickes, S. Ackler, Activity of the Bcl-2 family inhibitor ABT-263 in a panel of small cell lung cancer xenograft models, Clin. Cancer Res. 14 (11) (2008) 3268–3277.
[349] N. Tan, M. Malek, J. Zha, P. Yue, R. Kassees, L. Berry, et al., Navitoclax enhances the efficacy of taxanes in non-small cell lung cancer models, Clin. Cancer Res. 17 (6) (2011) 1394–1404.
[350] M. Wong, N. Tan, J. Zha, F.V. Peale, P. Yue, W.J. Fairbrother, et al., Navitoclax (ABT-263) reduces Bcl-x(L)-mediated chemoresistance in ovarian cancer models, Mol. Cancer Therapeut. 11 (4) (2012) 1026–1035.
[351] H. Dai, H. Ding, X.W. Meng, S.H. Lee, P.A. Schneider, S.H. Kaufmann, Contribution of Bcl-2 phosphorylation to Bak binding and drug resistance, Cancer Res. 73 (23) (2013) 6998–7008.
[352] S. Ackler, M.J. Mitten, K. Foster, A. Oleksijew, M. Refici, S.K. Tahir, et al., The Bcl-2 inhibitor ABT-263 enhances the response of multiple chemotherapeutic regimens in hematologic tumors in vivo, Cancer Chemother. Pharmacol. 66 (5) (2010) 869–880.
[353] O.S. Kwon, S.K. Hong, S.J. Kwon, Y.H. Go, E. Oh, H.J. Cha, BCL2 induced by LAMTOR3/MAPK is a druggable target of chemoradioresistance in mesenchymal lung cancer, Cancer Lett. 403 (2017) 48–58.
[354] L. Gandhi, D.R. Camidge, O.M. Ribeiro de, P. Bonomi, D. Gandara, D. Khaira, et al., Phase I study of navitoclax (ABT-263), a novel bcl-2 family inhibitor, in patients with small-cell lung cancer and other solid tumors, J. Clin. Oncol. 29 (7) (2011) 909–916.
[355] A.W. Roberts, J.F. Seymour, J.R. Brown, W.G. Wierda, T.J. Kipps, S.L. Khaw, et al., Substantial susceptibility of chronic lymphocytic leukemia to BCL2 inhibition: results of a phase I study of navitoclax in patients with relapsed or refractory disease, J. Clin. Oncol. 30 (5) (2012) 488–496.
[356] C.M. Rudin, C.L. Hann, E.B. Garon, O.M. Ribeiro de, P.D. Bonomi, D.R. Camidge, Phase II study of single-agent navitoclax (ABT-263) and biomarker correlates in patients with relapsed small cell lung cancer, Clin. Cancer Res. 18 (11) (2012) 3163–3169.
[357] A.W. Tolcher, P. LoRusso, J. Arzt, T.A. Busman, G. Lian, N.S. Rudersdorf, et al., Safety, efficacy, and pharmacokinetics of navitoclax (ABT-263) in combination with erlotinib in patients with advanced solid tumors, Cancer Chemother. Pharmacol. 76 (5) (2015) 1025–1032.
[358] T.J. Kipps, H. Eradat, S. Grosicki, J. Catalano, W. Cosolog, L.S. Dyagil, et al., A phase 2 study of the BH3 mimetic BCL2 inhibitor navitoclax (ABT-263) with or without rituximab, in previously untreated B-cell chronic lymphocytic leukemia, Leuk. Lymphoma 56 (10) (2015) 2826–2833.
[359] K.D. Mason, M.R. Carpinelli, J.I. Fletcher, J.E. Collinge, A.A. Hilton, S. Ellis, et al., Programmed anuclear cell death delimits platelet life span, Cell 128 (6) (2007) 1173–1186.

[360] E.C. Josefsson, C. James, K.J. Henley, M.A. Debrincat, K.L. Rogers, M.R. Dowling, et al., Megakaryocytes possess a functional intrinsic apoptosis pathway that must be restrained to survive and produce platelets, J. Exp. Med. 208 (10) (2011) 2017–2031.
[361] M. Vogler, D. Dinsdale, M.J. Dyer, G.M. Cohen, ABT-199 selectively inhibits BCL2 but not BCL2L1 and efficiently induces apoptosis of chronic lymphocytic leukaemic cells but not platelets, Br. J. Haematol. 163 (1) (2014) 139–142.
[362] T.L. Lochmann, K.V. Tloros, M. Naseri, K.M. Powell, W. Cook, R.J. March, et al., Venetoclax is effective in small-cell lung cancers with high BCL-2 expression, Clin. Cancer Res. 24 (2018) 360–369.
[363] F. Vaillant, D. Mérino, L. Lee, K. Breslin, B. Pal, M.E. Ritchie, et al., Targeting BCL-2 with the BH3 mimetic ABT-199 in estrogen receptor-positive breast cancer, Cancer Cell 24 (1) (2013) 120–129.
[364] S. Jilg, V. Reidel, C. Müller-Thomas, J. König, J. Schauwecker, U. Höckendorf, et al., Blockade of BCL-2 proteins efficiently induces apoptosis in progenitor cells of high-risk myelodysplastic syndromes patients, Leukemia 30 (1) (2016) 112–123.
[365] M. Klanova, L. Andera, J. Brazina, J. Svadlenka, S. Benesova, J. Soukup, et al., Targeting of BCL2 family proteins with ABT-199 and homoharringtonine reveals BCL2-and MCL1-dependent subgroups of diffuse large B-cell lymphoma, Clin. Cancer Res. 22 (5) (2016) 1138–1149.
[366] J. Montero, J. Stephansky, T. Cai, G.K. Griffin, L. Cabal-Hierro, K. Togama, et al., Blastic plasmacytoid dendritic cell neoplasm is dependent on BCL2 and sensitive to venetoclax, Cancer Discov. 7 (2) (2017) 156–164.
[367] F. Lucantoni, A.U. Lindner, N. O'Donovan, H. Düssmann, J.H.M. Prehn, Systems modeling accurately predicts responses to genotoxic agents and their synergism with BCL-2 inhibitors in triple negative breast cancer cells, Cell Death Dis. 9 (2) (2018) 42.
[368] K.L. Knorr, P.A. Schneider, X.W. Meng, H. Dai, B.D. Smith, A.D. Hess, et al., MLN4924 induces Noxa upregulation in acute myelogenous leukemia and synergizes with Bcl-2 inhibitors, Cell Death Differ. 22 (2015) 2133–2142.
[369] S.L. Khaw, D. Mérino, M.A. Anderson, S.P. Glaser, P. Bouillet, A.W. Roberts, et al., Both leukaemic and normal peripheral B lymphoid cells are highly sensitive to the selective pharmacological inhibition of prosurvival Bcl-2 with ABT-199, Leukemia 28 (6) (2014) 1207–1215.
[370] S. Stilgenbauer, B. Eichhorst, J. Schetelig, S. Coutre, J.F. Seymour, T. Munir, et al., Venetoclax in relapsed or refractory chronic lymphocytic leukaemia with 17p deletion: a multicentre, open-label, phase 2 study, Lancet Oncol. (2016) 768–778.
[371] J. Jones, M.Y. Choi, A.R. Mato, R.R. Furman, S.E. Coutre, Venetoclax (VEN) monotherapy for patients with chronic lymphocytic leukemia (CLL) who relapsed after or were refractory to ibrutinib or idelalisib, Blood 128 (22) (2016), 637.
[372] S. Kumar, J.L. Kaufman, C. Gasparetto, J. Mikhael, R. Vij, B. Peqourie, et al., Efficacy of venetoclax as targeted therapy for relapsed/refractory t(11;14) multiple myeloma, Blood 130 (22) (2017) 2401–2409.
[373] M.S. David, A.W. Roberts, J.F. Seymour, J.M. Pagel, B.S. Kahl, W.G. Wierda, et al., Phase I first-in-human study of venetoclax in patients with relapsed or refractory non-hodgkin lymphoma, J. Clin. Oncol. 35 (8) (2017) 826–833.
[374] C.S. Tam, M.A. Anderson, C. Pott, R. Agarwal, S. Handunnetti, R.J. Hicks, et al., Ibrutinib plus venetoclax for the treatment of mantle-cell lymphoma, N. Engl. J. Med. 378 (13) (2018) 1211–1223.
[375] G. Lessene, P.E. Czabotar, B.E. Sleebs, K. Zobel, K.N. Lowes, J.M. Adams, et al., Structure-guided design of a selective BCL-X(L) inhibitor, Nat. Chem. Biol. 9 (6) (2013) 390–397.

[376] B.E. Sleebs, W.J.A. Kersten, S. Kulasegaram, G. Nikolakopoulos, E. Hatzis, R.M. Moss, et al., Discovery of potent and selective benzothiazole hydrazone inhibitors of bcl-XL, J. Med. Chem. 56 (13) (2013) 5514–5540.
[377] D. Park, A.T. Magis, R. Li, T.K. Owonikoko, G.L. Sica, S.Y. Sun, et al., Novel small-molecule inhibitors of bcl-XL to treat lung cancer, Cancer Res. 73 (2013) 5485–5496.
[378] Z.F. Tao, L. Hasvold, L. Wang, X. Wang, A.M. Petros, C.H. Park, et al., Discovery of a potent and selective BCL-X-L inhibitor with in vivo activity, ACS Med. Chem. Lett. 5 (10) (2014) 1088–1093.
[379] J.D. Leverson, D.C. Phillips, M.J. Mitten, E.R. Boghaert, D. Diaz, S.K. Tahir, et al., Exploiting selective BCL-2 family inhibitors to dissect cell survival dependencies and define improved strategies for cancer therapy, Sci. Transl. Med. 7 (279) (2015), 279ra40.
[380] Z. Baranski, Y. de Jong, T. Ilkova, E.F. Peterse, A.M. Cleton-Jansen, B. van de Water, et al., Pharmacological inhibition of Bcl-xL sensitizes osteosarcoma to doxorubicin, Oncotarget 6 (34) (2015) 36113–36125.
[381] G. Karpel-Massler, C.T. Ishida, E. Bianchetti, C. Shu, R. Perez-Lorenzo, B. Horst, et al., Inhibition of mitochondrial matrix chaperones and antiapoptotic Bcl-2 family proteins empower antitumor therapeutic responses, Cancer Res. 77 (2017) 3513–3526.
[382] S. Khan, X. Zhang, D. Lv, Q. Zhang, Y. He, P. Zhang, et al., A selective BCL-XL PROTAC degrader achieves safe and potent antitumor activity, Nat. Med. 25 (12) (2019) 1938–1947.
[383] N.A. Cohen, M.L. Stewart, E. Gavathiotis, J.L. Tepper, S.R. Bruekner, B. Koss, et al., A competitive stapled peptide screen identifies a selective small molecule that overcomes MCL-1-dependent leukemia cell survival, Chem. Biol. 19 (9) (2012) 1175–1186.
[384] S. Varadarajan, M. Vogler, M. Butterworth, D. Dinsdale, L.D. Walensky, G.M. Cohen, Evaluation and critical assessment of putative MCL-1 inhibitors, Cell Death Differ. 20 (11) (2013) 1475–1484.
[385] D.J. Mallick, R.S. Soderquist, D. Bates, A. Eastman, Confounding off-target effects of BH3 mimetics at commonly used concentrations: MIM1, UMI-77, and A-1210477, Cell Death Dis. 10 (3) (2019) 185.
[386] A. Friberg, D. Vigil, B. Zhao, R.N. Daniels, J.P. Burke, P.M. Garcia-Barrantes, et al., Discovery of potent myeloid cell leukemia 1 (Mcl-1) inhibitors using fragment-based methods and structure-based design, J. Med. Chem. 56 (1) (2013) 15.
[387] J.P. Burke, Z. Bian, S. Shaw, B. Zhao, C.M. Goodwin, J. Belmar, et al., Discovery of tricyclic indoles that potently inhibit Mcl-1 using fragment-based methods and structure-based design, J. Med. Chem. 58 (9) (2015) 3794–3805.
[388] M. Bruncko, L. Wang, G. Sheppard, D.C. Phillips, S.K. Tahir, J. Xue, et al., Structure-guided design of a series of MCL-1 inhibitors with high affinity and selectivity, J. Med. Chem. 58 (5) (2015) 4089.
[389] J.D. Leverson, H. Zhang, J. Chen, S.K. Tahir, D.C. Phillips, J. Xue, et al., Potent and selective small-molecule MCL-1 inhibitors demonstrate on-target cancer cell killing activity as single agents and in combination with ABT-263 (navitoclax), Cell Death Dis. 6 (1) (2015) e1590.
[390] A. Kotschy, Z. Szlavik, J. Murray, J. Davidson, A.L. Maragno, G. Le Toumelin-Braizat, et al., The MCL1 inhibitor S63845 is tolerable and effective in diverse cancer models, Nature 538 (7626) (2016) 477–482.

[391] D. Merino, J.R. Whittle, F. Vaillant, A. Serrano, J.N. Gong, G. Giner, et al., Synergistic action of the MCL-1 inhibitor S63845 with current therapies in preclinical models of triple-negative and HER2-amplified breast cancer, Sci. Transl. Med. 9 (401) (2017).
[392] Z. Szlavik, M. Csekei, A. Paczal, Z.B. Szabo, A. Kotschy, Discovery of S64315, a potent and selective Mcl-1 inhibitor, J. Med. Chem. 63 (22) (2020) 13762–13795.
[393] M. Vogler, BCL2A1: the underdog in the BCL2 family, Cell Death Differ. 19 (1) (2012) 67–74.
[394] K.J. Kump, L. Miao, A.S.A. Mady, N.H. Ansari, U.K. Shrestha, Y. Yang, et al., Discovery and characterization of 2,5-substituted benzoic acid dual inhibitors of the anti-apoptotic Mcl-1 and Bfl-1 proteins, J. Med. Chem. 63 (5) (2020) 2489–2510.
[395] E.P. Harvey, Z.J. Hauseman, D.T. Cohen, T.J. Rettenmaier, S. Lee, A.J. Huhn, et al., Identification of a covalent molecular inhibitor of anti-apoptotic BFL-1 by disulfide tethering, Cell Chem. Biol. 27 (6) (2020) 647–656, e6.
[396] A.J. Huhn, R.M. Guerra, E.P. Harvey, G.H. Bird, L.D. Walensky, Selective covalent targeting of anti-apoptotic BFL-1 by cysteine-reactive stapled peptide inhibitors, Cell Chem. Biol. 23 (9) (2016) 1123–1134.
[397] E. Barile, G.D. Marconi, S.K. De, C. Baggio, L. Gambini, A.F. Salem, et al., hBfl-1/hNOXA interaction studies provide new insights on the role of Bfl-1 in cancer cell resistance and for the design of novel anticancer agents, ACS Chem. Biol. 12 (2) (2017) 444–455.
[398] J. Deng, N. Carlson, K. Takeyama, P. Dal Cin, M. Shipp, A. Letai, BH3 profiling identifies three distinct classes of apoptotic blocks to predict response to ABT-737 and conventional chemotherapeutic agents, Cancer Cell 12 (2) (2007) 171–185.
[399] J.M. Brouwer, P. Lan, A.D. Cowan, J.P. Bernardini, R.W. Birkinshaw, M.F. van Delft, et al., Conversion of Bim-BH3 from activator to inhibitor of bak through structure-based design, Mol. Cell 68 (4) (2017) 659–672.
[400] E. Jabbour, S. Paul, H. Kantarjian, The clinical development of antibody-drug conjugates - lessons from leukaemia, Nat. Rev. Clin. Oncol. (2021), https://doi.org/10.1038/s41571-021-00484-2.
[401] E. Gavathiotis, D.E. Reyna, J.A. Bellairs, E.S. Leshchiner, L.D. Walensky, Direct and selective small-molecule activation of proapoptotic BAX, Nat. Chem. Biol. 8 (7) (2012) 639–645.
[402] D.E. Reyna, T.P. Garner, A. Lopez, F. Kopp, G.S. Choudhary, A. Sridharan, et al., Direct activation of BAX by BTSA1 overcomes apoptosis resistance in acute myeloid leukemia, Cancer Cell 32 (4) (2017) 490–505.
[403] L.A. Barclay, T.E. Wales, T.P. Garner, F. Wachter, S. Lee, R.M. Guerra, et al., Inhibition of pro-apoptotic BAX by a noncanonical interaction mechanism, Mol. Cell 57 (5) (2015) 873–886.
[404] J. Ma, F. Edlich, G.A. Bermejo, K.L. Norris, R.J. Youle, N. Tjandra, Structural mechanism of Bax inhibition by cytomegalovirus protein Vmia, Proc. Natl. Acad. Sci. U.S.A. 109 (51) (2012) 20901–20906.
[405] J.R. Pritz, F. Wachter, S. Lee, J. Luccarelli, L.D. Walensky, Allosteric sensitization of pro-apoptotic BAX, Nat. Chem. Biol. 13 (9) (2017) 961.
[406] D. Amgalan, T.P. Garner, R. Pekson, X.F. Jia, M. Yanamandala, V. Paulino, et al., A small-molecule allosteric inhibitor of BAX protects against doxorubicin-induced cardiomyopathy, Nat. Cancer 1 (3) (2020) 315–328.
[407] J.F. Seymour, T.J. Kipps, B. Eichhorst, P. Hillmen, J. D'Rozario, S. Assouline, et al., Venetoclax-rituximab in relapsed or refractory chronic lymphocytic leukemia, N. Engl. J. Med. 378 (12) (2018) 1107–1120.

CHAPTER 8

Ferroptosis: lipids, iron, cellular defense mechanisms and opportunities for drug development

Daiqing Liao
Department of Anatomy and Cell Biology, College of Medicine, University of Florida, Gainesville, FL, United States

1. Introduction

In 2012, Dixon et al. described a form of caspase-independent but iron-dependent cell death; they termed this form of regulated cell death as ferroptosis [1]. Since this initial discovery, there is an exponential increase of research interest in understanding the mechanisms underlying ferroptosis and exploring the modulation of ferroptosis as therapeutic strategies to treat human diseases. Excessive oxidative stress can induce iron-dependent peroxidation of phospholipids (PLs) with polyunsaturated fatty acid (PUFA) tails. Accumulation of such lipid peroxides leads to membrane rapture and cell death. The glutathione peroxidase 4 (GPX4, also known as PL hydroperoxide glutathione peroxidase or PHGPx) is the only known enzyme that catalyzes the conversion of lipid peroxides into lipid alcohols [2]. Genetic or chemical inhibition of GPX4 promotes ferroptosis. Enzymes that synthesize PUFA-containing PLs potentiate ferroptosis. These pro-ferroptosis enzymes include ACSL4 (long-chain fatty acid CoA ligase 4, also known as acyl-CoA synthase 4) that catalyzes the esterification of FAs to generate CoA-conjugated FAs (acyl-CoA) and LPCAT3 (also known as lysophospholipid acyltransferase 5) that transfers FA from acyl-CoA to a lysophospholipid containing one FA chain [3,4]. Enzymes involved in the biosynthesis of various endogenous antioxidants such as coenzyme Q_{10} play critical roles in cellular defense against ferroptosis. Chemical biological approaches have been instrumental to the discovery of ferroptosis as a distinct mode of regulated cell death [1,5,6]. Chemical compounds that induce or inhibit ferroptosis are essential tools for deciphering the molecular mechanisms underlying ferroptosis. The known modulators of ferroptosis

Mechanisms of Cell Death and Opportunities for Therapeutic Development
ISBN 978-0-12-814208-0
https://doi.org/10.1016/B978-0-12-814208-0.00005-1

represent important starting compounds for developing therapeutics for treating cancer and other diseases. This chapter focuses on discussing key triggers of ferroptosis including PUFAs and irons and cellular mechanisms that inhibit ferroptotic cell death. Finally, it describes opportunities for developing therapeutics that target the ferroptotic pathway for treating cancer and diverse other diseases such as neurodegenerative diseases, stroke, kidney injury, diabetes, and osteoporosis.

2. Roles of PUFAs in ferroptosis

Lipids are the main component of cell membranes. Major lipids include (1) sterols such as cholesterol, (2) glycerophospholipids such as phosphatidylethanolamine (PE), phosphatidylserine (PS), and phosphatidylcholine (PC), (3) glycerolipids such as triacylglycerols (TAGs) and diacylglycerols (DAGs), and ether-linked glycerolipids, and (4) sphingolipids (e.g., sphingomyelin) [7]. The compositions of PLs with saturated and unsaturated fatty acid chains determine biophysical properties of cell membranes such as membrane fluidity. A recent study shows that the cytoplasmic leaflet contains approximately twofold more unsaturated FAs than the exoplasmic leaflet [8]. Fatty acids with one and more double bonds are collectively classified as monounsaturated and polyunsaturated fatty acids (MUFAs and PUFAs), respectively. In mammalian cells, MUFAs can be taken up from circulation or synthesized de novo from saturated FAs by desaturases. Stearyl-CoA desaturase (SCD, Δ-9 desaturase) introduces the first double bond into saturated fatty acyl-CoA substrates in a biochemical reaction using O_2 and electrons from reduced (Fe^{2+}) cytochrome b5. SCD protects cells from ferroptosis [9] and lipotoxicity [10,11]. FADS2 (fatty acid desaturase 2 or Δ-6 desaturase) desaturates palmitate to produce the MUFA sapienate (n-10) [12]. Both SCD and FADS2 are ubiquitously expressed; some tissues and organs such as brain, liver, and prostate exhibit elevated mRNA levels for both SCD and FADS2 with highest levels found in fat for SCD and adrenal glands for FADS2. Both SCD and FADS2 are highly expressed in cancer cells [9,13]. MUFAs generated by SCD (n-6) and FADS2 (n-10) constitute major FA classes in cancer cells [14]. It has been shown that sapienate biosynthesis is used in cancer cells to bypass the reliance on SCD [13]. Consistent with the findings that SCD protects cells against ferroptosis, MUFAs inhibit ferroptosis [9,15].

PUFAs can be acquired from dietary sources or synthesized from the basic building block acetyl CoA, through the action of acetyl CoA carboxylases (ACC1 and ACC2 encoded by the ACACA and ACACB gene, respectively). It has been shown that ACC inhibition due to phosphorylation by AMP-activated protein kinase on energy stress can inhibit ferroptosis [16]. Notably, two essential PUFAs for mammals, linoleic acid (18:2n-6, a polyunsaturated omega-6 FA) and α-linolenic acid (18:3n-3, a polyunsaturated omega-3 FA), cannot be synthesized de novo and must be obtained from dietary sources because the genes encoding the enzymes necessary to insert a cis double bond at the n-6 or the n-3 position of FA are absent in mammals [17,18]. FADS2 and FDAS1 (a Δ-5 desaturase) in conjunction with elongases synthesize PUFAs with ≥3 double bonds such as arachidonic acid and arachidonic acid-derived oxidized lipid molecules [17,18].

The peroxidation of PUFAs is a critical trigger of ferroptotic cell death [19]. Lipid peroxidation can occur spontaneously or be catalyzed by cellular enzymes. Compared to saturated FAs (SFAs) and MUFAs, bis-allylic hydrogen atoms in PUFAs are susceptible to free radical and enzyme-mediated peroxidation to generate PL hydroperoxides (PL-PUFA-OOH) and other reactive radicals [20]. Genetic depletion of two enzymes, ACSL4 and LPCAT3, inhibits ferroptosis [21–23]. This supports the concept that the presence of peroxidized lipids in cell membranes rather than their free forms is essential to the induction of ferroptosis. In mouse embryonic fibroblasts (MEFs), Acsl4 knockout blocks ferroptosis induced by RSL3, a potent and specific inhibitor of GPX4 [23], and Gpx4-Acsl4 double knockout cells are resistant to ferroptosis [22]. ACSL4 was shown to enrich PUFAs with omega-6 FAs in membranes and peroxidation of the arachidonoyl (20:4) and adrenoyl (22:4) PUFA chains in PEs induces ferroptosis [22,23]. Inhibition of lipoxygenases (LOXs/ALOXs) rather than other enzymes that oxidize PUFAs, such as cyclooxygenase (COX) and cytochrome P450 (CYP450), blocks RSL3-induced ferroptosis in MEFs [23]. In a recent study, cytochrome P450 oxidoreductase (POR, also known as CYPOR) was identified as a key mediator of intrinsic and PUFA-induced ferroptosis through a CRISPR screen [24]. POR along with cofactors cytochrome P450 (CYP)/heme b (with Fe^{2+}) was shown to peroxidize PUFAs [24]. Consistently, POR along with cytochrome b5 reductase (CYB5R1) has been shown to generate hydrogen peroxide (H_2O_2), which reacts with iron, resulting in reactive hydroxyl radicals that oxidize PUFA chains of membrane PLs [25].

Precisely how membranes containing oxidized PUFA-PEs and perhaps other lipid species lead to ferroptotic cell death remains to be determined [26]. Akin to necroptosis and pyroptosis, it is conceivable that proteinaceous pores on the plasma membranes could be formed for ferroptosis. Indeed, cell rupture during ferroptosis is mediated by plasma membrane pores, similarly to cell lysis in pyroptosis and necroptosis [27]. It is also possible that membranes with oxidized PUFA lipids could radically alter membrane structures, resulting in membrane rapture and subsequent cell death. This scenario is consistent with the findings that H_2O_2-induced and iron-dependent peroxidization of PUFA-PL in liposome in vitro is sufficient to cause membrane rapture [25]. Given the recent observation that unsaturated FAs are enriched in the cytosolic leaflet of the plasma membrane [8], it will be interesting to assess whether oxidized PUFA lipids would be flipflopped to the exoplasmic leaflet or whether the accumulation of such lipids in the cytosolic leaflet would markedly alter local membrane curvature to trigger membrane rapture. It has been observed that ferroptosis can spread through cell populations in a wave-like manner, resulting in spatiotemporal patches of dead cells. Interestingly, ferroptotic spreading whereby lipid peroxidation disseminates to neighboring cells appears to be a distinct step from the pore-forming event/cell lysis. The ferroptotic spreading requires lipid peroxidation and iron and occurs prior to cell lysis [27].

3. The requirement of labile iron in the induction ferroptosis

Limiting intracellular irons by iron chelators such as deferoxamine and ciclopirox inhibits ferroptosis [4,19]. Nonheme iron-containing dioxygenases known as lipoxygenases (LOXs or ALOXs) catalyze, mediated by iron, peroxidation of free and esterified PUFAs to yield diverse bioactive lipid mediators including PUFA-OOH. LOX-mediated lipid peroxidation was shown to activate cell death due to GPX4 inactivation [28,29]. In a nonenzymatic reaction, redox-active iron (ferrous iron, Fe^{2+}) catalyzes the conversion of PUFA-OOH to highly reactive lipid alkoxyl radicals (L-O•) via Fenton chemistry [20,30]. Likewise, POR was shown to use the reduced form of iron (Fe^{2+}) in heme b in cytochrome P450 to peroxidate PUFAs [24].

Transferrin (Tf) tightly binds extracellular iron, which renders Fe^{3+} (ferric iron) virtually redox-inactive and prevents its exchange with other metals. Extracellular iron is imported via receptor-mediated endocytosis of ferrotransferrin (Tf dimer carrying irons)-bound Tf receptor 1 (TfR1, or CD71 encoded by the TFRC gene). Fe^{3+} is released from the receptor-ferrotransferrin complex in the acidic environment in late endosome and the apo-ferrotransferrin-TfR1 complex recycles to the cell surface. TfR1 downregulation inhibits ferroptosis, indicating that this route of iron import is important for ferroptosis [31–33].

Intracellular labile iron is stored in ferritins. There are four different ferritin genes in mammals: FTH1 and FTL encoding cytosolic H- and L-chain, FTMT encoding the mitochondrial ferritin and FTHL17. Of note, the ferritin H chain (FHC) is named for its initial isolation from heart, whereas ferritin L chain (FLC) is named for its initial isolation from liver, although they are also referred to as ferritin heavy (21 kDa) and light (20 kDa) chain, respectively. FTH1 and FTL are ubiquitously expressed. Structurally, 24 subunits of ferritins, a mixture of both H- and L-chains, assemble to form a spherical structure of 12 nm in diameter containing a large hollow 8 nm cavity for iron storage. Each ferritin subunit is folded into a four-helix bundle of similar length, which binds two iron atoms (reviewed in Ref. [34]). The main function of ferritins is described to store iron in a soluble, nontoxic, readily available form. Ferrous irons are taken up by ferritins, in which irons are oxidized by the ferroxidase activity of FHC and deposited as ferric hydroxides after oxidation. In addition to intracellular iron storage, ferritins also circulate and FHC internalizes through TfR1 by endocytosis and subsequent trafficking through endosomes and lysosomes [35] as well as other mechanisms such as pinocytosis [36].

Both H- and L chains of ferritins can be degraded through autophagy dubbed ferritinophagy; nuclear receptor coactivator 4 (NCOA4) serves as a cargo receptor in this process [37]. It has been shown that NCOA4 depletion reduces intracellular levels of labile iron and hence sensitivity to ferroptosis, whereas NCOA4 overexpression increases ferritinophagy and consequently sensitizes cells to ferroptosis in studies based on cell culture models [32,38,39]. Interestingly, mice deficient of Ncoa4 on iron-enriched diets exhibits liver toxicity associated with elevated levels of Gpx and superoxide dismutase and premature death [40], suggesting that systemic NCOA4 deficiency perturbs iron homeostasis and might enhance ferroptosis in certain cell types. In the breast cancer MDA-MB-231 cell line, FTH1 expression is repressed by the ATM (Ataxia telangiectasia mutated) kinase; ATM depletion inhibits erastin and sorafenib-induced ferroptosis, as a consequence of increased expression of several genes including FTH1 involved in regulating intracellular iron levels [41]. Prominin2 was shown to inhibit ferroptosis by exporting iron through facilitating the formation of multivesicular bodies containing ferritin-bound iron [42]. Sorafenib treatment was shown to reduce FTH1 phosphorylation at Ser179 [43], but the functional impact of this event remains to be determined. Notably, arachidonic acid, a PUFA, was shown to bind an intersubunit hydrophobic pocket in a ferritin structure, which might shield this PUFA from oxidation [44]. It will be interesting to assess whether this contributes to protecting cells from ferroptosis.

The iron (ferrous) transporter DMT1 (also known as DCT1 and NRAMP2, encoded by the SLC11A2 gene) is also implicated in ferroptosis [45,46]. DMT1 is a transmembrane protein with 12 transmembrane domains. Four isoforms of DMT1 are identified, which are localized to endosome/lysosome or plasma membrane, depending on cell types [47]. High dietary iron increases DMT1 expression in vivo [48]. Sulfasalazine, a multikinase and xCT inhibitor [4] (Fig. 8.1), induces DMT1 expression in breast cancer cells [49]. DMT1 inhibitors block lysosomal iron export and increase local ROS and ferroptosis. Interestingly, these inhibitors selectively kill breast cancer stem cells [50]. Precisely how increased DMT1 levels lead to ferroptosis requires further investigation.

Other regulators of iron metabolism have also been implicated in regulating ferroptosis. ACO1 encoding the iron-sulfur cluster-containing iron sensor IRP1 was identified in an RNAi screen of ferroptosis inducers [32]. This protein binds to the TFRC Tf receptor (TfR1) mRNA and inhibits the degradation of this otherwise rapidly degraded mRNA. Therefore, it is possible that ACO1 potentiates ferroptosis through increasing TfR1 levels. The mitochondrial iron-sulfur cluster protein CISD1 was shown to protect cells from ferroptosis by limiting iron import into mitochondria and lipid peroxidation in this compartment [51]. Downregulation of NFS1, an essential enzyme for the biosynthesis of iron—sulfur clusters, sensitizes cancer cells to ferroptosis [52].

4. Cellular defense mechanisms against ferroptosis

GPX4-mediated detoxification of lipid peroxides. Excessive oxidative stress is a major trigger of lipid peroxidation and ensuing ferroptosis. There are several identified cellular mechanisms for protecting cells from ferroptotic cell death. Reduction of oxidized PUFA-PLs by GPX4, one of 25 known selenocysteine-containing proteins encoded in the human genome, to

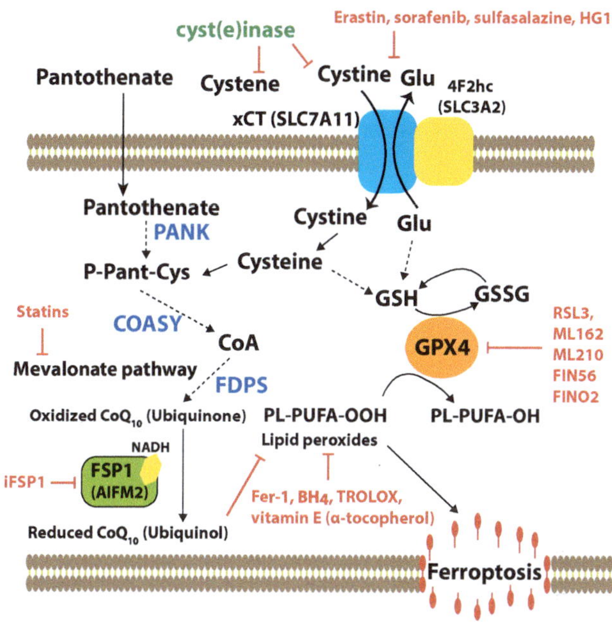

Figure 8.1 ***Regulators and pathways of ferroptosis and agents that modulate ferroptosis.*** The system x_c^- cystine/glutamate antiporter (a transmembrane protein complex containing the subunits xCT (the transporter subunit encoded by the SLC7A11 gene) and the chaperone subunit 4F2hc encoded by the SLC3A2 gene) imports extracellular cystine while exporting glutamate from the cytosol. The imported cystine is converted to cysteine for the production of glutathione and CoA. Glutathione is a cofactor for GPX4 to convert peroxidized PUFA-PLs to lipid alcohols. CoA is a precursor of the mevalonate pathway that provides metabolites for CoQ_{10} synthesis. Pantothenate along with cysteine is used for CoA synthesis. FSP1 reduces oxidized CoQ_{10} (ubiquinone) to the reduced form (ubiquinol) that serves as a lipophilic antioxidant to prevent lipid peroxidation. Therefore, both routes of cysteine metabolism are critical to inhibiting ferroptosis. Small molecule inhibitors of GPX4, xCT, and FSP1 are shown. These compounds can induce ferroptosis. Depletion of extracellular L-cystine and cysteine by a recombinant enzyme called cyst(e)inase can promote ferroptosis and suppress tumor growth. Statins that inhibit HMGCR (3-hydroxy-3-methylglutaryl-coenzyme A reductase), the rate-limiting enzyme in the mevalonate pathway, selectively inhibit the proliferation of cancer cells that depend on the GPX4-mediated lipid repair pathway for proliferation. Endogenous and synthetic antioxidant molecules such as BH_4 (5,6,7,8-tetrahydropterin), ferrostatin-1 (Fer-1), TROLOX (6-hydroxy-2,5,7,8-tetramethylchroman-2-carboxylic acid, a water-soluble analog of vitamin E) and α-tocopherol, a form of endogenous vitamin E are potent inhibitors of ferroptosis. Abbreviations: *CoA*, coenzyme A; *COASY*, CoA synthase; *CoQ_{10}*, coenzyme Q_{10}; *FDPS*, farnesyl pyrophosphate synthase; *FSP1*, ferroptosis suppressor protein 1; *GPX4*, glutathione peroxidase 4; *GSH*, glutathione; *GSSG*, glutathione disulfide; *NADH*, nicotinamide adenine dinucleotide; *PANK*, pantothenate kinase; *PL*, phospholipids; *PUFA*, polyunsaturated fatty acid.

lipid alcohols, is one of the best characterized mechanisms that prevent ferroptosis. Genetic or chemical GPX4 inhibition was shown to induce cell death [22,28,29]. GPX4 uses glutathione as a cofactor to catalyze the reduction of lipid hydroperoxides to their corresponding alcohols. As such, cellular glutathione depletion inhibits GPX4 activity, which is linked to reduced cystine import due to xCT/SCL7A11 inhibition (Fig. 8.1). GPX4 protein synthesis is regulated by the mevalonate pathway through isopentenylation of the selenocysteine-charged transfer RNA (tRNA) as part of selenocysteine–tRNA maturation. Statins that inhibit 3-hydroxy-3-methylglutaryl-coenzyme A reductase (HMGCR), a rate-liming enzyme in the mevalonate pathway, could induce ferroptosis in certain types of cancer cells partly through an impairment of GPX4 synthesis [53].

Despite its role in preventing ferroptosis, GPX4 has been considered as a tumor suppressor [54–56], probably through inhibiting the synthesis of various lipid-based signaling molecules such as eicosanoids (e.g., prostaglandin E2, PGE2) required for cell survival and proliferation or oxidation-induced oncogenic byproducts [56,57] as well as oncogenic cytokines such as IL-8 [58]. PGE2 is synthesized from arachidonic acid by cyclooxygenases (COXs), which require basal amounts of lipid peroxides for their activation. GPX4 overexpression reduces PGE2 level and impedes tumor growth and lung colonization of tumor cells [57]. GPX4 is downregulated in breast cancer clinical samples and cell lines [59,60]. However, cancer therapy could lead to a subpopulation of drug-resistant cells (the so-called persisters) with high levels of GPX4 and such cells have been shown to be sensitive to GPX4 inhibition [53,61].

Reduced coenzyme Q10 (CoQ10). Various coenzyme Qs are biosynthesized from isoprenoid metabolites generated in the mevalonate pathway, such as isopentenyl pyrophosphate [62,63]. CoQ10 is the most common coenzyme Q, which has 10 isoprene repeats. It exists in oxidized (ubiquinone) and reduced (ubiquinol) form. CoQ10 is embedded in the plasma and mitochondrial membranes. In mitochondria, CoQ10 is an essential electron carrier in the respiratory chain. It is well known that CoQ_{10} is one of the most significant lipid antioxidants to counteract oxidative stress.

Until recently, how CoQ_{10} functions in the plasma membranes are unclear. In screens aimed at identifying factors conferring resistance to ferroptosis independent of GPX4, two groups independently discovered ferroptosis suppressor protein 1 (FSP1, previously known as apoptosis-inducing factor mitochondrial 2 or AIFM2) is an inhibitor of ferroptosis

[64,65]. FSP1 is N-terminally myristoylated, which recruits FSP1 to the plasma membrane where it functions as a NADH-dependent oxidoreductase that reduces oxidized CoQ_{10} (ubiquinone) into reduced CoQ_{10} (ubiquinol). Ubiquinol acts as a lipophilic radical-trapping antioxidant and as such prevents ferroptosis. FSP1 is also localized to lipid droplets, although this localization is not involved in inhibiting ferroptosis [64]. However, despite previously being implicated in apoptosis, Bersuker et al. showed that FSP1 neither localizes to mitochondria nor promotes apoptosis [64]. Notably, The MDM2/MDMX oncoproteins were shown to potentiate ferroptosis independently of p53, possibly through inhibiting PPARα-mediated expression of antiferroptosis genes such as FSP1 [66]. In mitochondria, dihydroorotate dehydrogenase reduces ubiquinone to ubiquinol, acting in concert with mitochondrial GPX4 to prevent lipid peroxidation in mitochondria and ferroptosis [67].

The GCH1-BH_4 axis. Using a CRISPR/dCas9-mediated protein overexpression approach to identify factors implicated in regulating ferroptosis, Kraft et al. found that GCH1 overexpression confers resistance to ferroptosis [68]. GCH1 encodes GTP cyclohydrolase 1 that catalyzes the conversion of GTP to 7,8-dihydroneopterin 3′-triphosphate, which is the precursor for the biosynthesis of 7,8-dihydropterin (BH_2) and 5,6,7,8-tetrahydropterin (BH_4). BH_4 is an important cofactor for the production of aromatic amino acids, neurotransmitters, and nitric oxide (NO). GCH1 overexpression in mouse fibroblast or the human fibrosarcoma cell line HT1080 confers resistance to ferroptosis induced by the GPX4 inhibitor RLS3 or the xCT inhibitor imidazole ketone erastin (IKE). GCH1 overexpression increased levels of BH_2/BH_4 precursors, and BH_4 or BH_2 supplementation prevented RSL3 or IKE-induced ferroptosis. Increased GCH1 expression in cancer cell lines correlates with resistance to RLS3. Notably, BH_4 is a precursor of 4-OH-benzoate, which is an intermediate of CoQ_{10} biosynthesis. Therefore, the GCH1-BH_4 axis could protect cells from ferroptotic death through increased CoQ_{10}. In an independent study, the GCH1-BH_4 axis was also shown to be critical to protecting cells from GPX4 inhibition-induced ferroptosis [69]. Dihydrofolate reductase (DHFR)-mediated BH_4 regeneration is important for inhibiting ferroptosis. Notably, although water soluble, BH_4 appears to act directly to prevent lipid oxidation in lipid membranes similar to lipophilic antioxidants such as α-tocopherol (α-TOH), the most potent form of vitamin E and the most effective endogenous radical-trapping antioxidants [69].

5. Knowledge in the mechanisms of ferroptosis provides ample opportunities for developing therapeutics

Ferroptosis is triggered by increased levels of oxidized PUFA-PLs due to impaired cellular defense mechanisms against ferroptosis. The three major pathways that protect cells from ferroptosis may be hyperactive in cancer or inactivated in neurodegenerative diseases. As discussed above, cancer therapy can select cell populations that express high levels of GPX4. Notably, lymph nodes have been shown to protect metastasizing melanoma cells from ferroptosis, suggesting inhibition of ferroptosis plays an important role in the initial stage of metastasis [70]. Therefore, agents that inhibit these cellular defense mechanisms may be effective cancer therapeutics. Several classes of GPX4 inhibitors have been developed. RSL3, ML210, and ML162 analogs are covalent inhibitors of GPX4 [4,6,53,71]. Although some GPX4 inhibitors such as RSL3 and ML210 are not suitable for systemic in vivo dosing, an in vivo proof-of-principle that GPX4 activity is critical to tumor growth in tumor cells that persisted after cancer therapy has been established [61]. Several GPX4 inhibitors are in various stages of clinical trials [72].

Notably, xCT is a central hub of cellular regulation of ferroptosis (Fig. 8.1). Cystine import through xCT not only provides a starting material for glutathione synthesis required for GPX4 activity but also leads to acetyl-CoA synthesis that can be routed to CoQ_{10} synthesis (e.g., through the mevalonate pathway) [4,73]. Increased expression of SLC7A11 is found in many cancer types [74—76]. Importantly genetic inactivation of Slc7a11 in mice suppresses tumor growth through inducing ferroptosis [73]. Moreover, depletion of extracellular L-cystine and cysteine using an engineered recombinant enzyme known as cyst(e)inase exhibited in vivo anticancer efficacy in prostate and pancreatic cancer models [73,77] (Fig. 8.1).

Erastin and analogs are xCT inhibitors [4,78]. The erastin analog IKE, when formulated in PEG-PLGA (polyethylene glycol-poly(lactic-co-glycolic acid)) nanoparticles, has anticancer efficacy with reduced toxicity in a diffuse large B-cell lymphoma mouse xenograft model [79]. The FDA-approved drugs sulfasalazine and sorafenib inhibit xCT and thus deplete intracellular cystine/cysteine [74,78]. These drugs exhibit in vivo efficacy to inhibit tumor growth [74,75,80,81]. Novel xCT inhibitors have been discovered [76], among which HG106 shows potent in vivo anticancer efficacy in KRAS-mutated lung cancers [75]. Importantly cancer immunotherapy kills cancer cells by ferroptosis due to SLC7A11 down-regulation by CD8 T-cell-derived interferon-γ (IFN-γ) [82]. xCT inhibition also sensitizes tumors to radiotherapy [81,83]. Additionally, SLC7A11-based vaccines show potential anticancer effects [76].

In a screen of 10,000 drug-like compounds, the FSP1 inhibitor iFSP1 was discovered, which induces toxicity in cells with FSP1 overexpression when GPX4 is genetically depleted [65]. Further optimization and characterization of the compounds will be needed to demonstrate its therapeutic potential. Other important regulators of the ferroptosis pathways are also potential targets for therapeutic developments. As inhibitors of the mevalonate pathway (Fig. 8.1), the anticancer effects of statins have been investigated extensively in preclinical and clinical studies [84–88]. Whether or not statins have anticancer efficacy remain inconclusive in epidemiological analyses and clinical trials [88–92]. Cancer cells with mesenchymal-like phenotypes are more sensitive to statins and such cells are also more reliant on GPX4 activity for survival [53]. Thus a combination strategy of statins with GPX4 inhibition merits further investigation for cancer therapy.

Preventing excessive ferroptosis may have therapeutic potential for a number of diseases including neurodegeneration, stoke, diabetes, asthma, kidney injury, and osteoporosis [93–104]. Activating GPX4 expression via selenium (Se) dosing into the brain was shown to protect neurons and improves behavior in a hemorrhagic stroke model [102]. Sequestration of iron by iron chelators such as deferoxamine, deferiprone, and ciclopirox olamine can protect neurons after intracerebral hemorrhage [105].

6. Summary and perspectives

Ferroptosis is induced by excessive accumulation of peroxidized PUFA-containing PLs. The iron-fueled Fenton reaction is important for PUFA peroxidation. The molecular mechanisms and metabolic determinants underlying ferroptosis are subjects of intensive investigation in recent years [3,19,26,106]. A number of enzymes such as GPX4 and ACSL4 and pathways have been identified as key regulators of ferroptosis (Fig. 8.1). Nonetheless, the mechanistic details underlying ferroptosis are still being figured out [26]. For example, whether a protein pore is required for the execution of ferroptosis as in necroptosis and pyroptosis and whether the formation of such a pore is the final step (the point of no return) of ferroptotic cell death remains an open question [3,26]. Because of the involvement of GPX4 in other modes of cell death such as apoptosis and necroptosis [107,108], it will be important to delineate whether GPX4 uses distinct function(s) in preventing different forms of cell death. The demonstrated tumor suppressive roles for GPX4 will need to be reconciled with the requirement of its enzymatic activity for counteracting oxidative stress in highly proliferative tumor cells. The contribution of PUFA-mediated cell signaling events in ferroptosis is less understood, and further studies will be needed to delineate roles of PUFAs in ferroptosis signaling and structural alterations in membranes. Finally, although activating or inhibiting ferroptosis has clear implication for treating a wide range of human diseases, drug discovery targeting the ferroptotic pathways remain largely academic endeavors thus far. Efforts from the industry will be needed to drive drug discovery and clinical translation. A more complete understanding of molecular events during ferroptosis along with the availability of detectable biomarkers of ferroptosis will surely lead to robust efforts in drug development and eventually to clinical therapeutics.

References

[1] S.J. Dixon, et al., Ferroptosis: an iron-dependent form of nonapoptotic cell death, Cell 149 (5) (2012) 1060–1072.

[2] M. Maiorino, M. Conrad, F. Ursini, GPx4, lipid peroxidation, and cell death: discoveries, rediscoveries, and open issues, Antioxid. Redox Signal. 29 (1) (2018) 61–74.

[3] X. Jiang, B.R. Stockwell, M. Conrad, Ferroptosis: mechanisms, biology and role in disease, Nat. Rev. Mol. Cell Biol. 22 (4) (2021) 266–282, https://doi.org/10.1038/s41580-020-00324-8.

[4] B.R. Stockwell, et al., Ferroptosis: a regulated cell death nexus linking metabolism, redox biology, and disease, Cell 171 (2) (2017) 273–285.

[5] W.S. Yang, B.R. Stockwell, Synthetic lethal screening identifies compounds activating iron-dependent, nonapoptotic cell death in oncogenic-RAS-harboring cancer cells, Chem. Biol. 15 (3) (2008) 234–245.

[6] W.S. Yang, et al., Regulation of ferroptotic cancer cell death by GPX4, Cell 156 (1–2) (2014) 317–331.

[7] E. Fahy, et al., Update of the LIPID MAPS comprehensive classification system for lipids, J. Lipid Res. 50 (Suppl. l) (2009) S9–S14.

[8] J.H. Lorent, et al., Plasma membranes are asymmetric in lipid unsaturation, packing and protein shape, Nat. Chem. Biol. 16 (6) (2020) 644–652.

[9] L. Tesfay, et al., Stearoyl-CoA desaturase 1 protects ovarian cancer cells from ferroptotic cell death, Cancer Res. 79 (20) (2019) 5355–5366.

[10] K.J. Williams, et al., An essential requirement for the SCAP/SREBP signaling axis to protect cancer cells from lipotoxicity, Cancer Res. 73 (9) (2013) 2850–2862.

[11] M. Piccolis, et al., Probing the global cellular responses to lipotoxicity caused by saturated fatty acids, Mol. Cell 74 (1) (2019) 32e8–44e8.

[12] L. Ge, et al., Identification of the delta-6 desaturase of human sebaceous glands: expression and enzyme activity, J. Invest. Dermatol. 120 (5) (2003) 707–714.

[13] K. Vriens, et al., Evidence for an alternative fatty acid desaturation pathway increasing cancer plasticity, Nature 566 (7744) (2019) 403–406.
[14] C. Ferreri, et al., The n-10 fatty acids family in the lipidome of human prostatic adenocarcinoma cell membranes and extracellular vesicles, Cancers (Basel) 12 (4) (2020).
[15] L. Magtanong, et al., Exogenous monounsaturated fatty acids promote a ferroptosis-resistant cell state, Cell Chem. Biol. 26 (3) (2019) 420e9–432e9.
[16] H. Lee, et al., Energy-stress-mediated AMPK activation inhibits ferroptosis, Nat. Cell Biol. 22 (2) (2020) 225–234.
[17] H. Sprecher, Metabolism of highly unsaturated n-3 and n-6 fatty acids, Biochim. Biophys. Acta 1486 (2–3) (2000) 219–231.
[18] M.T. Nakamura, T.Y. Nara, Structure, function, and dietary regulation of delta6, delta5, and delta9 desaturases, Annu. Rev. Nutr. 24 (2004) 345–376.
[19] S.J. Dixon, B.R. Stockwell, The hallmarks of ferroptosis, Ann. Rev. Cancer Biol. 3 (2019) 35–54.
[20] G.C. Forcina, S.J. Dixon, GPX4 at the crossroads of lipid homeostasis and ferroptosis, Proteomics 19 (18) (2019) e1800311.
[21] S.J. Dixon, et al., Human haploid cell genetics reveals roles for lipid metabolism genes in nonapoptotic cell death, ACS Chem. Biol. 10 (7) (2015) 1604–1609.
[22] S. Doll, et al., ACSL4 dictates ferroptosis sensitivity by shaping cellular lipid composition, Nat. Chem. Biol. 13 (1) (2017) 91–98.
[23] V.E. Kagan, et al., Oxidized arachidonic and adrenic PEs navigate cells to ferroptosis, Nat. Chem. Biol. 13 (1) (2017) 81–90.
[24] Y. Zou, et al., Cytochrome P450 oxidoreductase contributes to phospholipid peroxidation in ferroptosis, Nat. Chem. Biol. 16 (3) (2020) 302–309.
[25] B. Yan, et al., Membrane damage during ferroptosis is caused by oxidation of phospholipids catalyzed by the oxidoreductases POR and CYB5R1, Mol. Cell 81 (2) (2021) 355–369.
[26] H. Feng, B.R. Stockwell, Unsolved mysteries: how does lipid peroxidation cause ferroptosis? PLoS Biol. 16 (5) (2018) e2006203.
[27] M. Riegman, et al., Ferroptosis occurs through an osmotic mechanism and propagates independently of cell rupture, Nat. Cell Biol. 22 (9) (2020) 1042–1048.
[28] A. Seiler, et al., Glutathione peroxidase 4 senses and translates oxidative stress into 12/15-lipoxygenase dependent- and AIF-mediated cell death, Cell Metabol. 8 (3) (2008) 237–248.
[29] W.S. Yang, et al., Peroxidation of polyunsaturated fatty acids by lipoxygenases drives ferroptosis, Proc. Natl. Acad. Sci. USA 113 (34) (2016) E4966–E4975.
[30] S. Doll, M. Conrad, Iron and ferroptosis: a still ill-defined liaison, IUBMB Life 69 (6) (2017) 423–434.
[31] M. Gao, et al., Glutaminolysis and transferrin regulate ferroptosis, Mol. Cell 59 (2) (2015) 298–308.
[32] M. Gao, et al., Ferroptosis is an autophagic cell death process, Cell Res. 26 (9) (2016) 1021–1032.
[33] S. Torii, et al., An essential role for functional lysosomes in ferroptosis of cancer cells, Biochem. J. 473 (6) (2016) 769–777.
[34] P. Arosio, et al., The importance of eukaryotic ferritins in iron handling and cytoprotection, Biochem. J. 472 (1) (2015) 1–15.
[35] L. Li, et al., Binding and uptake of H-ferritin are mediated by human transferrin receptor-1, Proc. Natl. Acad. Sci. USA 107 (8) (2010) 3505–3510.
[36] M.U. Muckenthaler, et al., A red carpet for iron metabolism, Cell 168 (3) (2017) 344–361.

[37] J.D. Mancias, et al., Quantitative proteomics identifies NCOA4 as the cargo receptor mediating ferritinophagy, Nature 509 (7498) (2014) 105–109.
[38] W. Hou, et al., Autophagy promotes ferroptosis by degradation of ferritin, Autophagy 12 (8) (2016) 1425–1428.
[39] M. Quiles Del Rey, J.D. Mancias, NCOA4-Mediated ferritinophagy: a potential link to neurodegeneration, Front. Neurosci. 13 (2019) 238.
[40] R. Bellelli, et al., NCOA4 deficiency impairs systemic iron homeostasis, Cell Rep. 14 (3) (2016) 411–421.
[41] P.H. Chen, et al., Kinome screen of ferroptosis reveals a novel role of ATM in regulating iron metabolism, Cell Death Differ. 27 (3) (2019) 1008–1022.
[42] C.W. Brown, et al., Prominin2 drives ferroptosis resistance by stimulating iron export, Dev. Cell 51 (5) (2019) 575–586.e4.
[43] E.G. Werth, et al., Time course of changes in sorafenib-treated hepatocellular carcinoma cells suggests involvement of phospho-regulated signaling in ferroptosis induction, Proteomics 20 (10) (2020) e2000006.
[44] W. Bu, et al., Ferritin couples iron and fatty acid metabolism, FASEB J. 26 (6) (2012) 2394–2400.
[45] Y. Song, et al., Human umbilical cord blood-derived MSCs exosome attenuate myocardial injury by inhibiting ferroptosis in acute myocardial infarction mice, Cell Biol. Toxicol. 37 (1) (2021) 51–64.
[46] X. Zeng, et al., Benefits of iron chelators in the treatment of Parkinson's disease, Neurochem. Res. 46 (5) (2021) 1239–1251.
[47] I. Yanatori, F. Kishi, DMT1 and iron transport, Free Radic. Biol. Med. 133 (2019) 55–63.
[48] L.B. Li, et al., Iron exposure and the cellular mechanisms linked to neuron degeneration in adult mice, Cells 8 (2) (2019).
[49] H. Yu, et al., Sulfasalazineinduced ferroptosis in breast cancer cells is reduced by the inhibitory effect of estrogen receptor on the transferrin receptor, Oncol. Rep. 42 (2) (2019) 826–838.
[50] A.L. Turcu, et al., DMT1 inhibitors kill cancer stem cells by blocking lysosomal iron translocation, Chemistry 26 (33) (2020) 7369–7373.
[51] H. Yuan, et al., CISD1 inhibits ferroptosis by protection against mitochondrial lipid peroxidation, Biochem. Biophys. Res. Commun. 478 (2) (2016) 838–844.
[52] S.W. Alvarez, et al., NFS1 undergoes positive selection in lung tumours and protects cells from ferroptosis, Nature 551 (7682) (2017) 639–643.
[53] V.S. Viswanathan, et al., Dependency of a therapy-resistant state of cancer cells on a lipid peroxidase pathway, Nature 547 (7664) (2017) 453–457.
[54] R. Brigelius-Flohe, A. Kipp, Glutathione peroxidases in different stages of carcinogenesis, Biochim. Biophys. Acta 1790 (11) (2009) 1555–1568.
[55] M.L. Zhang, et al., Involvement of glutathione peroxidases in the occurrence and development of breast cancers, J. Transl. Med. 18 (1) (2020) 247.
[56] E. Dai, et al., Ferroptotic damage promotes pancreatic tumorigenesis through a TMEM173/STING-dependent DNA sensor pathway, Nat. Commun. 11 (1) (2020) 6339.
[57] I. Heirman, et al., Blocking tumor cell eicosanoid synthesis by GP x 4 impedes tumor growth and malignancy, Free Radic. Biol. Med. 40 (2) (2006) 285–294.
[58] N. Rohr-Udilova, et al., Impact of glutathione peroxidase 4 on cell proliferation, angiogenesis and cytokine production in hepatocellular carcinoma, Oncotarget 9 (11) (2018) 10054–10068.
[59] P. Cejas, et al., Phospholipid hydroperoxide glutathione peroxidase (PHGPx) expression is downregulated in poorly differentiated breast invasive ductal carcinoma, Free Radic. Res. 41 (6) (2007) 681–687.

[60] F. Rusolo, et al., Comparison of the seleno-transcriptome expression between human non-cancerous mammary epithelial cells and two human breast cancer cell lines, Oncol. Lett. 13 (4) (2017) 2411–2417.
[61] M.J. Hangauer, et al., Drug-tolerant persister cancer cells are vulnerable to GPX4 inhibition, Nature 551 (7679) (2017) 247–250.
[62] J.A. Stefely, D.J. Pagliarini, Biochemistry of mitochondrial coenzyme Q biosynthesis, Trends Biochem. Sci. 42 (10) (2017) 824–843.
[63] S. Shukla, K.K. Dubey, CoQ10 a super-vitamin: review on application and biosynthesis, 3 Biotech 8 (5) (2018) 249.
[64] K. Bersuker, et al., The CoQ oxidoreductase FSP1 acts parallel to GPX4 to inhibit ferroptosis, Nature 575 (7784) (2019) 688–692.
[65] S. Doll, et al., FSP1 is a glutathione-independent ferroptosis suppressor, Nature 575 (7784) (2019) 693–698.
[66] D. Venkatesh, et al., MDM2 and MDMX promote ferroptosis by PPARalpha-mediated lipid remodeling, Genes Dev. 34 (7–8) (2020) 526–543.
[67] C. Mao, et al., DHODH-mediated ferroptosis defence is a targetable vulnerability in cancer, Nature 593 (7860) (2021) 586–590.
[68] V.A.N. Kraft, et al., GTP cyclohydrolase 1/tetrahydrobiopterin counteract ferroptosis through lipid remodeling, ACS Cent. Sci. 6 (1) (2020) 41–53.
[69] M. Soula, et al., Metabolic determinants of cancer cell sensitivity to canonical ferroptosis inducers, Nat. Chem. Biol. 16 (12) (2020) 1351–1360.
[70] J.M. Ubellacker, et al., Lymph protects metastasizing melanoma cells from ferroptosis, Nature 585 (7823) (2020) 113–118.
[71] J.K. Eaton, et al., Selective covalent targeting of GPX4 using masked nitrile-oxide electrophiles, Nat. Chem. Biol. 16 (5) (2020) 497–506.
[72] X. Chen, et al., Broadening horizons: the role of ferroptosis in cancer, Nat. Rev. Clin. Oncol. 18 (5) (2021) 280–296.
[73] M.A. Badgley, et al., Cysteine depletion induces pancreatic tumor ferroptosis in mice, Science 368 (6486) (2020) 85–89.
[74] L.A. Timmerman, et al., Glutamine sensitivity analysis identifies the xCT antiporter as a common triple-negative breast tumor therapeutic target, Cancer Cell 24 (4) (2013) 450–465.
[75] K. Hu, et al., Suppression of the SLC7A11/glutathione axis causes synthetic lethality in KRAS-mutant lung adenocarcinoma, J. Clin. Invest. 130 (4) (2020) 1752–1766.
[76] W. Lin, et al., SLC7A11/xCT in cancer: biological functions and therapeutic implications, Am. J. Cancer Res. 10 (10) (2020) 3106–3126.
[77] S.L. Cramer, et al., Systemic depletion of L-cyst(e)ine with cyst(e)inase increases reactive oxygen species and suppresses tumor growth, Nat. Med. 23 (1) (2017) 120–127.
[78] S.J. Dixon, et al., Pharmacological inhibition of cystine-glutamate exchange induces endoplasmic reticulum stress and ferroptosis, Elife 3 (2014) e02523.
[79] Y. Zhang, et al., Imidazole ketone erastin induces ferroptosis and slows tumor growth in a mouse lymphoma model, Cell Chem. Biol. 26 (5) (2019) 623e9–633e9.
[80] X. Sun, et al., Metallothionein-1G facilitates sorafenib resistance through inhibition of ferroptosis, Hepatology 64 (2) (2016) 488–500.
[81] L.F. Ye, et al., Radiation-induced lipid peroxidation triggers ferroptosis and synergizes with ferroptosis inducers, ACS Chem. Biol. 15 (2) (2020) 469–484.
[82] X. Lang, et al., Radiotherapy and immunotherapy promote tumoral lipid oxidation and ferroptosis via synergistic repression of SLC7A11, Cancer Discov. 9 (12) (2019) 1672–1686.
[83] G. Lei, et al., Ferroptosis, radiotherapy, and combination therapeutic strategies, Protein Cell (2021).

[84] K. Warita, et al., Statin-induced mevalonate pathway inhibition attenuates the growth of mesenchymal-like cancer cells that lack functional E-cadherin mediated cell cohesion, Sci. Rep. 4 (2014) 7593.
[85] T. Wang, et al., Simvastatin-induced breast cancer cell death and deactivation of PI3K/Akt and MAPK/ERK signalling are reversed by metabolic products of the mevalonate pathway, Oncotarget 7 (3) (2016) 2532–2544.
[86] C.H. Beckwitt, et al., Statins attenuate outgrowth of breast cancer metastases, Br. J. Cancer 119 (9) (2018) 1094–1105.
[87] C.H. Beckwitt, et al., Statin drugs to reduce breast cancer recurrence and mortality, Breast Cancer Res. 20 (1) (2018) 144.
[88] O. Bjarnadottir, et al., Targeting HMG-CoA reductase with statins in a window-of-opportunity breast cancer trial, Breast Cancer Res. Treat. 138 (2) (2013) 499–508.
[89] S. Woditschka, et al., Lipophilic statin use and risk of breast cancer subtypes, Cancer Epidemiol. Biomark. Prev. 19 (10) (2010) 2479–2487.
[90] K. Undela, V. Srikanth, D. Bansal, Statin use and risk of breast cancer: a meta-analysis of observational studies, Breast Cancer Res. Treat. 135 (1) (2012) 261–269.
[91] T.P. Ahern, et al., Statins and breast cancer prognosis: evidence and opportunities, Lancet Oncol. 15 (10) (2014) e461–e468.
[92] U. Ravnskov, P.J. Rosch, K.S. McCully, Statins do not protect against cancer: quite the opposite, J. Clin. Oncol. 33 (7) (2015) 810–811.
[93] L. Chen, et al., Lipid peroxidation up-regulates BACE1 expression in vivo: a possible early event of amyloidogenesis in Alzheimer's disease, J. Neurochem. 107 (1) (2008) 197–207.
[94] S.E. Yoo, et al., Gpx4 ablation in adult mice results in a lethal phenotype accompanied by neuronal loss in brain, Free Radic. Biol. Med. 52 (9) (2012) 1820–1827.
[95] K. Koulajian, et al., Overexpression of glutathione peroxidase 4 prevents beta-cell dysfunction induced by prolonged elevation of lipids in vivo, Am. J. Physiol. Endocrinol. Metab. 305 (2) (2013) E254–E262.
[96] J.P. Friedmann Angeli, et al., Inactivation of the ferroptosis regulator Gpx4 triggers acute renal failure in mice, Nat. Cell Biol. 16 (12) (2014) 1180–1191.
[97] L. Chen, et al., Ablation of the ferroptosis inhibitor glutathione peroxidase 4 in neurons results in rapid motor neuron degeneration and paralysis, J. Biol. Chem. 290 (47) (2015) 28097–28106.
[98] W.S. Hambright, et al., Ablation of ferroptosis regulator glutathione peroxidase 4 in forebrain neurons promotes cognitive impairment and neurodegeneration, Redox Biol. 12 (2017) 8–17.
[99] H. Imai, et al., Lipid peroxidation-dependent cell death regulated by GPx4 and ferroptosis, in: S. Nagata, H. Nakano (Eds.), Apoptotic and Non-apoptotic Cell Death, Springer International Publishing, Cham, 2017, pp. 143–170.
[100] B.R. Cardoso, et al., Glutathione peroxidase 4: a new player in neurodegeneration? Mol. Psychiatr. 22 (3) (2017) 328–335.
[101] I. Ingold, et al., Selenium utilization by GPX4 is required to prevent hydroperoxide-induced ferroptosis, Cell 172 (3) (2018) 409e21–422e21.
[102] I. Alim, et al., Selenium drives a transcriptional adaptive program to block ferroptosis and treat stroke, Cell 177 (5) (2019) 1262e25–1279e25.
[103] J. Lu, et al., Extracellular vesicles from endothelial progenitor cells prevent steroid-induced osteoporosis by suppressing the ferroptotic pathway in mouse osteoblasts based on bioinformatics evidence, Sci. Rep. 9 (1) (2019) 16130.
[104] M. Conrad, B. Proneth, Selenium: tracing another essential element of ferroptotic cell death, Cell Chem. Biol. 27 (4) (2020) 409–419.

[105] N. DeGregorio-Rocasolano, O. Marti-Sistac, T. Gasull, Deciphering the iron side of stroke: neurodegeneration at the crossroads between iron dyshomeostasis, excitotoxicity, and ferroptosis, Front. Neurosci. 13 (2019) 85.
[106] J. Zheng, M. Conrad, The metabolic underpinnings of ferroptosis, Cell Metab. 32 (6) (2020) 920–937.
[107] Q. Ran, et al., Transgenic mice overexpressing glutathione peroxidase 4 are protected against oxidative stress-induced apoptosis, J. Biol. Chem. 279 (53) (2004) 55137–55146.
[108] O. Canli, et al., Glutathione peroxidase 4 prevents necroptosis in mouse erythroid precursors, Blood 127 (1) (2016) 139–148.

Index

Note: 'Page numbers followed by "f" indicate figures and "t" indicate tables.'

D

FSC
www.fsc.org
MIX
Papier | Fördert gute Waldnutzung
FSC® C083411

Kolibri 360 GmbH
Ferdinand-Jühlke-Straße 7
99095 Erfurt, Deutschland
produktsicherheit@kolibri360.de